OHIO STATE UNIVERSITY BIOSCIENCES COLLOQUIA

Microbial Chemoautotrophy

THEMES, DATES, AND ORGANIZERS

Genetics and Biogenesis of
Mitochondria and Chloroplasts
5–7 September 1974
C. W. Birky, Jr., P. S. Perlman, and T. J. Byers

Regulatory Biology
4–6 September 1975
J. C. Copeland and G. A. Marzluf

Analysis of Ecological Systems
29 April–1 May 1976
D. J. Horn, G. R. Stairs, and R. D. Mitchell

Plant Cell and Tissue Culture:
Principles and Applications
6–9 September 1977
W. R. Sharp, P. O. Larsen, E. F. Paddock,
and V. Raghavan

Cellular Interactions in Symbiosis and Parasitism
7–9 September 1978
C. B. Cook, P. W. Pappas, and E. D. Rudolph

Gene Structure and Expression
6–8 September 1979
D. H. Dean, L. F. Johnson, P. C. Kimball,
and P. S. Perlman

Behavioral Energetics: Vertebrate Costs of Survival
30 October–1 November 1980
W. P. Aspey and S. I. Lustick

Microbial Chemoautotrophy
4–5 October 1982
W. R. Strohl and O. H. Tuovinen

EDITED BY W. R. STROHL AND O. H. TUOVINEN

Microbial Chemoautotrophy

OHIO STATE UNIVERSITY PRESS : COLUMBUS

Library of Congress Cataloging in Publication Data

Main entry under title:

Microbial chemoautotrophy.

(Ohio State University biosciences colloquia)

Papers presented at the Microbial Chemoautotrophy Colloquium held Oct. 4–5, 1982, at the Ohio State University.

Includes index.

1. Bacteria, Chemoautotrophic—Congresses. I. Strohl, W.R. II. Tuovinen, O.H. III. Microbial Chemoautotrophy Colloquium (1982 : Ohio State University) IV. Series.

QR88.7.M53 1984 589.9'013343 83–27344
ISBN 0-8142-0342-6

Contents

Preface

Microbial chemoautotrophy has emerged from its original definition of carbon dioxide fixation and inorganic substrate oxidation to encompass a continuing spectrum of chemolithotrophs, mixotrophs, methylotrophs, and methanogens whose obligate and facultative life styles are far from being resolved. Chemoautotrophs have been featured in many diverse publications, symposia, and meetings during the past several decades because of their impact on physiology, ecology, and biotechnology. Several recent symposia have convened to discuss specific topics such as C_1 utilization and biohydrometallurgy. We felt, however, that a unified approach comparing the salient characeristics of several different types of chemoautotrophs was warranted. Therefore, emphasis of the Microbial Chemoautotrophy Colloquium was on the comparative physiological and genetic aspects of chemoautotrophs. It was to our particular delight that chemoautotrophy genetics could be presented at a forefront stage at the Colloquium.

The Colloquium was convened on 4–5 October 1982 at the Ohio State University. In addition to the invited presentations included herein, a poster session highlighted the many different facets of chemoautotrophs. We thank all authors and speakers for their interesting contributions. Financial contributions were received from the Ohio State University (College of Biological Sciences, Graduate School, and Department of Microbiology), W. R. Grace & Co., Diamond Shamrock, Inc., Battelle Columbus Laboratories, industrial exhibitors, and Adria Laboratories, Inc. Special thanks are due to Richard O. Moore, associate dean of the College of Biological Sciences, and his secretary, Dianne Hardin, for helping with administrative matters. Technical support services by staff and graduate students too numerous to list were appreciated by everyone.

W. R. S.
O. H. T.

Introduction

In this symposium on chemoautotrophy, it occurs to me that we are once again coming full circle in our perspective of the topic. Allow me to explain.

The term *autotrophy* refers to the notion of self-nourishment and is specifically used in reference to carbon nourishment. Hence, it has come to mean the assimilation of carbon dioxide as the exclusive or at least the major source of cellular carbon. Autotrophy was expanded to the concept of *chemoautotrophy,* which incorporates the notion of chemical energy used for the assimilation of carbon dioxide.

In 1862 Pasteur suggested that microorganisms were responsible for the oxidation of ammonia to nitrate in the soil, and over the next 15 to 20 years, Schloesing and Muntz recognized that the process of nitrification took place in two stages: the conversion of ammonia to nitrite followed by the oxidation of nitrite to nitrate. It was during this period (1887–92) that the Winogradskys described several genera of autotrophic nitrifying bacteria as well as sulfur-utilizing bacteria and introduced the concept of *chemolithotrophy* to establish the linkage of energy release from inorganic chemical oxidations to the assimilation of carbon dioxide. Parenthetically, it is noted that sulfur-utilizing bacteria were studied earlier but with the misconception of many of the microbiologists during the 1870s who regarded the various morphological types, including filamentous sulfur bacteria, as stages in the life cycle of one or two bacteria.

S. Winogradsky went a step further in his concept of chemolithotrophy by inextricably linking inorganic oxidations (anorgoxydants) to CO_2 assimilation accompanied by the inhibitory effect of organic substrates, thereby creating the concept of *obligate chemolithotrophy*. This, of course, was contrasted to the more broadly understood concept of *chemoorganotrophy* or *heterotrophy*.

The word *chemosynthesis* was introduced in 1897 by Pfeffer, a plant physiologist who undoubtedly compared microbes that used inorganic oxidations while assimilating carbon dioxide to plant photosynthesis. By the turn of the century, both aerobic and anaerobic hydrogen-utilizing bacteria were recognized, and the study of chemolithotrophy proceeded in several directions. Those who embarked on the pioneering investigations of microbial CO_2 assimilation compared chemoautotrophy with photoautotrophs, and when they examined the chemoautotrophic utilization of H_2 gas, they were drawn to the organisms that: (1) reduced CO_2 to methane (e.g., *Methanobacterium*) and (2) reduced CO_2 to acetic acid (e.g., *Clostridium aceticum*). Beijerinck described mixed cultures of sulfate-reducing organisms in 1895, and pure cultures of *Spirillium desulfuricans* (now *Desulfovibrio desulfuricans*) were reported by Delden in 1903 (see Bisset and Grace, 1954).

As research progressed in the intervening years, it became evident that there was no clear line of demarcation between chemolithotrophy and heterotrophy. The aerobic hydrogen-oxidizers that assimilate CO_2 (e.g., *Hydrogenomonas,* which has recently been rejected as a genus) were recognized as facultative autotrophs, and some hydrogen-oxidizers also derive energy from the oxidation of carbon monoxide. The sulfate-reducer *Desulfovibrio desulfuricans* is an example of an organism that can use either hydrogen or an organic compound as its energy source and either CO_2 or organic C as its carbon source. The term *mixotroph* has been used to describe that kind of metabolic flexibility.

The point I wish to make is that many past symposia on chemoautotrophy were narrowly focused around the restrictive concepts of autotrophy espoused by Winogradsky. This symposium has accepted the suggestion of Werkman (1951) "that it is better to sense the differentiation between autotrophy and heterotrophy than to define it." That statement equally applies to obligate or facultative chemolithotrophy, chemolithoautotrophy, and mixotrophy. As others have suggested (Whittenbury and Kelly, 1977), autotrophy can be less restrictively viewed in the context of generating most cellular carbon from one-carbon compounds. The experts in this volume will discuss assimilation and metabolism of one-carbon compounds as well as lithotrophic and heterotrophic energy sources. The intrinsic fascination with the topic in my opinion has always resided with the metabolic and biochemical reactions and the ecological relationships among the organisms.

Although there has been intrinsic interest in chemoautotrophs since their discovery, the cadre of researchers has always been relatively small in relation to the general pool of microbiologists and biological chemists. Reflection leads to the observation that since virtually none of the chemoautotrophs are known animal or plant pathogens, the vast governmental research support base, being predominantly biomedical, has largely bypassed the world of chemoautotrophy. Most financial support for this kind of research in the United States has resulted from perceived applications, and in general the number of researchers in the subject has waxed and waned with the potential for application.

For example, the level of research activity on nitrogen fixation and nitrification follows the economics of agriculture. Interest in hydrogen-oxidizing bacteria peaked in the early 1960s with its potential use in closed-cycle ecosystems for space stations. Methanogenesis tends to receive emphasis in proportion to interest in biomass conversion—particularly with treatment of organic wastes and the economics of fossil fuels. Surges of research on sulfur- and iron-oxidizers parallel our concerns with environmental contamination by acid mine drainage and atmospheric contamination through combustion of sulfur-containing fossil fuels—particularly coal. Interest in the potential use of chemoautotrophs as agents for release of valuable minerals from ores fluctuates with the value and geographic location of the metals.

The point is that we as scientists while working on basic metabolic or biochemical aspects of chemoautotrophs should not overlook the importance of applications and potential commercialization because that has been and is likely to continue to be a source of our research support base. Chemoautotrophy has enormous potential in the emerging field of biotechnology.

Patrick R. Dugan

LITERATURE CITED

Bisset, K. A., and J. B. Grace. 1954. The nature and relationship of autotrophic bacteria. *In* B. L. Fry and J. L. Peel (eds.), Autotrophic micro-organisms, pp. 28–53. Cambridge University, Press, Cambridge.

Werkman, C. H. 1951. Assimilation of carbon dioxide by heterotrophic bacteria. *In* C. H. Werkman and P. W. Wilson (eds.), Bacterial physiology, p. 190. Academic Press, New York.

Whittenbury, R., and D. P. Kelly. 1977. Autotrophy: a conceptual phoenix. *In* B. A. Haddock and W. A. Hamilton (eds.), Microbial energetics, pp. 121–49. Cambridge University Press, Cambridge.

Eighth Annual Biosciences Colloquium
College of Biological Sciences
Ohio State University
4–5 October 1982

MICROBIAL CHEMOAUTOTROPHY

Organizers

W. R. Strohl, Department of Microbiology, Ohio State University, 484 West Twelfth Avenue, Columbus, Ohio 43210

O. H. Tuovinen, Department of Microbiology, Ohio State University, 484 West Twelfth Avenue, Columbus, Ohio 43210

Contributors

Aleem, M. I. H., Thomas Hunt Morgan School of Biological Sciences, University of Kentucky, Lexington, Kentucky 40506

Brand, M. D., Department of Biochemistry, University of Cambridge, Tennis Court Road, Cambridge CB2 1GW, England

Cobley, J. G., Department of Chemistry, University of San Francisco, 2130 Fulton Street, San Francisco, California 94117

Corder, R. E., Solar Energy Research Institute, 1617 Cole Boulevard, Golden, Colorado 80401

Cox, J. C., Martin Marietta Laboratories, 1450 South Rolling Road, Baltimore, Maryland 21227

DiSpirito, A. A., Department of Genetics and Cell Biology, University of Minnesota, St. Paul, Minnesota 55108

Dugan, P. R., College of Biological Sciences, Ohio State University, 484 West Twelfth Avenue, Columbus, Ohio 43210

Ehrlich, H. L., Department of Biology, Rensselaer Polytechnic Institute, Troy, New York 12181

Frea, J. I., Department of Microbiology, Ohio State University, 484 West Twelfth Avenue, Columbus, Ohio 43210

Gibson, J., Section of Biochemistry, Molecular and Cell Biology, Division of Biological Sciences, Cornell University, Ithaca, New York 14853

Hamilton, P. T., Department of Microbiology, Ohio State University, 484 West Twelfth Avenue, Columbus, Ohio 43210

Hegeman, G., Microbiology Group, Indiana University, Bloomington, Indiana 47405

Hook, L. A., Department of Microbiology, Ohio State University, 484 West Twelfth Avenue, Columbus, Ohio 43210

Hooper, A. B., Department of Genetics and Cell Biology, University of Minnesota, St. Paul, Minnesota 55108

Jarrell, K. F., Division of Biological Sciences, National Research Council of Canada, Ottawa, Ontario, Canada K1A 0R6

Lidstrom-O'Connor, M. E., Department of Microbiology and Immunology, University of Washington, Seattle, Washington 98185

Matin, A., Department of Medical Microbiology, Stanford University, Stanford, California 94305

Park, Y., Microbiology Group, Indiana University, Bloomington, Indiana 47405

Peck, Jr., H. D., Department of Biochemistry, University of Georgia, Athens, Georgia 30602

Radmer, R., Martin Marietta Laboratories, 1450 South Rolling Road, Baltimore, Maryland 21227

Reeve, J. N., Department of Microbiology, Ohio State University, 484 West Twelfth Avenue, Columbus, Ohio 43210

Schmidt, T. M., Environmental Biology Program, Ohio State University, 1735 Neil Avenue, Columbus, Ohio 43210

Sewell, D. A., Veterans Administration Medical Center, Portland, Oregon 97225

Sprott, G. D., Division of Biological Sciences, National Research Council of Canada, Ottawa, Ontario, Canada K1A 0R6

Strohl, W. R., Department of Microbiology, Ohio State University, 484 West Twelfth Avenue, Columbus, Ohio 43210

Tuovinen, O. H., Department of Microbiology, Ohio State University, 484 West Twelfth Avenue, Columbus, Ohio 43210

OHIO STATE UNIVERSITY BIOSCIENCES COLLOQUIA

Microbial Chemoautotrophy

PATRICK R. DUGAN

Desulfurization of Coal by Mixed Microbial Cultures

1

The potential utility of acidophilic thiobacilli and other microorganisms in the leach mining of metals and in the desulfurization of coal is related to (1) the diminishing reserves of high-grade ores and (2) the geographic location of strategic reserves that in turn is related to world geopolitics. Table 1 shows projected life expectancies of world reserves of several mineral commodities at two different demand-rate scenarios. Of particular interest is the 34-year "life expectancy" of our available sulfur reserves. One could argue that even if there is a 100 to 200% error in the estimate, we should be developing alternatives to utilize low-grade sources or otherwise recover these strategic minerals from dilute sources. Table 2 illustrates the United States' dependence on imports from other countries for several strategic metals.

We have previously reported on the microbial desulfurization of pyritic sulfur from high-sulfur coal in which acidophilic and acid-tolerant microbes function to oxidize the insoluble iron pyrite found in coal to soluble ferric sulfate and sulfuric acid in accordance with the following generalized relations:

$$FeS_2 + 3\ O_2 + 2\ H_2O \rightarrow 2\ H_2SO_4 + Fe^{3+}$$

$$3\ H_2O + Fe^{3+} \rightarrow Fe(OH)_3 + 3\ H^+ .$$

Loss of coal pyritic sulfur had previously been equated to formation of soluble sulfate in slurry suspensions, and we had shown that the rate of sulfur oxidation was related to: (1) coal mesh size (smaller size = greater rate), (2) initial pH (2.0–2.5), (3) nutritional supplements (in this coal, NH_4^+ and

TABLE 1

LIFE EXPECTANCIES OF 1976 WORLD RESERVES OF SELECTED MINERAL COMMODITIES AT TWO DIFFERENT RATES OF DEMAND

	1976 Reserves	1976 Primary Demand	Projected Demand Growth Rate (%)	Life Expectancy in Years[a]	
				Static at 1976 Level	Growing at Projected Rates
Fluorine (million s.t.)	37	2.1	4.58	18	13
Silver (million troy oz.)	6,100	305.0	2.33	20	17
Zinc (million s.t.)	166	6.4	3.05	26	19
Mercury (thousand flasks)	5,210	239.0	0.50	22	21
Sulfur (million l.t.)	1,700	50.0	3.16	34	23
Lead (million s.t.)	136	3.7	3.14	37	25
Tungsten (million lbs.)	4,200	81.0	3.26	52	31
Tin (thousand m.t.)	10,000	241.0	2.05	41	31
Copper (million s.t.)	503	8.0	2.94	63	36
Nickel (million s.t.)	60	0.7	2.94	86	43
Platinum (million troy oz.)	297	2.7	3.75	110	44
Phosphate rock (million m.t.)	25,732	107.0	5.17	240	51
Manganese (million s.t.)	1,800	11.0	3.36	164	56
Iron in ore (billion s.t.)	103	0.6	2.95	172	62
Aluminum in bauxite (million s.t.)	5,610	18.0	4.29	312	63
Chromium (million s.t.)	829	2.2	3.27	377	80
Potash (million s.t.)	12,230	26.0	3.27	470	86

Note: Corresponding data for helium and industrial diamonds not available.
[a]Assumes no increase to 1976 reserves.
Source: After *Global 2000* Technical Report, Table 12-4, but with updated and corrected entries. Updated reserves and demand data from U.S. Bureau of Mines, *Mineral Trends and Forecasts, 1979.* Projected demand growth rates are from *Global 2000* Technical Report.

SOURCE: Adapted from ERIC Bulletin No. 1 (1982).

CO_2), (4) the optimum temperature range (between 25°C and 35°C), and (5) the preconditioning of mixed cultures (Dugan and Apel, 1977, 1978; Dugan, 1980).

MATERIALS AND METHODS

Thiobacillus ferrooxidans was grown under forced aeration at 23 ± 2°C in "9K" medium. After 5 days' incubation, the cells were harvested, washed, suspended in pH 3 H_2SO_4 solution, and stored as previously described. Cell suspensions were used as inoculum in 1 week or less after harvesting.

T. thiooxidans was grown on ATCC medium #450, harvested, and stored using the same method described for *T. ferrooxidans*.

T. acidophilus was cultured on "9K" glucose (0.2%) at pH 2.8, harvested, and stored as described.

TABLE 2

U.S. Import Dependence for Some Strategic Metals

Metal	Principal Industrial Uses	1980–81 Price Range	Import Dependence	Major Sources
Antimony	Constituent in alloys for type metals and metal bearings	$1.50–$1.70/lb.	53%	China, South Africa, Bolivia
Chromium	Alloyed with nickel in heat-resistant metals; alloyed with iron and nickel in stainless steel; also used in corrosion-resistant plating	$3.25–$4.50/lb.	91%	Soviet bloc, South Africa, Philippines
Cobalt	Alloyed with iron in ferroalloys; alloyed with chromium and iron in cobalt-chromium steel, which is used for valves for internal combustion engines	$13.50–$50.00/lb.	93%	Zaire, Zambia
Germanium	Component in manufacture of solid rectifiers or diodes in microwave detectors; used pure in transistors	$522.00–$1450.00/kg	11%	U.S., Soviet bloc, Belgium, Zaire
Indium	Ingredient in plating of lead-coated silver airplane bearings	$7.00–$20.00/troy oz.	NA	Soviet bloc, Canada, Japan
Manganese	Component of high-grade steel alloys and alloys of other metals	$0.65–$0.75/lb.	97%	U.S.S.R., South Africa, Gabon, Brazil
Rhodium	Component in silver plating and in thermocouples	$550.00–$900.00/ troy oz.	87%	South Africa, U.S.S.R.
Tantalum	Component in manufacture of corrosion-resistant apparatus for laboratories and in electronic equipment	$80.00–$120.00/lb.	97%	Canada, Brazil, Australia
Titanium	Alloying metal used extensively in aircraft construction	$16.00–$19.00/kg	14%	U.S.S.R., China, U.S., Australia

Data from Bache Halsey Stuart Shields, Sinclair Group, and ''Exploring Strategic Metals,'' by Paul Sarnoff, in *Moneymaker*, April/May 1981, pp. 41–45.

NA, information not available.

Source: ERIC Bulletin No. 1 (1982).

Pyrite enrichment cultures were obtained by inoculating slurries of pyrite adjusted to pH 3 with H_2SO_4 with acid mine drainage and incubating under the conditions described previously.

Coal slurries were prepared with a commercially pulverized coal (table 3) provided by the Columbus and Southern Ohio Electric Company. The slurries were adjusted to a pH of 2.5 with 10 N H_2SO_4, and 5 ml aliquots were dispensed in 250 ml Erlenmeyer flasks and sterilized by autoclaving at 20 p.s.i. for 25 min. After cooling to ambient temperature, duplicate flasks were inoculated with 1% inoculum (w/v) from individual preconditioned culture isolates. Duplicate flasks from each of the above pH slurries remained uninoculated and served as sterile controls. Both inoculated and sterile flasks were placed on a gyratory shaker (200 rpm) and incubated at 23 ± 2°C. At intervals during the incubation period, 1 ml aliquots were aseptically removed from each of the flasks and used to determine pH and sulfate concentration. The pH of the samples was determined utilizing a Corning model 12 expanded scale pH meter.

The procedure used for determination of sulfate was a modification of a barium sulfate turbidimetric procedure in which samples were collected, diluted in H_2O (if necessary), and treated with excess $BaCl_2$. The turbidity of each sample was determined in a Shimadzu model 14PS-50L spectrophotometer set at a wavelength of 450 nm. Percent transmission of each sample was recorded and compared with a standard sulfate curve prepared in the range of 0 to 600 mg l^{-1} of sulfate.

RESULTS

The sulfur content of various mesh-size fractions of pulverized coal before and after 14 days' treatment with enrichment cultures is shown in table 3. Caution must be exercised to compare the after-treatment values with the appropriate control mesh-size because the sulfur content of coal is not homogeneously distributed and therefore varies with mesh-size. Data are also presented for 20, 30, 40, and 50% slurries (weight of sub-200-mesh coal per volume of water). Rates of removal have been reported elsewhere (Dugan and Apel, 1977, 1978), and table 4 is a tabulation of calculated rates of pyritic sulfur removal in grams of sulfur removed per gram of coal per day from various mesh-sizes, and percentages slurries treated for various durations of time.

We have been able to substantiate that mixtures of pure cultures of acidophilic *Thiobacillus* species plus acidophilic heterotrophic species are capable of achieving rates of desulfurization that approach those of our enrichment cultures. Figure 1 illustrates the rate of SO_4^{2-} accumulation in the

TABLE 3

PERCENTAGE OF SULFUR BEFORE AND AFTER BIOLOGICAL TREATMENT
(Values are percent S by weight based on analysis of coal)

	Total S	Pyritic S	Organic S	SO_4^{2-}
CONTROL COAL BLEND				
(prior to treatment)	4.6	3.1	1.4	0.2
10% 50–100 mesh (0.297–0.149 mm)	–	–	–	–
45% 100–200 mesh (0.149–0.074 mm)	4.1	2.9	1.2	0.1
45% sub-200 mesh (< 0.074 mm)	5.4	4.2	1.0	0.2
AFTER TREATMENT				
100–200 (20% slurry)	1.8	0.1	1.6	0.1
sub-200 (20%)	2.0	0.5	1.3	0.2
sub-200 (20%) (sterile control)	5.2	–	–	–
BLEND (20%)				
BLEND (30%)	2.0	0.6	1.2	0.2
BLEND (40%)	2.6	1.0	1.4	0.2
BLEND (50%)	3.0	1.4	1.5	0.2
BLEND 50% + $(NH_4)_2SO_4$	2.9	1.1	1.5	0.3

SOURCE: Dugan and Apel (1977).

TABLE 4

CALCULATED RATES OF TOTAL SULFUR REMOVAL FROM SO_4^{2-} RELEASE DATA

	g S/g coal/day	Lb. S/ton coal/day
*20% slurry sub-200 (14 d)	0.0024	5.3
*20% slurry sub-200 + salts (18 d)	0.0021	4.45
20% slurry 100–200 (20 d)	0.0012	2.66
30% slurry blend (14 d)	0.0018	4.0
50% slurry blend + NH_4^+ (14 d)	0.0012	2.66
**If calculated during max release*		
20% slurry sub-200 (8 d)	0.0042	9.44
20% slurry sub-200 + salts (7 d)	0.0053	11.73

All removal was FeS_2 from coal which contained 84 lb S per ton coal (4.2% S).

SOURCE: Dugan (1980).

presence of pure cultures alone and as mixtures. These data verify that mixed cultures of autotrophic and heterotrophic acidophilic microbes are much more effective in oxidizing pyritic sulfur than are pure cultures alone. This is likely due to the heterotrophic removal of autotoxic organic by-products formed by the autotrophic bacteria.

Although most of our work has been carried out in either shaken flasks or 10-liter aerated batch fermenters in the laboratory, we believe that several

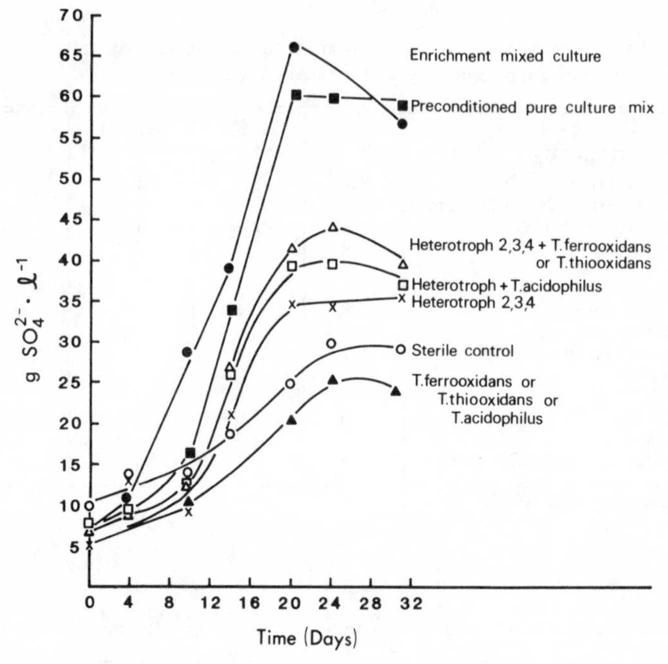

Fig. 1. Formation of sulfate ion from pyritic sulfur in a pulverized coal sample in the presence of various acidophilic microorganisms.

possibilities exist for scale-up that are operationally analogous to those employed for aerobic domestic or industrial waste treatment processes (for example, slurry lagoons or a counterpart to the activated sludge process).

Figure 2 is a schematic diagram of an aerated slurry process. Our data suggest that we should be able to hold up to 40% coal in suspension with aeration and agitation. The coal will settle out rapidly when agitation and aeration are terminated, providing for separation of the treated coal. The settled coal contains a high number of adsorbed microorganisms (in excess of 10^8 cells per gram) and in addition approximately 10^8 cells remain suspended per ml of supernatant. Return of 5 to 10% of the coal solids would provide an active preconditioned inoculum to the continuous process, and return of a percentage of the supernatant would provide additional inoculum as well as initial acidification of the untreated coal slurry.

The waste supernatant would have a high content of H_2SO_4 (approximately pH 2.2), Fe^{3+}, Ca^{2+}, Mg^{2+}, Al^{3+} and other cations leached from mineral

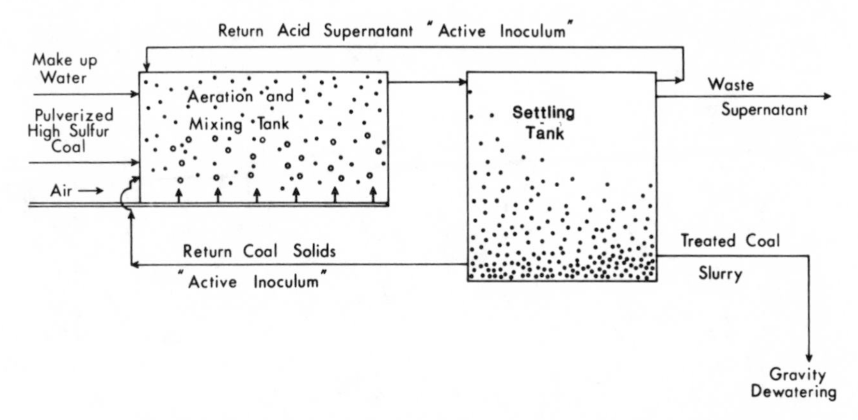

Fig. 2. Schematic microbial process for desulfurization of pulverized coal.

impurities in the coal. Total sulfate could reach as high as 7 to 8% as combined metal sulfate and sulfuric acid. This, of course, necessitates further treatment of the effluent prior to release into a receiving stream or recycle as make-up water.

In conclusion, the desulfurization of high-pyrite coal by a microbiological process is feasible, and in contrast to post-combustion techniques, its potential is greater with increasing sulfur content in the coal because the sulfur is the nutritional energy supply for the microorganisms that solubilize it. This method also has the advantage of leach removal of metals, thereby lowering the ash content and potential atmospheric particulate emissions.

LITERATURE CITED

Dugan, P. R. 1980. Microbial desulfurization of coal. Presented at 2d Chemical Congress of North American Continent, Division of Microbial and Biochemical Technology, Las Vegas, Nevada, August 26, 1980.

Dugan, P. R., and W. A. Apel. 1977. Microbiological removal of sulfur from a pulverized coal blend. *In* Proc. 3d symposium on coal preparation, pp. 11–21. National Coal Association/ Bituminous Coal Research Coal Conference and Expo IV, Louisville, Kentucky.

Dugan, P. R., and W. A. Apel. 1978. Microbiological desulfurization of coal. *In* L. E. Murr, A. E. Torma, and J. A. Brierley (eds.), Metallurgical applications of bacterial leaching and related microbiological phenomena, pp. 223–50. Academic Press, New York.

ERIC Bulletin No. 1. 1982. ERIC Clearinghouse for Science, Mathematics, and Environmental Education, Ohio State University, 1200 Chambers Road, Columbus, Ohio 43212.

ALAN A. DISPIRITO AND OLLI H. TUOVINEN

Oxidations of Nonferrous Metals by Thiobacilli

2

INTRODUCTION

Current commercial applications of the microbiological leaching of metals involve large-scale operations for the recovery of copper and uranium from ore materials. The predominant role of *Thiobacillus ferrooxidans* in the solubilization of metal sulfides has long been recognized (Tuovinen and Kelly, 1974a; Torma and Bosecker, 1982), but the mechanism through which this catalysis is carried out is still largely unknown in metal-leaching systems. Recently there also have been attempts to characterize the heterogeneous microbial populations and bacterial and chemical interactions in these acid processes (Norris and Kelly, 1982), but the role of chemoorganotrophs has not been elucidated in metal-leaching systems.

In the bacterial oxidation of iron sulfides, ferric iron is produced by oxidation of ferrous minerals. Soluble ferric iron acts as an electron acceptor for the chemical oxidation of sulfide minerals. The role of iron-oxidizing thiobacilli is then to reoxidize the reduced electron carrier, ferrous iron. Sulfuric acid is produced during the bacterial oxidation of sulfides, thereby accelerating the rate of metal solubilization. The acid also neutralizes carbonate materials in the mineral matrix. Incomplete oxidation of the sulfide entity commonly occurs in the acid leaching. This results in the formation of polythionates and the precipitation of elemental sulfur. The latter effectively coats metal sulfides and prevents their further oxidation until the sulfur film is removed by bacterial oxidation.

Studies with nonferrous metal sulfides have shown that *T. ferrooxidans* oxidizes the sulfide entity of ZnS, NiS, CoS, and other metal sulfides. Theoretically, a number of transition metals in addition to Fe may be biologically oxidizable, as also are actinides that occur in reduced form in minerals. If nonferrous metals are present in sulfides, it is extremely difficult to demonstrate whether bacteria oxidize also the metal entity parallel with sulfide oxidation. Natural sulfide minerals invariably contain at least trace amounts of iron, which cycles as a redox carrier in the presence of *T. ferrooxidans*. This can be exemplified with the following equations:

$$4\ Fe^{2+} + O_2 + 4\ H^+ \rightarrow 4\ Fe^{3+} + 2\ H_2O \qquad (1)$$

$$FeS_2 + 2\ Fe^{3+} \rightarrow 3\ Fe^{2+} + 2\ S \qquad (2)$$

$$2\ S + 3\ O_2 + 2\ H_2O \rightarrow 2\ SO_4^{2-} + 4\ H^+ \qquad (3)$$

$$FeS_2 + 3\frac{1}{2}\ O_2 + H_2O \rightarrow Fe^{2+} + 2\ SO_4^{2-} + 2\ H^+ \qquad (4)$$

$$2\ FeS_2 + 7\frac{1}{2}\ O_2 + H_2O \rightarrow 2\ Fe^{3+} + 4\ SO_4^{2-} + 2\ H^+ . \qquad (5)$$

Equation 2 describes the ferric-iron-dependent oxidation of pyrite; for other equations, the bacteria catalyze the reactions at much faster rates than observed in a solely chemical system.

Ferric iron also mediates the oxidation of other metal sulfides:

$$CuFeS_2 + 4\ Fe^{3+} \rightarrow Cu^{2+} + 5\ Fe^{2+} + 2\ S \qquad (6)$$

$$ZnS + 2\ Fe^{3+} \rightarrow Zn^{2+} + 2\ Fe^{2+} + S \qquad (7)$$

$$CuS + 2\ Fe^{3+} \rightarrow Cu^{2+} + 2\ Fe^{2+} + S . \qquad (8)$$

Oxygen does not have a direct role in these redox reactions because they involve only the redox carrier (i.e., Fe^{3+}/Fe^{2+}). The combined role of bacteria and oxygen is to regenerate ferric iron (equation 1) and to remove the precipitated sulfur (equation 3), as summarized in the following equations (the cyclic role of iron is not indicated):

$$CuFeS_2 + 4\frac{1}{2}\ O_2 + H^+ \rightarrow Cu^{2+} + Fe^{3+} + 2\ SO_4^{2-} + H_2O \qquad (9)$$

$$ZnS + 2\frac{1}{2}\ O_2 + 2\ H^+ \rightarrow Zn^{2+} + SO_4^{2-} + H_2O \qquad (10)$$

$$CuS + 2\frac{1}{2}\ O_2 + 2\ H^+ \rightarrow Cu^{2+} + SO_4^{2-} + H_2O . \qquad (11)$$

Unlike the transition metals, uranium occurs mainly as oxide minerals and not as sulfides. The oxidation of uraninite is relatively slow in sulfuric acid (equation 12), and ferric iron is commonly used to accelerate the oxidative dissolution (equation 13) as follows:

$$UO_2 + \tfrac{1}{2} O_2 + 2\,H^+ \rightarrow UO_2^{2+} + H_2O \tag{12}$$

$$UO_2 + 2\,Fe^{3+} \rightarrow UO_2^{2+} + 2\,Fe^{2+}\,. \tag{13}$$

Ferrous iron is sufficiently stable in acid solutions for use as a substrate in microbiological studies. The oxidation of Fe^{2+} by *T. ferrooxidans* has been recently summarized by Ingledew (1982) and Cox and Brand (1984). *T. ferrooxidans* avails itself to studies in acid solutions because of its ability to function at low pH values that are prohibitive to most other microorganisms. Many other reduced metal ions are extremely rapidly oxidized by oxygen in acid solutions, and hence their microbiological oxidations are not well characterized. Evidence has been presented in the literature for the oxidation of Cu(I), Sn(II), and U(IV) catalyzed by *T. ferrooxidans*. In addition, other reduced elements—for example, selenide (Torma and Habashi, 1972)—may be able to be oxidized by *T. ferrooxidans,* but their biological transformations need to be further examined for verification. Possible redox reactions that *T. ferrooxidans* may mediate include the oxidations of the anions of Mo and As, and of trivalent chromium, but no information is presently available on their transformations by *T. ferrooxidans*. In the present paper, emphasis is given to the oxidation of Cu(I) and U(IV), both of which are currently of commercial significance in microbiological leaching process. The oxidation of stannous tin was reported by Lewis and Miller (1977), but it has not been verified since then. Manganous ion, which is oxidized by many different microorganisms (Ehrlich, 1984), is not an electron donor for *T. ferrooxidans*. The lack of an active manganese oxidation system has been confirmed in manometric studies using manganous sulfate as the substrate for washed cell suspensions of *T. ferrooxidans* (A. A. DiSpirito and O. H. Tuovinen, unpublished results).

MICROBIOLOGICAL OXIDATION OF CU(I)

The monovalent copper (as CuCl) is soluble in acid solutions, but its chemical oxidation is extremely rapid, thereby preventing the use of Cu^+ in manometric studies at the pH range suitable for *T. ferrooxidans*. Cuprous ion oxidation was tentatively demonstrated by Lewis and Miller (1977) by using it as an electron donor for cytochrome *c*. No oxygen uptake measurements coupled with biological Cu^+ oxidation have been reported because of the rapid chemical oxidation. Theoretically, the oxidation may be represented by the following equation:

$$2\ Cu^+ + \frac{1}{2}\ O_2 + 2\ H^+ \rightarrow 2\ Cu^{2+} + H_2O\ . \quad (14)$$

Copper occurs as Cu(I) in sulfide minerals such as chalcocite. Partial oxidation of chalcocite results in the formation of covellite:

$$2\ Cu_2S + \frac{1}{2}\ O_2 + 2\ H^+ \rightarrow Cu^{2+} + 2\ CuS + H_2O\ . \quad (15)$$

Covellite is oxidized directly by *T. ferrooxidans*:

$$2\ CuS + 5\ O_2 + 4\ H^+ \rightarrow 2\ Cu^{2+} + 2\ SO_4^{2-} + 2\ H_2O\ . \quad (16)$$

Nielsen and Beck (1972) demonstrated the incomplete oxidation of chalcocite and suggested that only the copper entity was oxidized by *T. ferrooxidans* since no increase in sulfate concentration could be detected. Conversely, no net change in the sulfur balance in the solid phase was observed. The data also indicated that carbon dioxide was assimilated by *T. ferrooxidans* during chalcocite oxidation. Both digenite (Cu_9S_5) and covellite were detected as oxidation products in the solid phase. Iron, which constituted 0.17% of the initial mineral sample, was not analyzed in the acid reaction mixture and thus the cyclic role of iron, as schematically presented in figure 1, could not be ruled out.

Imai and colleagues (1973) studied the biological oxidation of chalcocite using a sample that was washed free of acid-soluble Fe and acetone-soluble S. Intact cells of *T. ferrooxidans* oxidized chalcocite with a stoichiometry of about 1 O_2 for each 2 Cu^{2+} detected in the reaction mixture. The same

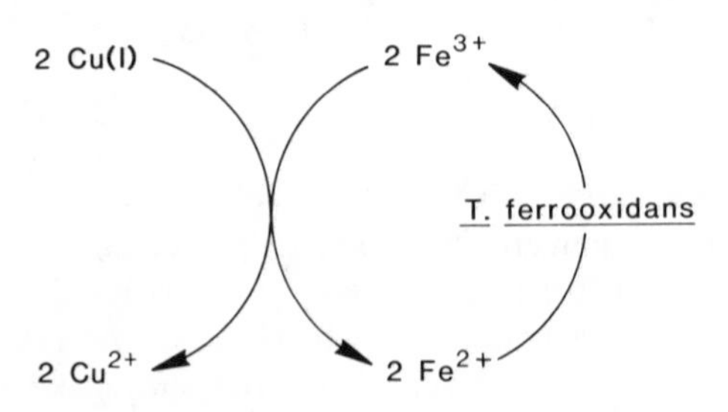

Fig. 1. Indirect oxidation of cuprous copper by *T. ferrooxidans* with iron as an electron carrier.

stoichiometry prevailed in the presence of added ferric iron (0.25–60 mM Fe^{3+}), but the oxidation rate was doubled, the largest rate increase being observed in the lower range of iron concentration. No change in the stoichiometry would be expected because added ferric iron serves as a redox carrier between Cu(I) and *T. ferrooxidans* while the net number of electrons transferred in the reaction remains the same.

Equation 17 was used by Nielsen and Beck (1972) to explain the observed stoichiometry of 2 Cu^{2+}/1 O_2:

$$2\ Cu_2S + 2\ H_2SO_4 + O_2 \rightarrow 2\ CuS + 2\ CuSO_4 + 2\ H_2O\ . \qquad (17)$$

Both copper atoms in Cu_2S are oxidized, but the solution and solid phase both contain CuII, and thus it is of doubtful value to monitor the stoichiometry of oxidation by measuring soluble Cu^{2+}. Should all copper remain soluble, then the determination of Cu^{2+} would serve to measure the solubilization based on the stoichiometry of 4 Cu^{2+}/1 O_2 (equation 14).

Imai and colleagues (1973) presented the following equation to describe incomplete oxidation of chalcocite:

$$Cu_2S + 2\ H_2SO_4 + O_2 \rightarrow 2\ CuSO_4 + S + 2\ H_2O\ . \qquad (18)$$

This reaction has a stoichiometry of 2 Cu^{2+}/1 O_2. This is an apparent ratio with respect to copper because it also involves the oxidation of the sulfide entity to elemental sulfur. If sulfur were completely oxidized (equation 3), the stoichiometric oxygen consumption would be decreased to a ratio of 0.8 Cu^{2+}/1 O_2. The solubilization of Cu_2O by *T. ferrooxidans* was reported by Imai and colleagues (1973), but no data were given on the parallel oxygen uptake. Assuming that 1 O_2 is required for the oxidation of 4 Cu(I) (i.e., 2 Cu_2O), the copper solubilization rates reported by Imai and colleagues (1973) represent an oxygen uptake rate of 3.25 μmol O_2 h^{-1} mg $protein^{-1}$. This compares well with the estimated rate of 3.75 μmol O_2 h^{-1} mg $protein^{-1}$ for Cu_2S oxidation (Imai et al., 1973); for the latter calculation, 50% of the observed uptake was presumed to be due to the oxidation of Cu(I), the other 50% accounting for the parallel oxidation of sulfide to elemental sulfur.

Golding and colleagues (1974) used a Cu(I)-bromine analogue of bis(piperidylidithiocarbamato)Cu(II) $[(CuS_2CNC_5H_{10})_2(CuBr)_4]$ in growth studies of *T. ferrooxidans*. Soluble Cu^{2+} was detected in the culture medium, but no experimental values or other details were presented. Thus, it is not possible to evaluate the significance of the release of copper from the substrate complex. Golding and colleagues (1974) suggested that organic Cu-compounds, such as the Cu(I)-Cu(II) mixture that they studied, may find use in microbiological studies of copper oxidation because of the chemical stability of organic copper complexes as opposed to the extremely rapid chemical oxidation of Cu^+.

Ehrlich (1978) discussed chalcocite oxidation by *T. ferrooxidans*. The results were inconclusive because the role of iron contamination could not be excluded. The H^+/Cu^{2+} ratios varied from 1.6 to 2.7, indicating a mixed-type oxidation that cannot be evaluated from the data summarized by Ehrlich (1978).

OXIDATION OF U(IV)

The oxidation of uranium by *T. ferrooxidans* was believed to be indirect with soluble iron acting as a redox carrier (see fig. 2). The catalytic role of *T. ferrooxidans* in regenerating ferric iron in uranium leaching systems has been demonstrated by several authors (Harrison et al., 1966; Derry et al., 1976; Manchee, 1977; Tuovinen et al., 1983).

The oxidation of uranous compounds in the absence of added iron has been investigated by microcalorimetric and manometric techniques (Soljanto and Tuovinen, 1980; DiSpirito and Tuovinen, 1981, 1982a). The sequence of reactions was proposed to be the following:

$$2\ U^{4+} + O_2 + 4\ H^+ \rightarrow 2\ U^{6+} + 2\ H_2O \quad (19)$$

$$2\ U^{6+} + 4\ H_2O \rightarrow 2\ UO_2^{2+} + 8\ H^+ \quad (20)$$

$$2\ U^{4+} + O_2 + 2\ H_2O \rightarrow 2\ UO_2^{2+} + 4\ H^+ . \quad (21)$$

Equation 19 describes the direct oxidation of U^{4+}; in acid solutions U^{6+} forms a hydrolysis product (equation 20), and therefore equation 21 is used to describe the net reaction.

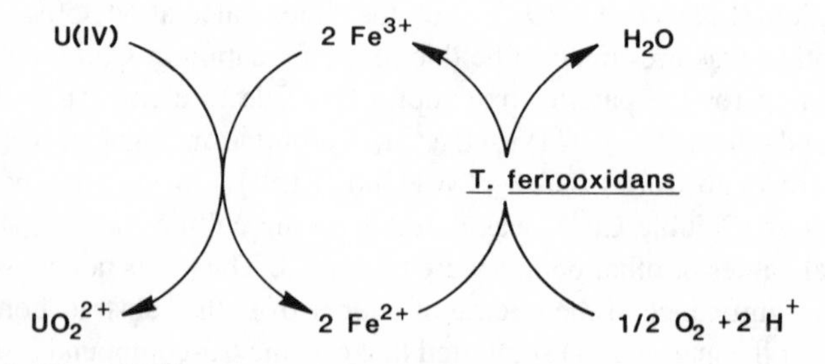

Fig. 2. Indirect oxidation of uranous uranium by *T. ferrooxidans* with iron as an electron carrier.

The first attempt to demonstrate the direct oxidation of uranium was based on microcalorimetric measurement of heat liberated during oxidation of UO_2 by *T. ferrooxidans* (Soljanto and Tuovinen, 1980). Heat evolution was not observed with UO_2 alone nor was it detected upon addition of boiled cells of *T. ferrooxidans*. The reaction mixture contained 1.72 μmol total Fe ml^{-1} of which 0.32 μmol Fe ml^{-1} was soluble, the rest being bound and incorporated into cellular material. Thus, the results may have represented a combination of the direct and indirect oxidation of UO_2 by *T. ferrooxidans* and Fe^{3+}. Similarly, Ivarson's (1980) demonstration of uranous sulfate oxidation was inconclusive because the contaminating iron may have recycled as an electron carrier between U^{4+} and the bacteria.

In order to demonstrate the direct oxidation of U(IV) by *T. ferrooxidans*, the presence of iron must be minimized in the reaction mixture. Iron is an important component in several enzymes and electron transport carriers and can never be completely excluded from a reaction mixture containing viable cells. In the case of *T. ferrooxidans*, the cell-bound iron can be reduced to approximately 8 nmol total Fe mg cell protein^{-1} for studying the Fe-independent oxidation of uranous ion.

The rates of oxygen uptake by washed cell suspensions of *T. ferrooxidans* were dependent on the concentration of U^{4+}, previous growth history, and cell density of the organism (DiSpirito and Tuovinen, 1981). Strains of *T. ferrooxidans* that had developed resistance to high levels of uranium oxidized U^{4+} faster than strains that had not previously been exposed to uranium. Subculturing the uranium-resistant strains in the absence of uranium resulted in a loss of resistance and a decreased rate of oxidation (DiSpirito and Tuovinen, 1982b). These relationships were indicative of biological activity, but did not exclude the possibility of iron recycling as the major route of uranium oxidation.

Manometric studies of uranous ion oxidation by *T. ferrooxidans* indicated a stoichiometry of 2.15 U^{4+}/1 O_2, which was 107% of the theoretical value of 2 U^{4+}/1 O_2 as indicated in equations 19 and 21 (DiSpirito and Tuovinen, 1982a). However, the stoichiometric measurements of uranous ion oxidation do not distinguish between the direct and Fe-dependent oxidation because the number of electrons transferred to *T. ferrooxidans* to reduce oxygen to water remains the same in both cases.

On the other hand, kinetic studies of uranous ion oxidation indicated that *T. ferrooxidans* may directly oxidize U^{4+}. The V_{max} for iron oxidation was shown to be approximately thirty times greater than that determined for uranous ion oxidation (DiSpirito and Tuovinen, 1982b). Similarly, the K_m^{app} values were approximately ten times higher for Fe^{2+} compared with U^{4+}.

Such large differences in the kinetic parameters suggested the presence of separate systems for Fe^{2+} and U^{4+}. Mixed substrate studies indicated that uranous ion was predominantly a competitive inhibitor of iron oxidation (DiSpirito and Tuovinen, 1982b).

If the differences in the kinetic parameters of uranous ion and ferrous iron oxidation were due to limiting concentrations of iron, there should be a proportional increase in the rate of O_2 uptake with increasing concentrations of iron. However, no rate increase was observed for uranous ion oxidation in the presence of $\leq 0.5\ \mu M\ Fe^{2+}$ in the reaction mixture (DiSpirito and Tuovinen, 1982b). The reaction mixtures used in these studies contained less than 50 nmol total Fe ml^{-1} as impurity. Thus, the results indicated that an increase in the iron concentration of approximately 20–100 times the contaminating iron was required before an increase in the oxygen uptake rate was observed. These results also indicated that the recycling of trace levels of iron does not appear to be a major factor in the microbiological oxidation of U^{4+} in these reaction mixtures.

The fixation of $^{14}CO_2$ was monitored during U^{4+} oxidation, and the results illustrated that energy was derived by *T. ferrooxidans* from the oxidation of U^{4+} to support growth-related functions (DiSpirito and Tuovinen, 1982a). The amount of $^{14}CO_2$ incorporated into cell material per number of electrons was approximately the same with either U^{4+} of Fe^{2+} as electron donors, but on an equimolar substrate basis, twice the amount of $^{14}CO_2$ was fixed with U^{4+} as the substrate when compared with Fe^{2+}. Since both the oxygen uptake and carbon dioxide fixation were coupled with uranium oxidation by *T. ferrooxidans*, this also suggested the forward electron flow to O_2 and the reverse electron flow to $NAD(P)^+$ in the electron transport chain.

Rusticyanin has been proposed to be the initial electron acceptor for Fe^{2+} oxidation (Ingledew et al., 1977; Cox and Boxer, 1978). The spectrophotometric assay of rusticyanin is based on the reversal of the reaction, i.e., the oxidation of rusticyanin by Fe^{3+}. As with ferric iron, uranyl ion reduction by rusticyanin in cell-free extracts was demonstrated (DiSpirito and Tuovinen, 1982a). The absorption peak of the oxidized form of rusticyanin was abolished upon the addition of Fe^{2+}, whereas with U^{4+} it could not be demonstrated because of the absorbance of uranous ion in the 400–680 nm range of the spectrum.

The use of electron transport inhibitors indicated the entry of electrons from Fe^{2+} and U^{4+} at the cytochrome *c* level (DiSpirito and Tuovinen, 1982a). With U^{4+} as the electron donor, many of the inhibitors indicated an uncoupling pattern that was much less pronounced with Fe^{2+}. At low concentrations, uranium may act as a weak ionophore and thus promote the uncoupling effects of respiratory inhibitors. Previously, it had been shown that at high concentrations (> 10 mM) UO_2^{2+} uncoupled $^{14}CO_2$ fixation from the Fe^{2+}-dependent O_2 uptake in a manner similar to that of 2,4-dinitrophenol (Tuovinen and Kelly, 1974b). Complete inhibition of both O_2 uptake and $^{14}CO_2$ fixation coupled with uranium and iron oxidation was observed with KCN and NaN_3 (DiSpirito and Tuovinen, 1982a).

The direct oxidation of uranous ion by *T. ferrooxidans* was proposed in view of the kinetic parameters of uranium and iron oxidation, the oxidation of rusticyanin by UO_2^{2+}, and the lack of increase in the rate of uranium oxidation upon the addition of trace levels of iron (DiSpirito and Tuovinen, 1982b). A direct reduction of membrane-bound Fe^{3+} by uranous ion cannot be ruled out because components of the uranium oxidation system have not been isolated from *T. ferrooxidans* for redox studies with U^{4+}.

Thiobacillus acidophilus, a related bacterium but incapable of growing with Fe^{2+} as an energy source, was able to oxidize uranous ion (DiSpirito and Tuovinen, 1981). After growth on tetrathionate, the cells oxidized both iron and uranium; after growth on glucose, the cells lost the ability to directly oxidize U^{4+} but the Fe^{2+}-oxidation activity was retained. This lends support for U^{4+}-oxidation system that is not associated with Fe^{2+}-oxidation. On the other hand, the electron transport chain is probably similar for both U^{4+}- and Fe^{2+}-oxidation. In *T. ferrooxidans* the electron transport is coupled with energy transduction, whereas in *T. acidophilus* the iron oxidation is not coupled to the growth of the bacteria.

In the present paper, the oxidation of uranous ion is examined further. In this work, the concentration of iron was minimized by either growing *T. ferrooxidans* with tetrathionate or by extensive washings of iron-grown cells in 0.11 N H_2SO_4. The level of iron was less than 50 nmol Fe ml^{-1} of reaction mixtures containing 1 mg cell protein ml^{-1}. Ferrous iron was used in some experiments as a standard reference substrate.

Comparison of the Chemical and Bacterial Oxidation of U^{4+}

Figure 3 shows that above 1.75 mM U^{4+} concentrations the chemical oxidation rate was concentration-dependent at pH 1.5. The addition of washed cell suspensions of *T. ferrooxidans* (1 mg cell protein ml reaction $mixture^{-1}$) increased the rate of U^{4+}-oxidation, and a saturation effect was observed. Below 1 mM U^{4+} the chemical oxidation was negligible whereas the rate of bacterial oxidation was almost linearly dependent on the substrate concentration. The increased rate of oxygen uptake between 1.75 and 4 mM U^{4+} was primarily attributed to the accelerated rate of chemical oxidation.

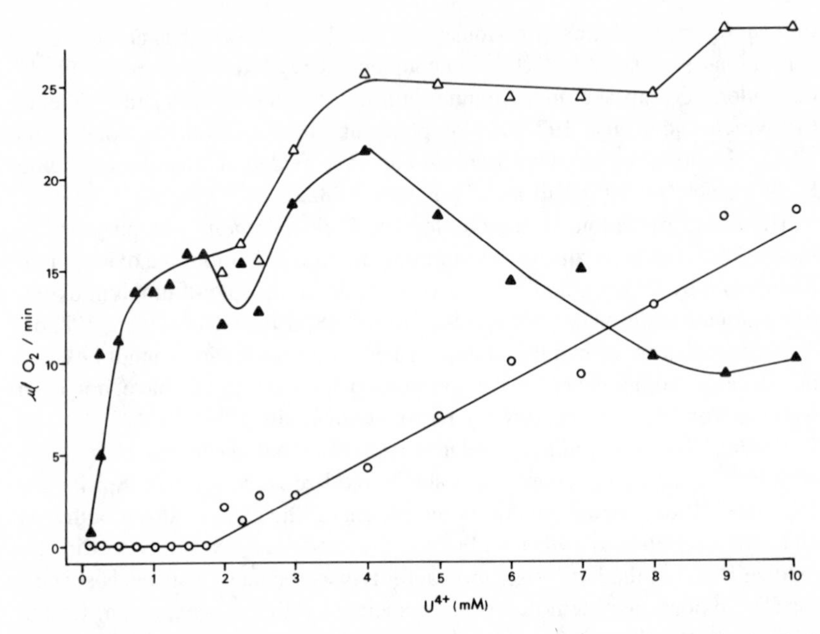

Fig. 3. Effect of U^{4+}-concentration on the rate of oxygen uptake at pH 1.5; ○, chemical oxidation; △, oxidation by *T. ferrooxidans* strain TFI-35 (1 mg protein ml^{-1}); ▲, differential rate (bacterial *minus* chemical oxidation).

Kinetics of U^{4+}- and Fe^{2+}-Oxidation by T. ferrooxidans

Silver (1978) suggested that *T. ferrooxidans* may not be a distinct species but rather a group of metabolically similar bacteria. Ferrous iron oxidation is the key diagnostic character for all isolates, but several differences have been reported in morphological characteristics (DiSpirito et al., 1982) and in the guanine and cytosine content of DNA of different strains (Guay et al., 1976). Genetic heterogeneity among iron-oxidizing thiobacilli was also reported by Harrison (1982). Variation in the resistance to metals in different strains of *T. ferrooxidans* was reported by DiSpirito and Tuovinen (1982b). Based on the resistance to uranyl ion, three strains were selected for kinetic analysis of U^{4+}-oxidation; these included a sensitive and a resistant strain and one with intermediate resistance. The Lineweaver-Burk plots of uranium and iron oxidation by the uranium-sensitive strain TFI-35 have been published previously (DiSpirito and Tuovinen, 1982b). The V_{max} of Fe^{2+}-oxidation was 2.23 μmol Fe^{2+} oxidized min^{-1} mg $protein^{-1}$ with an apparent K_m of 1.51 mM. The rate of uranium oxidation showed a V_{max} and K_m^{app} of 0.077 μmol

U^{4+} oxidized min^{-1} mg protein $^{-1}$ and 0.132 mM, respectively. Thus, the rate of Fe^{2+}-oxidation was 29 times faster than the rate of uranium oxidation. The Lineweaver-Burk plots of uranium and iron oxidation by the resistant strain TFI-1 (resistant to 3.0 mM UO_2^{2+}) and strain TFI-29 (resistant to 1 mM UO_2^{2+}) are shown in figures 4 and 5, respectively. The V_{max} and K_m^{app} of iron oxidation by TFI-1 were 2.47 μmol Fe^{2+} min^{-1} mg $protein^{-1}$ and 1.58 mM, respectively, and for uranium oxidation 0.056 μmol U^{4+} min^{-1} mg $protein^{-1}$ and 0.031 mM. The V_{max} and K_m^{app} of iron oxidation by TFI-29 were 2.13 μmol Fe^{2+} min^{-1} mg $protein^{-1}$ and 1.41 mM; the respective values for uranium oxidation by *T. ferrooxidans* were 0.069 μmol U^{4+} min^{-1} mg $protein^{-1}$ and 0.126 mM. As for TFI-35, the rate of iron oxidation by TFI-1 and TFI-29 was 31 to 44 times faster than the rate of uranium oxidation. Except for the low K_m^{app} for uranium oxidation by TFI-1, there was little difference in the kinetic parameters between the three strains examined.

Mixed Substrate Studies

Below 2.5 mM concentrations, uranous ion acts as a competitive inhibitor of iron oxidation (DiSpirito and Tuovinen, 1982b). At higher concentrations, a mixed type of inhibition is observed as illustrated in figure 6. At high iron concentrations the inhibition by U^{4+} of Fe^{2+}-oxidation was proportional to the concentration of U^{4+} (fig. 6A,B). A mixed type of inhibition was observed as the concentration of Fe^{2+} decreased and that of U^{4+} increased (fig. 6C). At low concentrations of Fe^{2+}, the oxygen uptake rate was similar to that determined for U^{4+} alone (fig. 6D,E).

Spectrophotometric Determination of Uranium Oxidation

In addition to manometric measurements of oxygen uptake, the oxidation of U^{4+} could also be monitored spectrophotometrically. Figure 7A shows the absorption spectrum of U^{4+} and UO_2^{2+} at the 350–700 nm range. The oxidation of U^{4+} could be monitored by the decrease in the absorption spectrum above 475 nm; the peak of the absorption spectrum was detected at 650.2 nm. Figure 7B shows an example of the spectrophotometric assay to monitor the oxidation of U^{4+} by *T. ferrooxidans* at pH 1.5. The rate-limiting factor in this experiment was the aeration; in the experiment illustrated in figure 7B, oxygen was introduced between the scans by mixing the contents of the cuvette by means of a Pasteur pipette. Based on the absorption spectrum of U^{4+}, the spectrophotometric method may be developed to a continuous assay system, but continuous aeration by means of a stirrer appears to

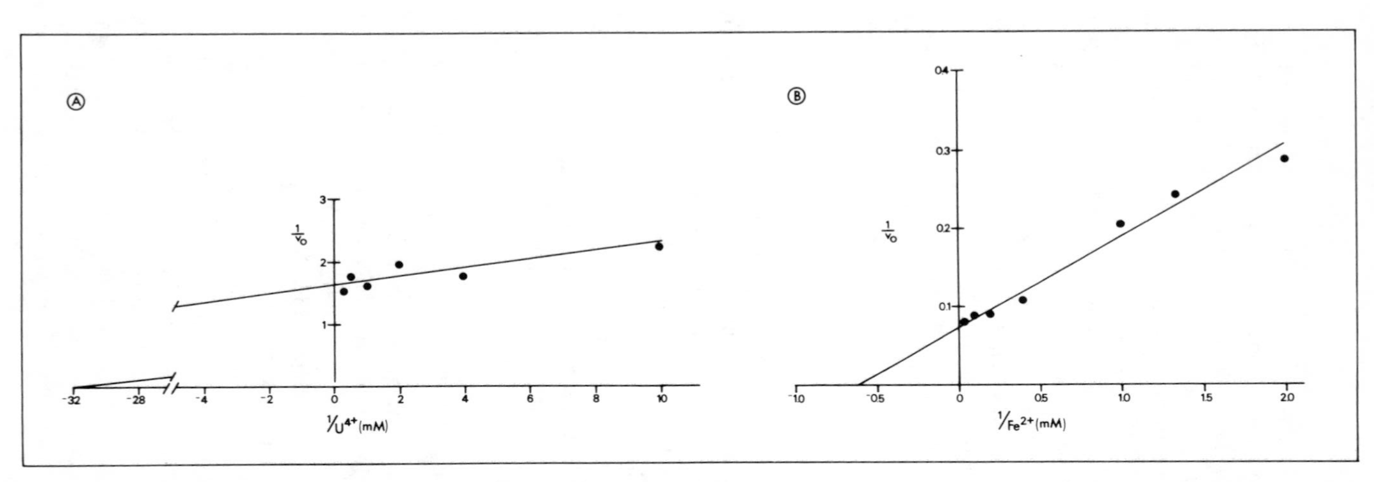

Fig. 4. Lineweaver-Burk plot of uranium oxidation (A) and iron oxidation (B) by *T. ferrooxidans* strain TFI-1. S is given as mM concentration of the substrate and v_o as μmol substrate oxidized min^{-1} mg $protein^{-1}$.

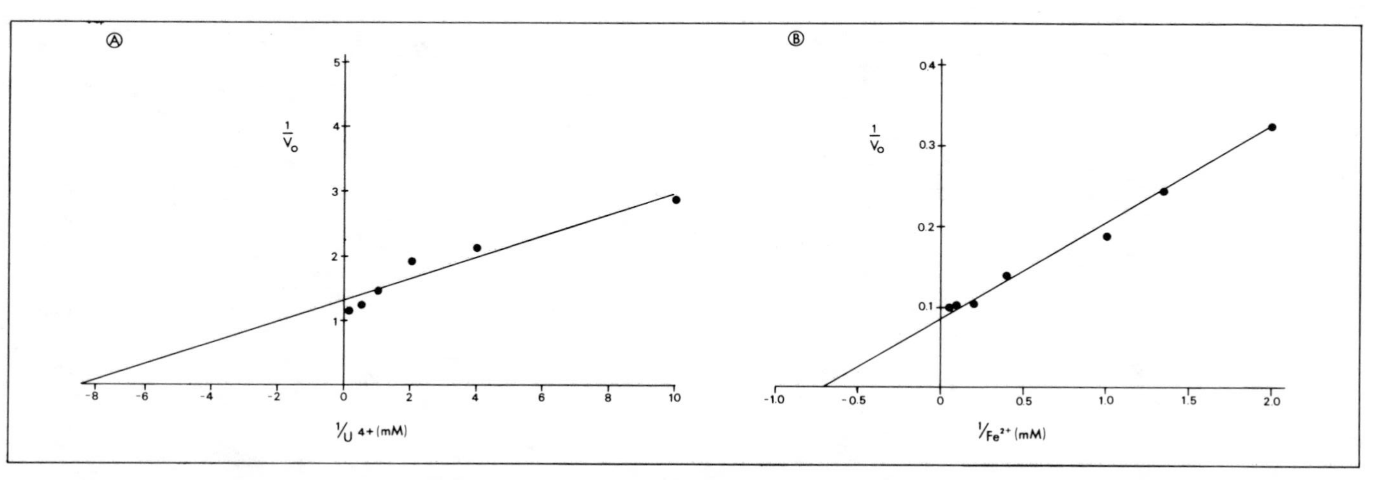

Fig. 5. Lineweaver-Burk plot of uranium oxidation (A) and iron oxidation (B) by *T. ferrooxidans* strain TFI-29. S and v_o as in fig. 4.

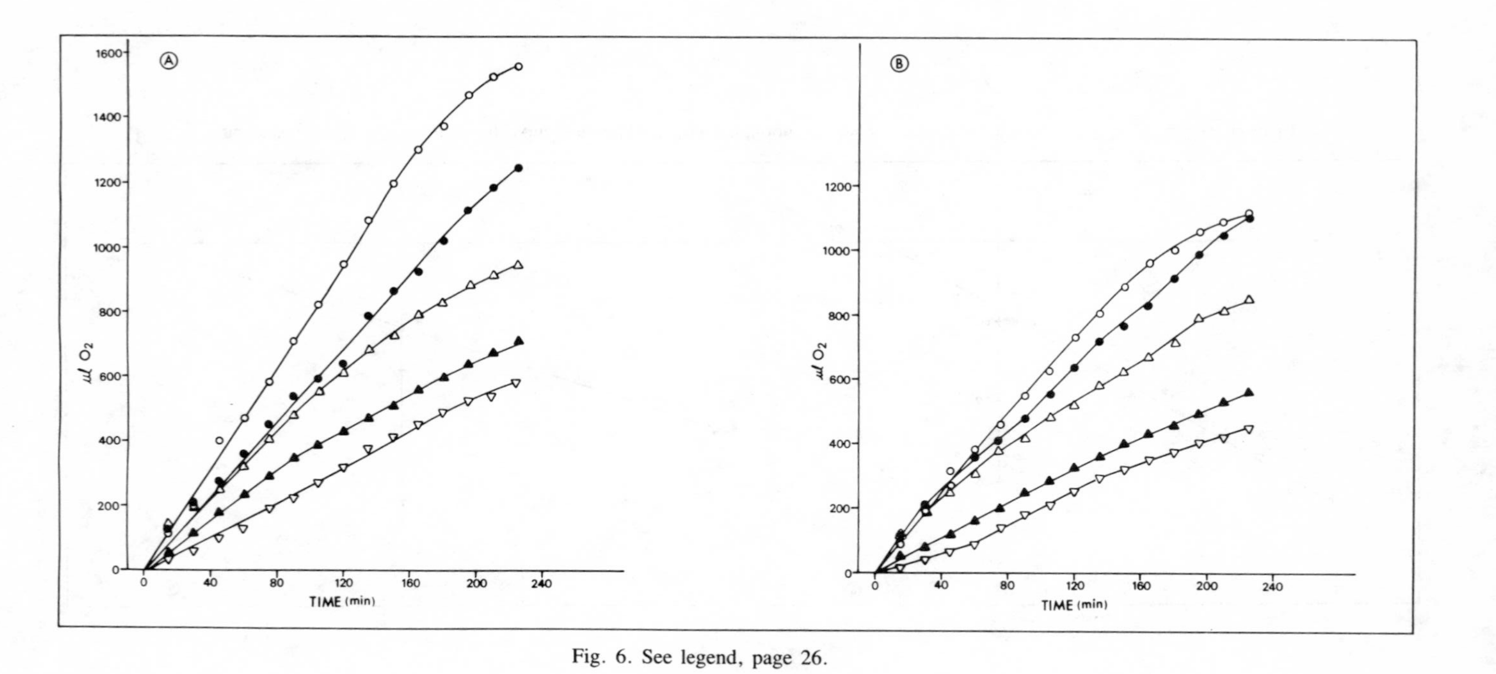

Fig. 6. See legend, page 26.

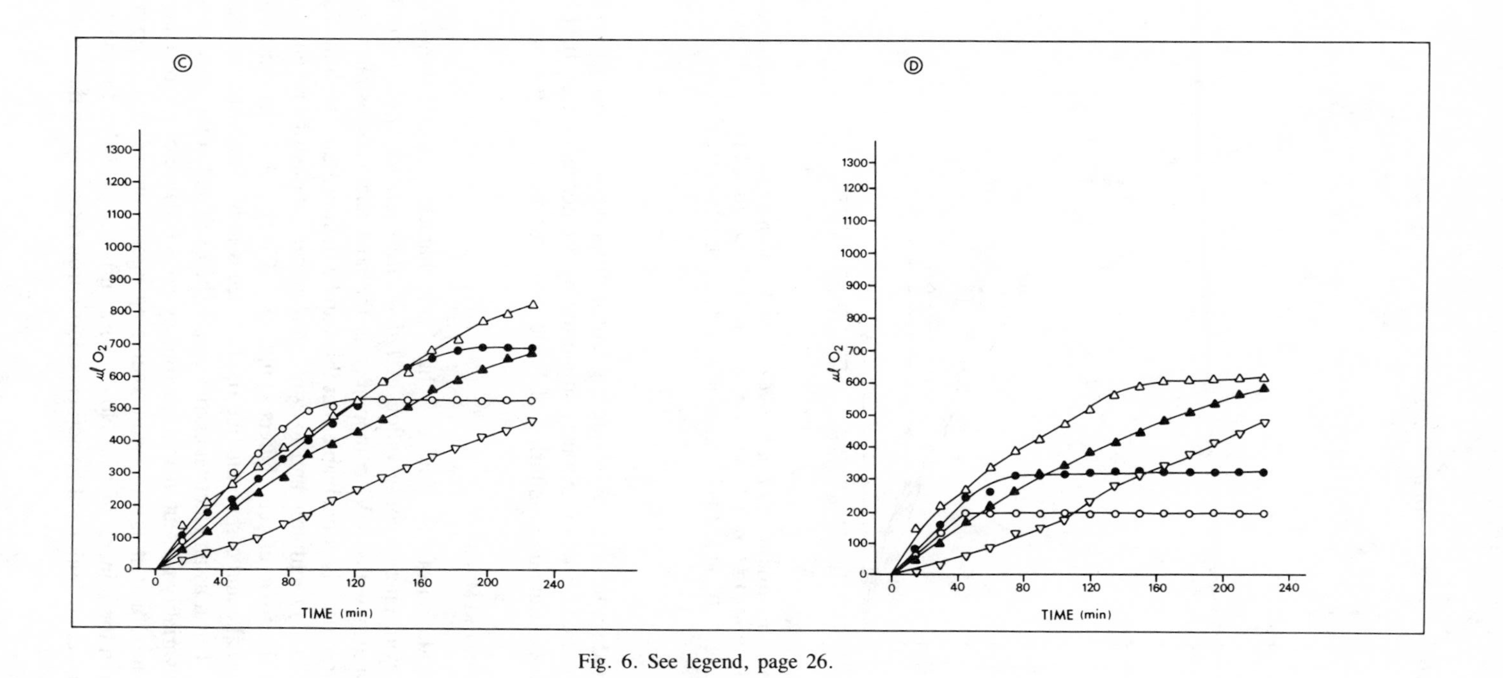

Fig. 6. See legend, page 26.

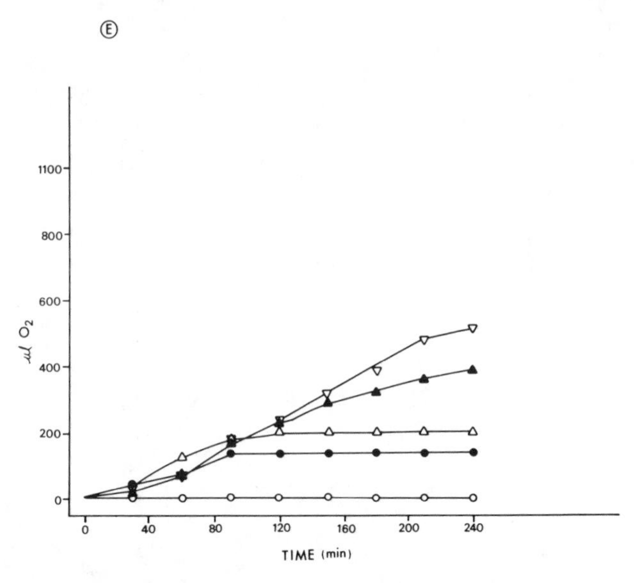

Fig. 6. Ferrous iron oxidation by *T. ferrooxidans* strain TFI-35 in the presence of uranous sulfate at pH 1.5; ○, 0 mM U^{4+}; ●, 0.25 mM U^{4+}; △, 5 mM U^{4+}; ▲, 10 mM U^{4+}; ▽, 15 mM U^{4+}. A, 75 mM Fe^{2+}; B, 50 mM Fe^{2+}; C, 25 mM Fe^{2+}; D, 10 mM Fe^{2+}; E, 0 mM Fe^{2+}.

be required for this type of monitoring. A spectrophotometric assay technique for iron oxidation was previously suggested by Steiner and Lazaroff (1974), based on continuous monitoring of ferric iron absorbance at 304 nm.

CONCLUDING REMARKS

The oxidation of Fe^{2+} and U^{4+} has been characterized and shown to be coupled with energy conservation in *Thiobacillus ferrooxidans*. Growth of this organism with U(IV)-compounds as the sole energy substrate has not been reported. It is conceivable that this can be demonstrated with the use of continuous-flow cultures and low-influent concentrations of uranous ion to alleviate the toxic effects. Direct coupling of Cu(I) and Sn(II) oxidation to oxygen uptake and carbon dioxide fixation needs to be demonstrated because the role of iron as an electron carrier cannot be excluded at the present time. Iron is invariably present in acid leaching systems because of its abundance in sulfide ores, and it is the prime oxidizing agent for reduced copper and uranium in the mineral matrix. No data are currently available to resolve the individual contributions of bacteria and ferric iron recycling on the solubiliza-

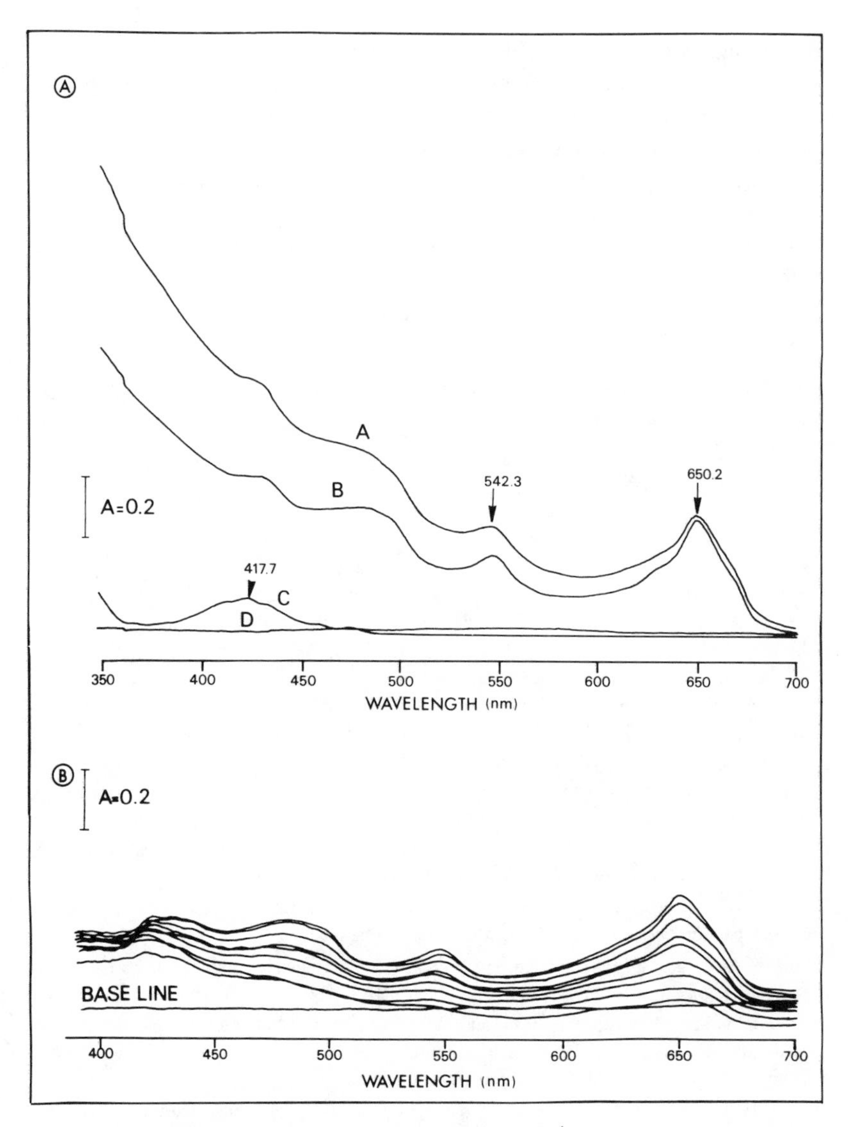

Fig. 7A. Absorption spectrum of uranium sulfate. A, 2.5 mM U^{4+} at pH 1.5; B, 2.5 mM U^{4+} at pH 2.5; C, 2.5 mM UO_2^{2+}.

Fig. 7B. Absorption spectrum of 2.5 mM uranous sulfate at pH 1.5 in the presence of *T. ferrooxidans* strain TFI-35 (1 mg protein ml^{-1}). The spectra were recorded at approximately 2.5 min intervals and show a successive decrease at 650.2 nm of the spectrum.

tion of metal sulfides in acid leaching processes. For the examination of individual metal oxidations in systems free of iron and reduced S-compounds and because of the low pH requirement for the stability of these metal cations, synthetic lipid membranes may be useful as matrix of isolated electron transport components. To better understand the chemical and microbiological processes in metal leaching systems, more data are required on the kinetics of both mixed (e.g., FeS_2, $CuFeS_2$, Cu_2S) and single (e.g., UO_2, CuCl) substrates.

ACKNOWLEDGEMENT

Partial support (O.H.T.) of the work was received from the Office of Water Research and Technology (Grant No. 14-34-0001-2137).

LITERATURE CITED

Cox, J. C., and D. H. Boxer. 1978. The purification and some properties of rusticyanin, a blue copper protein involved in iron (II) oxidation from *Thiobacillus ferro-oxidans*. Biochem. J. 174:497–502.

Cox, J. C., and M. D. Brand. 1984. Iron oxidation and energy conservation in the chemautotroph *Thiobacillus ferrooxidans*. *In* W. R. Strohl and O. H. Tuovinen (eds.), Microbial chemoautotrophy, pp. 31–46. Ohio State University Press, Columbus.

Derry, R., K. H. Garrett, N. W. Le Roux, and S. E. Smith. 1976. Bacterially assisted plant process for leaching uranium ores. *In* M. J. Jones (ed.), Geology, mining and extractive processing of uranium, pp. 56–62. Institution of Mining and Metallurgy, London.

DiSpirito, A. A., and O. H. Tuovinen. 1981. Oxygen uptake coupled with uranous sulfate oxidation by *Thiobacillus ferrooxidans* and *T. acidophilus*. Geomicrobiol. J. 2:275–91.

DiSpirito, A. A., and O. H. Tuovinen. 1982a. Uranous ion oxidation and carbon dioxide fixation by *Thiobacillus ferrooxidans*. Arch. Microbiol. 133:28–32.

DiSpirito, A. A., and O. H. Tuovinen. 1982b. Kinetics of uranous ion and ferrous iron oxidation by *Thiobacillus ferrooxidans*. Arch. Microbiol. 122:33–37.

DiSpirito, A. A., M. Silver, L. Voss, and O. H. Tuovinen. 1982. Flagella and pili of iron-oxidizing thiobacilli isolated from a uranium mine in northern Ontario, Canada. Appl. Environ. Microbiol. 43:1196–1200.

Ehrlich, H. L. 1978. Inorganic energy sources for chemolithotrophic and mixotrophic bacteria. Geomicrobiol. J. 1:65–83.

Ehrlich, H. L. 1984. Different forms of bacterial manganese oxidation. *In* W. R. Strohl and O. H. Tuovinen (eds.), Microbial chemoautotrophy, pp. 47–56. Ohio State University Press, Columbus.

Golding, R. M., A. D. Rae, B. J. Ralph, and L. Sulligoi. 1974. A new series of polynuclear copper dithiocarbamate-copper halide polymers. Inorg. Chem. 13:2499–2504.

Guay, R., M. Silver, and A. E. Torma. 1976. Base composition of DNA isolated from *Thiobacillus ferrooxidans* grown on different substrates. Rev. Can. Biol. 35:61–67.

Harrison, V. F., W. A. Gow, and M. R. Hughson. 1966. I. Factors influencing the application of bacterial leaching to a Canadian uranium ore. J. Metals 18:1189–94.

Harrison, A. P. 1982. Genomic and physiological diversity amongst strains of *Thiobacillus ferrooxidans,* and genomic comparison with *Thiobacillus thiooxidans*. Arch. Microbiol. 131:68–76.

Imai, K., H. Sakaguchi, T. Sugio, and T. Tano. 1973. On the mechanism of chalcocite oxidation by *Thiobacillus ferrooxidans*. J. Ferment. Technol. 51:865–70.

Ingledew, W. J., J. C. Cox, and P. J. Halling. 1977. A proposed mechanism for energy conservation during Fe^{2+} oxidation by *Thiobacillus ferro-oxidans:* chemiosmotic coupling to net H^{+} influx. FEMS Microbiol. Lett. 2:193–97.

Ingledew, J. W. 1982. *Thiobacillus ferrooxidans*: the bioenergetics of an acidophilic chemolithotroph. Biochim. Biophys. Acta 683:89–117.

Ivarson, K. C. 1980. Enhancement of uranous-ion oxidation by *Thiobacillus ferrooxidans*. Curr. Microbiol. 3:253–54.

Lewis, A. J., and J. D. A. Miller. 1977. Stannous and cuprous ion oxidation by *Thiobacillus ferrooxidans*. Can. J. Microbiol. 23:319–24.

Manchee, R. J. 1977. Laboratory-scale bacterially assisted leaching of Canadian uranium ores. Trans. Instn Min. Metall. C80:126–33.

Nielsen, A. M., and J. V. Beck. 1972. Chalcocite oxidation and coupled carbon dioxide fixation by *Thiobacillus ferrooxidans*. Science 175:1124–26.

Norris, P. R., and D. P. Kelly. 1982. The use of mixed microbial cultures in metal recovery. *In* A. T. Bull and J. H. Slater (eds.), Microbial interactions and communities, pp. 443–74. Academic Press, New York.

Silver, M. 1978. Metabolic mechanisms of iron-oxidizing thiobacilli. *In* L. E. Murr, A. E. Torma, and J. A. Brierley (eds.), Metallurgical applications of bacterial leaching and related microbiological phenomena, pp. 3–17. Academic Press, New York.

Soljanto, P., and O. H. Tuovinen. 1980. A microcalorimetric study of U(IV)-oxidation by *Thiobacillus ferrooxidans* and ferric-iron. *In* P. A. Trudinger, M. R. Walter, and B. J. Ralph (eds.), Biogeochemistry of ancient and modern environments, pp. 469–75. Springer Verlag, New York.

Steiner, M., and N. Lazaroff. 1974. Direct method for continuous determination of iron oxidation by autotrophic bacteria. Appl. Microbiol. 28:872–80.

Torma, A. E., and F. Habashi. 1972. Oxidation of copper (II) selenide by *Thiobacillus ferrooxidans*. Can. J. Microbiol. 18:1780–81.

Torma, A. E., and K. Bosecker. 1982. Bacterial leaching. Progr. Ind. Microbiol. 16:77–118.

Tuovinen, O. H., and D. P. Kelly. 1974a. Use of micro-organisms for the recovery of metals. Int. Metall. Rev. 19:21–31.

Tuovinen, O. H., and D. P. Kelly. 1974b. Studies on the growth of *Thiobacillus ferrooxidans*. III. Influence of uranium, other metal ions, and 2:4-dinitrophenol on ferrous iron oxidation and carbon dioxide fixation by cell suspensions. Arch. Microbiol. 95:165–80.

Tuovinen, O. H., P. Hiltunen, and A. Vuorinen. 1983. Solubilization of phosphate, uranium, and iron from apatite- and uranium-containing rock samples in synthetic and microbiologically produced acid leach solutions. Eur. J. Appl. Microbiol. Biotechnol. 17:327–33.

JOHN C. COX AND MARTIN D. BRAND

Iron Oxidation and Energy Conservation in the Chemoautotroph *Thiobacillus ferrooxidans*

3

INTRODUCTION

In experiments designed to explain the mechanism by which living cells synthesize ATP from ADP and inorganic phosphate, unique microorganisms such as the chemoautotrophic bacteria have been studied only to a small extent (Peck, 1968; Trudinger, 1969; Kelly, 1971; Suzuki, 1974; Schlegel, 1975; Aleem, 1977). Yet such microorganisms are worthy of consideration because they provide a stringent test of any general hypothesis for the mechanism of energy conservation.

The purpose of this article is to summarize briefly the literature on energy conservation by the chemoautotrophic acidophile *Thiobacillus ferrooxidans* and to suggest how ADP phosphorylation is driven by the small amount of energy available to the organism during Fe^{2+} oxidation.

T. FERROOXIDANS: AN ACIDOPHILIC CHEMOAUTOTROPH

T. ferrooxidans has been isolated from various acidic, mineral habitats and is one of the most widely studied examples of chemoautotrophic bacteria (e.g., Leathen et al., 1953, 1956; Silverman and Lundgren, 1959; Beck, 1960; Silverman and Ehrlich, 1964; Dugan and Lundgren, 1965; Apel and Dugan, 1978). The organism is able to grow using Fe^{2+} or reduced sulfur compounds as the sole source of energy, and CO_2 as the sole source of cell carbon (Lundgren et al., 1964, 1972; Peck, 1968). The optimal pH for growth is in the range pH 1.5–3.5, except during growth on elemental sulfur when the pH optimum is pH 5.5 (McGoren et al., 1969). It has also been reported that *T. ferrooxidans* grows heterotrophically, at pH 7.0, with glucose serving as the carbon and energy source (Doelle, 1975). However, this finding has been disputed, and it has been suggested that *T. ferrooxidans* exists in very close association with the facultative chemoautotroph *Thiobacillus acidophilus,* which would be selected for at neutral pH (Guay and Silver, 1975; Tuovinen et al., 1978; Arkesteyn and de Bont, 1980; Harrison et al., 1980). Generally, *T. ferrooxidans* is considered to be an obligate autotroph.

CHARACTERIZATION OF THE IRON-OXIDASE OF *T. FERROOXIDANS*

The iron-oxidase system of *T. ferrooxidans* is a membrane-bound enzyme complex (Peck, 1968) that spans the cytoplasmic membrane of the organism (Ingledew et al., 1977, 1978; Ingledew and Cobley, 1980). The Fe^{2+}-oxidizing portion of the respiratory chain is short, consisting of only four redox proteins, namely, rusticyanin (a copper-containing protein) and three cytochromes (Cobley and Haddock, 1975; Ingledew et al., 1977; Ingledew and Cobley, 1980). It is envisioned that these redox proteins form a single span across the cytoplasmic membrane (fig. 1), and that electrons produced by the oxidation of Fe^{2+} on the periplasmic face of the cytoplasmic membrane (at pH 2.0) are transferred across the membrane and consumed on the inner face by the action of cytochrome oxidase (at pH 7.0; Ingledew et al., 1977, 1978; Ingledew and Cobley, 1980). Quinones, *b*-type cytochromes, and iron-sulfur proteins occur in these bacteria, but it is assumed that they are

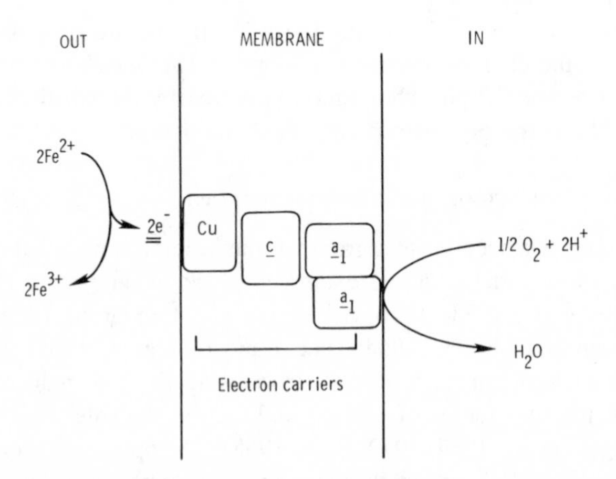

Fig. 1. Components of the iron oxidase of *T. ferrooxidans*. The components are identified by their prosthetic groups and are arranged from left to right in order of increasing redox potential (Cobley and Haddock, 1975; Ingledew and Cobley, 1980). "Out" and "in" refer to the bulk phase and cytoplasm, respectively.

involved in reverse electron flow for the purpose of NAD^+ reduction (Cobley and Haddock, 1975; Aleem, 1977; Ingledew and Cobley, 1980).

Regarding the components of the iron-oxidase complex, the copper protein rusticyanin is thought to be the initial acceptor of electrons from Fe^{2+} (Ingledew et al., 1977; Cox and Boxer, 1978). The midpoint potential (E_m) of rusticyanin has been measured to be 680 mV at pH 3.2 (Ingledew and Cobley, 1980). During growth of *T. ferrooxidans*, the electrical potential (E_h) of the Fe^{2+}/Fe^{3+} couple increases from approximately 500 mV to 800 mV as Fe^{2+} is oxidized by the cells.

When the electrode potential of the growth medium is greater than 750 mV, the concentration of Fe^{2+} in the medium is less than 5 mM. Yet during continuous culture of *T. ferrooxidans*, the steady-state E_h of the growth medium is 770 mV (Ingledew et al., 1977). Two theories have been advanced to explain how Fe^{2+} oxidation, and hence growth, can occur at such low substrate concentrations: (1) the presence of binding proteins in the membrane, or periplasmic space, which interact with rusticyanin and effectively scavenge Fe^{2+} from the growth medium; and (2) the following reaction between iron and rusticyanin has been observed by monitoring EPR signals of the copper center (Cox et al., 1978) and by monitoring oxidation/reduction of rusticyanin spectrophotometrically (J. C. Cox and D. H. Boxer, unpublished observations):

$$\text{rusticyanin } (Cu^{2+}) + Fe^{2+} \rightleftharpoons \text{rusticyanin } (Cu^{+}) + Fe^{3+} \ .$$

The K_m for Fe^{2+} during rusticyanin reduction is 0.5 mM, whereas the K_m for Fe^{3+} during reoxidation of rusticyanin is approximately 30 mM (using $FeCl_3$ in solution as the oxidant). This means that rusticyanin itself is an extremely efficient scavenger of Fe^{2+}. Growth of *T. ferrooxidans* is probably limited when the Fe^{2+} concentration of the growth medium falls below the level that reacts sufficiently rapidly with rusticyanin.

Bearing in mind the kinetic data above, one can conclude that reoxidation of rusticyanin by Fe^{3+} in the growth vessel would be slow. We propose that reoxidation is driven by an Fe^{3+}-protein complex that increases the rate of the reverse reaction. Ingledew and Cobley (1980) observed in *T. ferrooxidans* a protein with properties very similar to other Fe^{3+}-binding proteins.

Regarding the other components of the Fe^{2+}-oxidase system, Cobley and Haddock (1975) demonstrated that Fe^{2+} reduces cytochromes of *c* and *a* types in membranes of *T. ferrooxidans*. On the basis of kinetic evidence, a pathway of electron flow from Fe^{2+} to O_2 similar to that shown in figure 1 has been suggested by Ingledew and Cobley (1980).

OXIDATIVE PHOSPHORYLATION: A GENERAL MECHANISM FOR ATP FORMATION FROM ADP + PI

The chemiosmotic hypothesis of Peter Mitchell (Mitchell, 1968, 1976, 1977, 1979, 1981) has been largely accepted as a universal mechanism accounting for oxidative phosphorylation and other membrane-linked energetic functions (Harold, 1972, 1977; Haddock and Hamilton, 1977; Haddock and Jones, 1977). According to this hypothesis, the phosphorylation of ADP requires a coupling membrane containing a series of oxido-reductase enzymes that catalyze the transfer of electrons from an oxidizable substrate of low redox potential to a terminal electron acceptor of high redox potential (fig. 2). As a result of the electron flow, a gradient of H^+ is set up across the cytoplasmic membrane such that the cell matrix becomes alkaline with respect to the external medium. Thus a proton electrochemical gradient ($\Delta\bar{\mu}_{H^+}$) is created, and the energy available in this system is used to drive ADP phosphorylation by the action of a reversible H^+-translocating ATPase (fig.

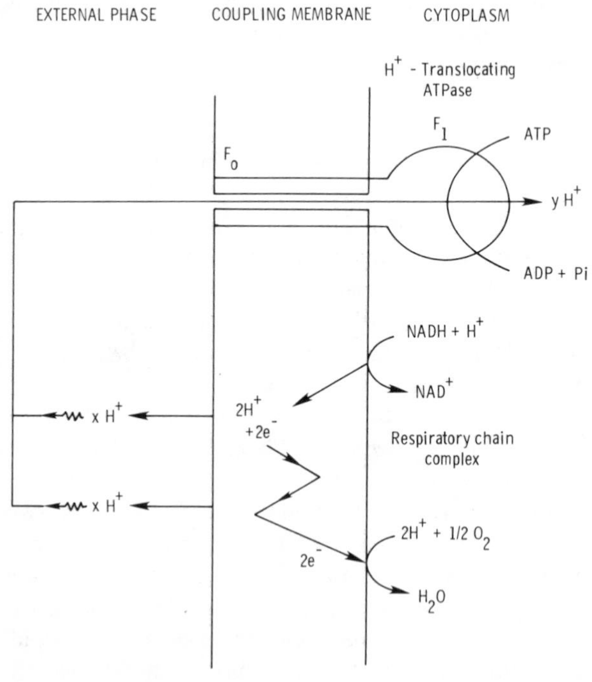

Fig. 2. Scheme for oxidative phosphorylation driven by electron flow in a bacterial cytoplasmic membrane.

2). Hence the transmembrane $\Delta\bar{\mu}_{H^+}$ is the coupling factor between electron flow and ATP synthesis.

The energy available from respiratory chain driven electron transport can be quantitated in the following way:

$$\Delta G = -nF\Delta E \tag{1}$$

where n = the number of electrons transferred between two redox couples of potential difference ΔE; F = the Faraday constant (23.06 kcal mol^{-1}) and ΔG = the Gibbs free energy. This electromotive force drives H^+ transloca-tion producing the $\Delta\bar{\mu}_{H^+}$. Thus, at equilibrium the $\Delta\bar{\mu}_{H^+}$ is poised against the electromotive force:

$$x\Delta\bar{\mu}_{H^+} = -nF\Delta E \tag{2}$$

where x is the number of H^+ translocated as n electrons pass between a redox couple of potential difference ΔE ($\rightarrow H^+/e^-$ ratio).

The amount of energy required to synthesize ATP from ADP + Pi is called the phosphorylation potential, ΔGp:

$$\Delta Gp = \Delta G^{\circ\prime} + 2.30RT \log_{10} \frac{[ATP]}{[ADP]\ [Pi]} . \tag{3}$$

During electron transport, the $\Delta\bar{\mu}_{H^+}$ is poised against the force required to phosphorylate ADP. Thus, the energy available from electron transport must be greater than or equal to the phosphorylation potential, i.e.,

$$y\Delta\bar{\mu}_{H^+} \geq \Delta Gp \tag{4}$$

where y is the number of H^+ translocated per ADP phosphorylated to ATP ($\rightarrow H^+/{\sim}P$ ratio). Assuming a value of +455 mV for ΔGp (Alberty, 1969), then for the synthesis of ATP by oxidative phosphorylation, $y\Delta\bar{\mu}_{H^+}$ must be ≥ -455 mV (equation 4). Measurements of $\Delta\bar{\mu}_{H^+}$ have been made on several types of bacteria, and estimates have usually fallen in the range of 120–230 mV (Haddock and Jones, 1977; Harold, 1972, 1977; Jones, 1979; Thauer et al., 1977). Therefore, if $\Delta\bar{\mu}_{H^+}$ were 230 mV, an $H^+/{\sim}P$ ratio of two would be required to synthesize ATP from ADP + Pi.

$\Delta\bar{\mu}_{H^+}$ is the coupling factor between electron flow and ATP synthesis. Clearly, because electron flow is ultimately the source of energy for oxidative phosphorylation, sufficient electromotive force (emf) must be generated by the respiratory chain to satisfy the phosphorylation potential.

ENERGY TRANSDUCTION IN HETEROTROPHIC ORGANISMS

Typically, the respiratory chains of heterotrophic organisms obtain electrons at low redox potential from primary dehydrogenases in which $NADH_2$ or $FADH_2$ serve as coenzymes. There is a large amount of potential energy available from redox reactions involving primary dehydrogenases when O_2 serves as the terminal electron acceptor (table 1). The chemiosmotic hypothesis explains how the energy available from intramembrane redox reactions can be used to drive the vectorial translocation of H^+ across the cytoplasmic membrane in order to create a $\Delta\bar{\mu}_{H^+}$ (Garland, 1977; Harold, 1972; Mitchell, 1968). At pH 7.0, in the resting state, the pH gradient (ΔpH) across the cytoplasmic membrane is approximately zero.

ENERGY TRANSDUCTION IN *T. FERROOXIDANS*

Fe^{2+}/Fe^{3+} is an electropositive redox couple (Ingledew et al., 1977; Lazaroff, 1963, 1977), and since O_2 reduction occurs inside the cell (pH 7.0), the potential energy available for energy conservation in *T. ferrooxidans* is small (table 1). *T. ferrooxidans* grows optimally at pH 2.0 during Fe^{2+} oxidation, yet the internal pH of this organism is close to neutrality (Ingledew et al., 1977; Cox et al., 1979; Apel et al., 1980). Therefore, the organism faces a pH gradient (ΔpH) of 4–5 units across its plasma membrane. It has

TABLE 1

OXIDATION/REDUCTION POTENTIALS ($E^{\circ\prime}$) AND $\Delta G^{\circ\prime}$ FOR SELECTED ENERGY SOURCES

System	$E^{\circ\prime}$ (volts)	$\Delta G^{\circ\prime} = -nF\Delta E^{\circ\prime}$[a] (kcal/mole)
½ O_2/H_2O	0.82	.....
$NAD^+/NADH_2$	−0.32	52.53
$FAD/FADH_2$	−0.22	47.96
Pyruvate/lactate	−0.18	46.12
Oxaloacetate/malate	−0.17	45.65
Fumarate/succinate	0.03	36.43
$2NO_2^-/2NO$	0.35[b]	21.67
$2NO_3^-/2NO_2^-$	0.42	18.45
$S_2O_4^{2-}/S_2O_3^{2-}$	0.48	15.67
$S_2O_3^{2-}/S^o$	0.50	14.76
$2Fe^{2+}/2Fe^{3+}$	0.77	2.31
$2Fe^{2+}/2Fe^{3+}$	0.65[c]	7.84

SOURCE: Data obtained from CRC Handbook of Biochemistry, 2d ed. (1970), pp. J34–J39.
[a]n = 2. Oxygen is terminal electron acceptor.
[b]Thauer et al. (1977).
[c]At pH 2.0 in 200 mM SO_4^{2-} (Ingledew et al., 1977; Lazaroff, 1963, 1977).

been suggested that this gradient could be used by the organism to drive oxidative phosphorylation; the H^+ entry required by the H^+-translocating ATPase is driven by the ΔpH that exists as a result of the acidic environment in which *T. ferrooxidans* lives (Ingledew et al., 1977; Cox et al., 1979; Apel et al., 1980). If the organism were permeable to H^+ in this fashion, the internal pH would quickly become acidic (≃ pH 2.0), and the organism would die.

Cox and colleagues (1979) have measured the components of $\Delta\bar{\mu}_{H^+}$ in *T. ferrooxidans* (table 2), and their data indicate (in keeping with results from experiments with other acidophiles) that the internal pH remains near neutrality as the external pH is varied (Hsung and Haug, 1975; Krulwich et al., 1978; Matin et al., 1982). In addition, Apel and colleagues (1980) have shown that pH gradients of this magnitude can drive ATP synthesis in *T. ferrooxidans* membrane vesicles. It was also observed that metabolic poisons did not cause the internal pH of the cell to rise, though they did decrease the magnitude $\Delta\bar{\mu}_{H^+}$ (Cox et al., 1979). These results have been interpreted in the following way in the mechanism proposed by Ingledew and colleagues (1977). There are two basic questions to be considered:

1. How does *T. ferrooxidans* remove H^+ from the cytoplasm in order to maintain an internal pH of near neutrality?

2. How does the organism regulate H^+ influx?

Fe^{2+} oxidation in *T. ferrooxidans* involves two half reactions (Ingledew et al., 1977; Apel and Dugan, 1978):

$$\text{(a)} \quad 2\,Fe^{2+} \rightarrow 2\,Fe^{3+} + 2\,e^- \qquad (5)$$

$$\text{(b)} \quad \tfrac{1}{2}\,O_2 + 2\,H^+ + 2\,e^- \rightarrow H_2O \qquad (6)$$

$$\text{Sum} \quad 2\,Fe^{2+} + \tfrac{1}{2}\,O_2 + 2\,H^+ \rightarrow 2\,Fe^{3+} + H_2O \qquad (7)$$

$$\Delta G^{\circ\prime} = -2.31 \text{ kcal mol}^{-1} .$$

TABLE 2

$\Delta\bar{\mu}_{H^+}$ IN *T. FERROOXIDANS*

External pH	E°′ Internal pH	Δψ (mV)	ΔpH (mV)	$\Delta\bar{\mu}_{H^+}$ (mV)
1.0	6.8	−70	342.2	272.2
2.0	6.5	−10	265.5	255.5
3.0	6.7	0	218.3	218.3

SOURCE: Data obtained from Cox et al. (1979).

Therefore, Fe^{2+} oxidation results in net H^+ consumption. The ΔpH provides 265 mV of $\Delta\bar{\mu}_{H^+}$ (table 2), and electron flow provides approximately 120 mV if 2 electrons are transferred from Fe^{2+} to O_2 (table 1). All the energy required for ATP synthesis in the chemoautotroph *T. ferrooxidans* is obtained from the pH gradient that is imposed upon the organism by the nature of its environment. Protons enter the cell by the H^+-translocating ATPase (driven by a ΔpH of 256 mV), thus catalyzing ADP phosphorylation. These protons are then consumed in the oxidase reaction (equation 6). The oxidase reaction is driven in the forward direction by the ΔE_h (120 mV) from the respiratory chain, thereby accounting for the H^+ consumption. In addition, because metabolic poisons do not collapse the ΔpH, there must be a passive (nonmetabolic) mechanism for regulation of H^+ entry, possibly through a Donnan potential (Hsung and Haug, 1975, 1977a, 1977b; Cox et al., 1979). In such a case, the influx of H^+ would quickly result in the buildup of an opposing charge within the cell, thereby preventing further H^+ entry. In *T. ferrooxidans*, a redox potential of approximately 120 mV is available when two electrons pass from Fe^{2+} to O_2 (Ingledew et al., 1977), whereas the pH gradient supplies a potential of 266 mV. These authors suggest that because electron flow is not the driving force required to set up the pH gradient by H^+ translocation, the energy produced from electron flow does not necessarily have to balance the energy available from the pH gradient (Ingledew et al., 1977).

THE THERMODYNAMICS OF FE^{2+} OXIDATION BY *T. FERROOXIDANS*

Initially the above mechanism appears to disagree both with the chemiosmotic hypothesis and with the laws of thermodynamics. However, certain minor modifications to this mechanism would satisfy both problems. Ingledew and colleagues (1977) showed that during growth of *T. ferrooxidans* the steady-state electrode potential of the Fe^{2+}/Fe^{3+} couple was 770 mV, and they used this value to calculate the potential difference across the respiratory chain (ΔE_h). If one uses the E_h of the growth medium to calculate the ΔE_h for the respiratory chain, then the emf available from the respiratory chain would decrease as the organism grows and oxidizes Fe^{2+}. Table 3 shows the results of an experiment where the medium E_h was monitored as a function of cell growth and how the respiratory chain emf would vary if indeed the bulk phase E_h were used in its calculation. The results show that under normal growth conditions the ΔE_h (respiratory chain) would decrease as the E_h (Fe^{2+}/Fe^{3+}) becomes more positive.

The chemiosmotic theory postulates that in steady-state oxidative phosphorylation, $\Delta E_h \geq x\Delta\bar{\mu}_{H^+}$ (Mitchell, 1968); therefore, $\Delta\bar{\mu}_{H^+}$ should

TABLE 3

VARIATION OF E_h OF THE MEDIUM DURING GROWTH OF *T. FERROOXIDANS*

Time (h)	Dry wt. of cells (mg/100 ml)	E_h (mV)	Calculated ΔE_h for respiratory chain[b]
0	.....	535	...
24	0.462	562	258
48	1.201	575	245
72	2.769	625	195
96	3.693	655	165
120	4.300	675	145
144	7.077	760	60
155	8.120	790	30
168	8.250	800	20
196	8.160	805	15

[a]The cells were grown in a 5-liter batch culture using the medium described in Cobley and Haddock (1975), and were aerated vigorously (1.5–2.0 ℓ min^{-1}). Redox potentials were measured by means of a platinum electrode with a calomel reference. One hundred ml samples were taken, filtered through dried, preweighed Millipore filters and dry weights taken after 2 days at 80°C.

[b]Using a value of 820 mV for ½ O_2/H_2O.

change as ΔE_h (respiratory chain) changes. Notably, when Cox and colleagues (1979) measured the parameters of $\Delta\bar{\mu}_{H^+}$ in *T. ferrooxidans*, they used a freshly made solution of $FeSO_4$ ($E_h \simeq 550$ mV). Thus, under these experimental conditions the ΔE_h (respiratory chain) would have been 280 mV when the observed $\Delta\bar{\mu}_{H^+}$ was 260 mV.

An experiment was performed in which the parameters of $\Delta\bar{\mu}_{H^+}$ in *T. ferrooxidans* were measured as a function of bulk phase E_h, and the results are summarized in table 4. The range of bulk phase E_h values spanned the actual E_m of the Fe^{2+}/Fe^{3+} couple in H_2SO_4 (650 mV) and included the value used by Ingledew and colleagues (1977) in their model for Fe^{2+} oxidation by *T. ferrooxidans* (770 mV). Neither the ΔpH nor the $\Delta\psi$ varied significantly over this range of bulk phase E_h.

According to the mechanism outlined in figure 3, $2H^+$ enter the cell matrix through the H^+-translocating ATPase per $2e^-$ released by Fe^{2+} oxidation. The formation of ATP from ADP + Pi requires 455 mV (equation 4, and see Alberty, 1969). Because $\Delta\bar{\mu}_{H^+}$ during Fe^{2+} oxidation is 260 mV, $2H^+$ are required if the $\Delta\bar{\mu}_{H^+}$ of 260 mV is to balance the requirement of 455 mV for ADP phosphorylation ($2 \times 260 = 520$ mV; equation 4). Also, under the experimental conditions used by Cox and colleagues (1979) a ΔE_h of 280 mV is available, so transfer of 2 electrons down the respiratory chain would result

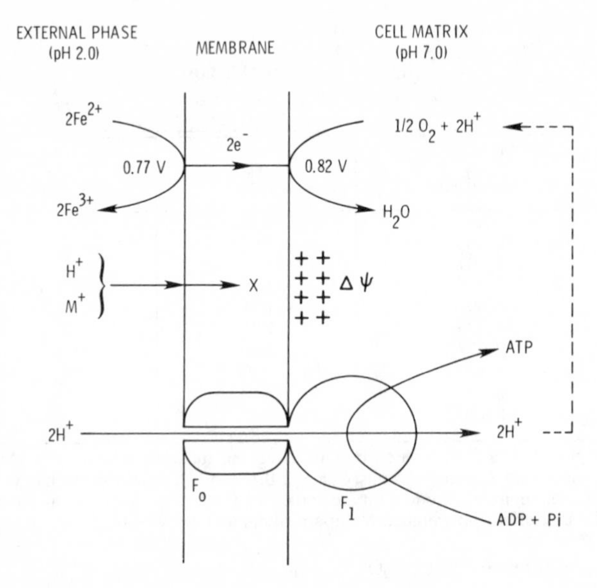

Fig. 3. Scheme for ATP synthesis linked to Fe^{+} oxidation in *T. ferrooxidans*.

in an emf of 560 mV (i.e., $-n\Delta E_h \geq x\ \Delta\bar{\mu}_{H^+}$), and x = 1 (i.e., $1H^+$ equivalent transferred per e^- flowing).

Under growth conditions where the steady-state E_h of the growth medium is 770 mV, the ΔE_h (respiratory chain) is only 60 mV (Ingledew et al., 1977), although $\Delta\bar{\mu}_{H^+}$ is about 260 mV (table 4). Because $\Delta E_h \geq x\ \Delta\bar{\mu}_{H^+}$ (equation 2), x is approximately 0.25. Hence $1H^+$ equivalent is transferred per 4 e^- flowing. In order to maintain electroneutrality, there must also be an inwardly directed flow of H^+ associated with the respiratory chain so that overall for every $4e^-$ available from Fe^{2+} oxidation, $1H^+$ enters the cytoplasm through the ATPase and $3H^+$ enter through the respiratory chain. Thus the stoichiometry of Fe^{2+} oxidation is as follows:

$$4\ Fe^{2+} + 4\ H^+ + O_2 \xrightarrow{4e^-} 4\ Fe^{3+} + 2\ H_2O\ . \qquad (8)$$

This means that *T. ferrooxidans* growing on Fe^{2+} would consume 4 mol of Fe^{2+} per mol of oxygen, which has been observed experimentally (Lees et al., 1969; Apel and Dugan, 1978). The theoretical $\rightarrow H^+/{\sim}P$ and ${\sim}P/O$ ratios would be 4 and 0.5, respectively, through these parameters have not been estimated experimentally for *T. ferrooxidans*.

It is possible that this H^+ entry may be a universal mechanism whereby chemoautotrophs are able to balance the electromotive force generated by substrate oxidation with the phosphorylation potential (Ingledew et al., 1974; Cobley, 1976). The electron transport chains of most chemolithotrophs are short, and large amounts of their electron donor must be oxidized so that ΔG from the respiratory chain balances ΔGp. This produces a surfeit of negative charge in the organism that can be consumed by the oxidase reaction provided there is an influx of H^+.

THE MECHANISM OF FE^{2+} OXIDATIVE PHOSPHORYLATION IN *T. FERROOXIDANS*

Two mechanisms have been considered in this review, both based on the model proposed by Ingledew and colleagues (1977, fig. 3). In most respects this original model accounts adequately for the experimental observations in that the protons that enter the cell are removed in the oxidase reaction (Ingledew et al., 1977; Apel et al., 1980).

We have sought to explain how energy is conserved by *T. ferrooxidans* under normal growth conditions in which the ΔE_h (respiratory chain) decreases as growth proceeds (table 3). The first possibility to be considered is that $\Delta\bar{\mu}_{H^+}$ decreases as the bulk phase E_h becomes more positive. The data reported in table 4 seem to discount this theory, since $\Delta\bar{\mu}_{H^+}$ apparently does not vary throughout the growth cycle of a culture of *T. ferrooxidans*.

A second explanation is that there may be a variable stoichiometry of charge movements by the ATPase and respiratory chain. Thus, as ΔE_h de-

TABLE 4

$\Delta\bar{\mu}_{H^+}$ IN *T. FERROOXIDANS* STRAIN T.F.3 AS THE E_h OF THE INCUBATION MEDIUM IS VARIED[a]

E_h	pH_o	pH_i	ΔpH	(mV)	$\Delta\bar{\mu}_{H^+}$ ($\Delta\psi - 59\ \Delta pH$) (mV)
600	2.0	6.6	4.6	−10	261
650	2.0	6.8	4.8	−15	268
700	2.0	6.7	4.7	0	277
750	2.0	6.4	4.4	− 5	255
770	2.0	6.4	4.4	−12	248

[a]*T. ferrooxidans* was grown in batch culture in 5-liter carboy flasks to a final E_h of 770 mV. Cells were harvested and the internal pH and membrane potential determined by acetate and thiocyanate distribution as described by Cox et al. (1979). Incubation mixtures were prepared by mixing quantities of 50 mM of $FeSO_4$ and $Fe_2(SO_4)_3$ until the desired E_h was obtained. Intracellular volume was assumed to be 2.4 μl mg protein^{-1}.

creases more H^+ per electron are transferred in order that $\Delta\bar{\mu}_{H^+}$ and ΔE_h may balance.

The mechanism by which H^+ entry into the cell is controlled is unclear. Under conditions that would tend to collapse the ΔpH (uncouplers; inhibitors of cytochrome oxidase) a $\Delta\psi$, positive inside, arises in response to H^+ entry, and so further entry of H^+ is quickly restricted. Hsung and Haug (1975, 1977a,b) ascribe this effect to the presence of a Donnan potential, but this is doubtful in view of the large concentrations of positively charged macromolecules that this would require in the cell (Cobley, 1984). In addition (as observed by Matin et al., 1982), large changes in $\Delta\psi$ result in response to addition of uncouplers and respiratory inhibitors (see also Cox et al., 1979). Such changes would be unlikely if the internal potential were maintained by a fixed positive charge.

Matin and colleagues (1982) discuss the possibility that entry of H^+ into the cell is important in generating a $\Delta\psi$ that then opposes further H^+ entry, and they suggest that movement of some other ion (probably K^+) is also involved. Interestingly, Krulwich and colleagues (1978) and Guffanti and colleagues (1979) observed that ΔpH in *Bacillus acidocaldarius* was completely collapsed by nigericin, which would facilitate K^+ efflux in response to H^+ entry. Therefore in *T. ferrooxidans*, $\Delta\psi$ may also play a prominent role in regulating H^+ entry into metabolically poisoned or uncoupled cells. Whether a similar mechanism is prominent in regulating H^+ entry into untreated cells remains to be elucidated (Cobley, 1983). Protons that do actually enter the cell (to drive ADP phosphorylation) are consumed by cytochrome oxidase. In order for the flow of electrons through the respiratory chain to balance the potential energy available from the $\Delta\bar{\mu}_{H^+}$, the most reasonable hypothesis is that the number of H^+ per electron crossing the respiratory chain varies in response to changes in ΔE_h.

SUMMARY

1. The potential difference between Fe^{2+}/Fe^{3+} and ½ O_2/H_2O varies as the Fe^{2+} of the growth medium is oxidized by the organism. Therefore the emf from electron transport decreases during the oxidation of Fe^{2+}.

2. A $\Delta\bar{\mu}_{H^+}$ of 260 mV is observed during Fe^{2+} oxidation by *T. ferrooxidans*. Therefore $2H^+$ must traverse the ATPase in order to satisfy the phosphorylation potential (455 mV; 10.49 kcal).

3. Two possibilities are presented to account for the mechanism by which *T. ferrooxidans* couples electron transport to oxidative phosphorylation; either variation in $\Delta\bar{\mu}_{H^+}$, or alteration in the stoichiometry of charge transfer during growth.

4. The mechanism by which cells control the influx of H^+ into the cytoplasm is unknown, though ion movements over than H^+ may be involved.

5. We propose that H^+ entry associated with the respiratory chain is responsible for providing the remaining H^+ to be consumed in the oxidase reaction. An analogous system has been observed in *Nitrobacter winogradskyi* (Cobley, 1976), and it is possible that such H^+ movements occur in all chemoautotrophs as a response to the production of large amounts of electrons from substrate oxidation.

ACKNOWLEDGEMENTS

We thank J. G. Cobley for helpful discussions and access to unpublished information.

LITERATURE CITED

Arkesteyn, G. J. M. W., and J. A. M. de Bont. 1980. *Thiobacillus acidophilus*: a study of its presence in *Thiobacillus ferrooxidans* cultures. Can. J. Microbiol. 26:1057–65.

Alberty, R. A. 1969. Standard Gibbs free energy, enthalpy, and entropy changes as a function of pH and pMg for several reactions involving adenosine phosphates. J. Biol. Chem. 244:3290–3302.

Aleem, M. I. H. 1977. Coupling of energy with electron transfer reactions in chemolithotrophic bacteria. *In* B. A. Haddock and W. A. Hamilton (eds.), Microbial energetics, pp. 351–82. Cambridge University Press, Cambridge.

Apel, W. A., and P. R. Dugan. 1978. Hydrogen ion utilization by iron-grown *Thiobacillus ferrooxidans*. *In* L. E. Murr, A. E. Torma, and J. A. Brierley (eds.), Metallurgical applications of bacterial leaching and related microbiological phenomena, pp. 45–59. Academic Press, New York.

Apel, W. A., P. R. Dugan, and J. H. Tuttle. 1980. Adenosine 5′-triphosphate formation in *Thiobacillus ferrooxidans* vesicles by H^+ ion gradients comparable to those of environmental conditions. J. Bacteriol. 142:295–301.

Beck, J. V. 1960. A ferrous-ion-oxidizing bacterium. I. Isolation and some general physiological characteristics. J. Bacteriol. 79:502–9.

Cobley, J. G. 1976. Reduction of cytochromes by nitrite in electron transport particles from *Nitrobacter winogradskyi*. Biochem. J. 156:493–99.

Cobley, J. G. 1984. The maintenance of pH gradients in acidophilic and alkalophilic bacteria: Gibbs-Donnan equilibrium calculations. *In* W. R. Strohl and O. H. Tuovinen (eds.), Microbial chemoautotrophy, pp. 121–32. Ohio State University Press, Columbus.

Cobley, J. G., and B. A. Haddock. 1975. The respiratory chain of *Thiobacillus ferrooxidans*: the reduction of cytochromes by Fe^{2+} and the preliminary characterization of rusticyanin, a novel 'blue' copper protein. FEBS Lett. 60:29–33.

Cox, J. C., and D. H. Boxer. 1978. The purification and some properties of rusticyanin, a blue copper protein involved in iron (II) oxidation from *Thiobacillus ferro-oxidans*. Biochem. J. 174:497–502.

Cox, J. C., R. Aasa, and B. G. Malmström. 1978. EPR studies on the blue copper protein, rusticyanin. A protein involved in Fe^{2+} oxidation at pH 2.0 in *Thiobacillus ferro-oxidans*. FEBS Lett. 93:157–60.

Cox, J. C., D. G. Nicholls, and W. J. Ingledew. 1979. Transmembrane electrical potential and transmembrane pH gradient in the acidophile *Thiobacillus ferro-oxidans*. Biochem. J. 178:195–200.

Doelle, H. W. 1975. Bacterial metabolism, 2d ed., pp. 322–32. Academic Press, New York.

Dugan, P. R., and D. G. Lundgren. 1965. Energy supply for the chemoautotroph *Ferrobacillus ferrooxidans*. J. Bacteriol. 89:825–34.

Garland, P. B. 1977. Energy transduction and transmission in microbial systems. *In* B. A. Haddock and W. A. Hamilton (eds.), Microbial energetics, pp. 1–21. Cambridge University Press, Cambridge.

Guay, P. and M. Silver. 1975. *Thiobacillus acidophilus* sp. nov.: isolation and some physical characteristics. Can. J. Microbiol. 21:281–88.

Guffanti, A. A., L. F. Davidson, T. M. Mann, and T. A. Krulwich. 1979. Nigericin induced death of an acidophilic bacterium. J. Gen. Microbiol. 114:201–6.

Haddock, B. A. and W. A. Hamilton, eds. 1977. Microbial energetics. Cambridge University Press, Cambridge.

Haddock, B. A., and C. W. Jones. 1977. Bacterial respiration. Bacteriol. Rev. 41:47–99.

Harold, F. M. 1972. Conservation and transformation of energy by bacterial membranes. Bacteriol. Rev. 36:172–230.

Harold, F. M. 1977. Membranes and energy transduction in bacteria. Curr. Top. Bioenerg. 6:83–149.

Harrison, A. P., B. W. Jarvis, and J. L. Johnson. 1980. Heterotrophic bacteria from cultures of autotrophic *Thiobacillus ferro-oxidans*: relationships as studied by means of DNA homology. J. Bacteriol. 143:448–54.

Hsung, J. C., and A. Haug. 1975. Intracellular pH of *Thermoplasma acidophila*. Biochim. Biophys. Acta 389:447–82.

Hsung, J. C., and A. Haug. 1977a. Membrane potential of *Thermoplasma acidophila*. FEBS Lett. 73:47–50.

Hsung, J. C., and A. Haug. 1977b. Zeta potential and surface charge of *Thermoplasma acidophila*. Biochim. Biophys. Acta 461:124–30.

Ingledew, W. J., and J. G. Cobley. 1980. A potentiometric and kinetic study on the respiratory chain of ferrous-iron-grown *Thiobacillus ferro-oxidans*. Biochim. Biophys. Acta 590:141–58.

Ingledew, W. J., J. G. Cobley, and J. B. Chappell. 1974. Cytochromes of the *Nitrobacter* respiratory chain. Biochem. Soc. Trans. 2:149–51.

Ingledew, W. J., J. C. Cox, and P. J. Halling. 1977. A proposed mechanism for energy conservation during Fe^{2+} oxidation by *Thiobacillus ferro-oxidans:* chemiosmotic coupling to net H^{+} influx. FEMS Microbiol. Lett. 2:193–97.

Ingledew, W. J., J. C. Cox, R. W. Jones, and P. B. Garland. 1978. Vectorial oxido-reductions: the ferrous iron oxidase complex of *Thiobacillus ferro-oxidans* and the nitrate reductase complex of *Escherichia coli*. *In* P. L. Dutton, J. S. Leigh and A. Scarpa (eds.), Frontiers of biological energetics, pp. 334–41. Academic Press, New York.

Jones, C. W. 1979. Energy metabolism in aerobes. Int. Rev. Biochem. 21:49–84.

Kelly, D. P. 1971. Autotrophy: concepts of lithotrophic bacteria and their organic metabolism. Ann. Rev. Microbiol. 25:177–210.

Krulwich, T. A., L. F. Davidson, S. J. Filip, R. S. Zuckerman, and A. A. Guffanti. 1978. The proton motive force and β-galactoside transport in *Bacillus acidocaldarius*. J. Biol. Chem. 253:4599–4603.

Lazaroff, N. 1963. Sulfate requirement for iron oxidation by *Thiobacillus ferrooxidans*. J. Bacteriol. 85:78–83.

Lazaroff, N. 1977. The specificity of the anion requirement for iron oxidation by *Thiobacillus ferrooxidans*. J. Gen. Microbiol. 101:85–91.

Leathen, W. W., S. A. Brayley, and L. D. McIntyre. 1953. The role of bacteria in the formation of acid from certain sulfuritic constituents associated with bituminous coal. II. Ferrous iron oxidizing bacteria. Appl. Microbiol. 1:65–68.

Leathen, W. W., N. A. Kinsel, and S. A. Braley. 1956. *Ferrobacillus ferrooxidans*: a chemosynthetic autotrophic bacterium. J. Bacteriol. 72:700–704.

Lees, H., S. C. Kwok, and I. Suzuki. 1969. The thermodynamics of iron oxidation by the ferrobacilli. Can. J. Microbiol. 15:43–46.

Lundgren, D. G., K. J. Anderson, C. C. Remsen, and R. P. Maloney. 1964. Culture, structure, and physiology of the chemoautotrophic *Ferrobacillus ferrooxidans*. Dev. Ind. Microbiol. 6:250–59.

Lundgren, D. G., J. R. Vestal, and F. R. Tabita. 1972. The microbiology of mine drainage pollution. *In* R. Mitchell (ed.), Water pollution microbiology, pp. 100–114. Academic Press, New York.

Matin, A., B. Wilson, E. Zychlinsky, and M. Matin. 1982. Proton motive force and the physiological basis of delta pH maintenance in *Thiobacillus acidophilus*. J. Bacteriol. 150:582–91.

McGoren, C. J. M., D. W. Duncan, and C. C. Walden. 1969. Growth of *Thiobacillus ferrooxidans* on various substrates. Can. J. Microbiol. 15:135–38.

Mitchell, P. 1968. Chemiosmotic coupling and energy transduction. Glynn Research, Bodmin, U.K.

Mitchell, P. 1976. Vectorial chemistry and the molecular mechanisms of chemiosmotic coupling: power transmission by proticity. Biochem. Soc. Trans. 4:399–430.

Mitchell, P. 1977. From energetic abstraction to biochemical mechanisms. *In* B. A. Haddock and W. A. Hamilton (eds.), Microbial energetics, pp. 383–432. Cambridge University Press, Cambridge.

Mitchell, P. 1979. Direct chemiosmotic ligand conduction mechanisms in protonmotive complexes. *In* C. P. Lee, G. Schatz, and L. Ernster (eds.), Membrane bioenergetics, pp. 361–72. Addison-Wesley, Reading, Mass.

Mitchell, P. 1981. From black box bioenergetics to molecular mechanisms: vectorial ligand-conduction mechanisms in biochemistry. *In* V. P. Skulachev and P. C. Hinkle (eds.), Chemiosmotic proton circuits in biological membranes, pp. 611–32. Addison-Wesley, Reading, Mass.

Peck, H. D. 1968. Energy coupling mechanisms in chemolithotrophic bacteria. Ann. Rev. Microbiol. 22:489–518.

Schlegel, H. G. 1975. Mechanisms of chemo-autotrophy. *In* O. Kinne (ed.), Marine ecology, 2:9–60. John Wiley & Sons, New York.

Silverman, M. P., and H. L. Ehrlich. 1964. Microbial formation and degradation of minerals. Adv. Appl. Microbiol. 6:153–206.

Silverman, M. P., and D. G. Lundgren. 1959. Studies on the chemoautotrophic iron bacterium *Ferrobacillus ferrooxidans*. I. An improved medium and harvesting procedure for securing high cell yields. J. Bacteriol. 77:642–47.

Suzuki, I. 1974. Mechanisms of inorganic oxidation and energy coupling. Ann. Rev. Microbiol. 28:85–101.

Thauer, R. K., K. Jungermann, and K. Decker. 1977. Energy conservation in chemotrophic anaerobic bacteria. Bacteriol. Rev. 41:100–180.

Trudinger, P. A. 1969. Assimilatory and dissimilatory metabolism of inorganic sulphur compounds by micro-organisms. Adv. Microb. Physiol. 3:111–58.

Tuovinen, O. H., D. P. Kelly, C. S. Dow, and M. Eccleston. 1978. Metabolic transitions with cultures of acidophilic thiobacilli. *In* L. E. Murr, A. E. Torma, and J. A. Brierley (eds.), Metallurgical applications of bacterial leaching and related microbiological phenomena, pp. 61–81. Academic Press, New York.

HENRY L. EHRLICH

Different Forms of Bacterial Manganese Oxidation

4

Bacterial precipitation of manganese oxide was reported as far back as 1901 by Jackson, who observed the phenomenon with *Crenothrix manganifera*. Beijerinck (1913) is more commonly credited with the first observation of Mn^{2+} oxidation by bacteria and fungi. It is now realized that microbial manganese oxidation may be enzymatic or nonenzymatic. One of the clearest demonstrations of nonenzymatic manganese oxidation was a recent observation by Bromfield (1978, 1979). He isolated an actinomycete, *Streptomyces* sp., from soil, which in laboratory culture produced a water-soluble, secondary metabolite of low-molecular weight that oxidized Mn^{2+} at pH 5–5.6, a pH range in which Mn^{2+} is not oxidized by oxygen. Another example is *Bacillus* SG-1, whose spores but not vegetative cells oxidize Mn^{2+} nonenzymatically (Rosson and Nealson, 1982). All manganese-oxidizing fungi probably oxidize manganese nonenzymatically, in part, at least, by producing hydroxycarboxylic acids that promote oxidation (Ehrlich, 1981, p. 217). The oxides are frequently precipitated on the fungal mycelia (Schweisfurth, 1971), as was already noted by Beijerinck (1913).

Enzymatic manganese oxidation was first shown by Bromfield (1956) with intact cell suspensions of *Corynebacterium* (now *Arthrobacter*) strain B from soil. Ehrlich (1966, 1968) demonstrated enzymatic manganese oxidation with intact cells and cell extracts of a marine *Arthrobacter* strain 37 and later with *Oceanospirillum* and *Vibrio* strains (Ehrlich et al., 1972; Ehrlich, 1976, 1978). Other observations of enzymatic manganese oxidation have been made with intact *Sphaerotilus discophorus* (now *Leptothrix discophora*) by Johnson and Stokes (1966) and confirmed by Mills and Randles (1979) with intact cells and cell extracts. Enzymatic manganese oxidation has also been noted with intact cells of *Leptothrix pseudoochracea, Metallogenium personatum,* and *Arthrobacter siderocapsulatus* (formerly *Siderocapsa*) (Dubinina,

1978a); with *Citrobacter freundii* strain E_4 and *Pseudomonas* strain E_1 from soil concretions (Douka, 1977); with intact cells and cell extracts from two unidentified Gram-negative marine bacteria from the Galapagos Rift zone (Ehrlich, 1983); and with intact cells and cell extracts of a marine *Vibrio* (Schuett, 1983). Other observations of manganese oxidation by bacteria have been made, but whether the oxidations are enzymatic or nonenzymatic remains to be clarified (Ehrlich, 1981; Marshall, 1979).

In general, bacteria that oxidize Mn^{2+} enzymatically can be assigned to one of the three groups according to their mode of attack on Mn^{2+}, as follows:

$$\text{Group I:}\quad Mn^{2+} + \tfrac{1}{2}\,O_2 + H_2O \xrightarrow{\text{Mn oxidase}} MnO_2 + 2\,H^+ \,. \tag{1}$$

$$\text{Group II:}\quad MnMnO_3 + \tfrac{1}{2}\,O_2 + 2H_2O \xrightarrow{\text{Mn oxidase}} 2\,HO_2MnO_3 \,. \tag{2}$$

$$\text{Group III:}\quad Mn^{2+} + H_2O_2 \xrightarrow{\text{catalase}} MnO_2 + 2\,H^+ \,. \tag{3}$$

The following is a summary of the evidence on which this grouping is based.

The Group I mechanism is illustrated by the action of *Arthrobacter* strain B, *Leptothrix discophora, Citrobacter freundii* strain E_4, *Pseudomonas* strain E_1, two unidentified gram-negative marine bacteria, and a marine *Vibrio*. Bromfield (1956) showed that resting cells from a stationary phase culture of *Arthrobacter* strain B oxidized Mn^{2+} at an initial concentration of 0.005% as $MnSO_4{\cdot}4H_2O$. Oxidation occurred at 27, 35, 40 and 45°C but not at 100°C. The activity was heat-labile and was inhibited by various poisons such as $HgCl_2$, $CuCl_2$, KCN, NaN_3, and benzene. The activity was measured semi-quantitatively with benzidine reagent, which turns blue on reaction with manganese at valence states greater than +2. Some of the manganese oxide was precipitated on the cells. Bromfield and David (1976) quantified the Mn^{2+}-oxidizing activity of *Arthrobacter* strain B by measuring either residual unoxidized manganese after first displacing adsorbed but unoxidized Mn^{2+} from the cells by treatment with 10 mM $CuSO_4$ solution, or by measuring the oxidized manganese after separating it from the culture medium, washing it, and then converting it to Mn^{2+} with H_2O_2 in acid solution. They then measured the Mn^{2+} by atomic absorption spectroscopy.

Johnson and Stokes (1966) showed that suspensions of *Sphaerotilus discophorus (Leptothrix discophora)* strain 35 consumed significantly greater amounts of oxygen in the presence than in the absence of $MnSO_4$. Only cells that had been induced by growth in a $MnSO_4$-containing medium exhibited this heat-labile activity. The formation of Mn(IV) oxide was demonstrated qualitatively by browning of cultures on incubation. The oxide that caused the browning could be dissolved in HCl accompanied with evolution of Cl_2. Mills and Randles (1979) confirmed these observations with a newly isolated strain of *Leptothrix discophora*. In addition, they demonstrated Mn^{2+}-oxidizing activity with cell-free preparations from $MnSO_4$-induced but not uninduced cultures. The oxidation of Mn^{2+} by cell-free preparations was inhibited by KCN, NaN_3, and o-phenanthroline. Like Johnson and Stokes, Mills and Randles measured the Mn^{2+}-oxidizing activity by oxygen consumption.

Douka (1977) demonstrated oxidation of Mn^{2+} by *Citrobacter freundii* strain E_4 and *Pseudomonas* strain E_1 from soil concretions by showing that pure cultures browned $MnSO_4$-containing agar. The brown deposit in the agar gave a positive test for manganese oxide with benzidine reagent. Manganese-oxidizing activity was followed quantitatively with intact cells and cell extracts by measuring the residual Mn^{2+} in solution after first removing oxidized manganese from suspension by centrifugation. The manganese determinations were made by atomic absorption or neutron activation. The Mn^{2+}-oxidizing activity was heat-labile and was inhibited by $HgCl_2$. The cells did not adsorb Mn^{2+} in measurable amounts. Sterile concretion material in the reaction mixture did not affect the rate of oxidation.

Ehrlich (1983) demonstrated oxidation by two isolates, strains SSW_{22} and S_{13} from a hydrothermal discharge area in the Galàpagos Rift zone, by showing Mn^{2+} oxidation by intact cells and cell extracts from induced but not uninduced cultures. The oxidation was not dependent on an initial presence of MnO_2, nor was the rate of oxidation increased by the presence of MnO_2 except in the case of cell extract from strain SSW_{22}. The extent of oxidation was determined colorimetrically in terms of residual Mn^{2+} in solution as measured by persulfate oxidation after centrifuging each sample to remove oxidized manganese. Mn^{2+} was not adsorbed by the cells or cell extracts as shown by $CuSO_4$ treatment. The oxidation was heat-labile, oxygen-dependent, and inhibited by antimycin A, 2-n-nonyl-4-hydroxyquinoline-N-oxide (NOQNO), NaN_3, and KCN. In strain SSW_{22}, the Mn^{2+}-oxidizing activity required a soluble and particulate component. Cytochrome of the *c*-type in the particulate fraction was reduced by the Mn^{2+} in the presence of the soluble factor. Adenosine 5′-triphosphate (ATP) synthesis was shown to be coupled to Mn^{2+} oxidation.

Schuett (1983; personal communication) isolated a *Vibrio* from a deep-sea sediment sample. Whole cells and cell extracts from this strain oxidized Mn^{2+}. Manganese-oxidizing activity was followed quantitatively by an *o*-tolidine reaction that measures manganese of valence states of +3 and higher. Schuett found that the activity was inhibited by 1 mM $HgCl_2$ and was destroyed by boiling the cell extract.

All the foregoing Group I cultures could initiate Mn^{2+} oxidation in the absence of added Mn(IV) oxide, which is required by Group II organisms, and in the absence of any other externally added, oxidizable organic substrate. Therefore, unlike Group III organisms, H_2O_2 is not used as an oxidant by Group I organisms. Based on the various assay methods used, *C. freundii, Pseudomonas* sp., and strains SSW_{22} and S_{13} did not adsorb measurable amounts of Mn^{2+}. Therefore, it is probable that adsorption of Mn^{2+} is not a prerequisite for its oxidation by Group I organisms. The initial adsorption of Mn^{2+} by *Arthrobacter* strain B measured by Bromfield and David (1976) may not be an absolute prerequisite for Mn^{2+} oxidation by that organism but may, at least in part, be an independent process. The oxidase systems of these organisms may be either constitutive (Bromfield, 1956; Douka, 1977; Schuett, 1983) or inducible (Johnson and Stokes, 1966; Ehrlich, 1983).

The reaction that is common to members of Group II has been found thus far only in a limited number of unrelated bacterial isolates from deep-sea ferromanganese nodules, sediments, and seawater. The isolates include among others *Arthrobacter* strain 37, *Oceanospirillum* strains BIII 45 and BIII 82, and *Vibrio* strains BIII 39 and BIII 47 (Ehrlich 1966, 1968, 1974, 1976, 1978). Intact cells or cell extracts of these organisms oxidize Mn^{2+} only if MnO_2 is initially present. In the initial studies, Mn-Fe oxide, which adsorbs Mn^{2+} very strongly even in the absence of bacteria or bacterial cell extracts, was used (Ehrlich, 1966, 1968, 1971). More recent experiments were performed with a reagent-grade MnO_2 (Baker) (e.g., Ehrlich, 1976), which, in the absence of cells or cell extract of the test cultures, adsorbs immeasurably small amounts of Mn^{2+} (table 1). Neither intact cells nor cell extracts of Group II organisms so far tested bind Mn^{2+} in amounts measurable by persulfate oxidation (see Ehrlich, 1963, for the assay method). Table 2 shows that with all organisms tested, the residual Mn^{2+} in solution had not changed in 4 h in the absence of MnO_2. In the presence of MnO_2, intact cells and cell extracts removed 0.4–0.9 μmol Mn^{2+} per ml from solution (table 2). Thus, the added MnO_2 played a role in Mn^{2+} removal from solution when intact cells or cell extracts were present. The function of MnO_2 or other Mn(IV) oxide is to adsorb the Mn^{2+} that the bacteria oxidize. The adsorption reaction may be written

$$Mn^{2+} + H_2MnO_3 \rightleftharpoons MnMnO_3 + 2\,H^+ . \qquad (4)$$

It does not require catalysis. Within limits, the ratio of MnO_2 initially avail-

TABLE 1

EFFECT OF MnO_2 ON THE CONCENTRATION OF DISSOLVED Mn^{2+} IN SEAWATER OVER A 4-HOUR INCUBATION PERIOD

Length of Incubation	Mn^{2+} Concentration (μmol ml^{-1})
5 min	3.57
10 min	3.58
15 min	3.58
1 h	3.57
2 h	3.57
3 h	3.57
4 h	3.57

The reaction mixture contained 0.1 g reagent grade MnO_2 (Baker), 8 ml of seawater (pH 7.5), 1 ml of 0.114 M $NaHCO_3$ (pH 7.5), and 1 ml of 35.7 mM $MnSO_4$ solution. Mn^{2+} in solution was measured by persulfate oxidation (Ehrlich, 1966).

TABLE 2

BACTERIAL Mn^{2+} REMOVAL FROM SEAWATER SOLUTION IN THE PRESENCE AND ABSENCE OF MnO_2[a]

			Dissolved manganese after 4 h (μmol)					
			with MnO_2			without MnO_2		
Organism[b]	Preparation	Cell OD[c] or protein	−Cells[d] or extract	+Cells or extract	ΔMn[e]	−Cells[d] or extract	+Cells or extract	ΔMn[e]
BIII 39	intact cells	0.435	33.3	32.9	0.4	33.4	33.4	0
	cell extract[f]	2.4 mg	33.3	32.9	0.4	33.4	33.4	0
BIII 45	intact cells	0.242	34.7	34.0	0.7	34.8	34.9	−0.1
	cell extract	2.2 mg	33.3	32.4	0.9	33.4	33.4	0
BIII 47	intact cells	0.340	33.3	32.8	0.5	33.4	33.4	0
	cell extract	2.2 mg	33.3	32.4	0.9	33.4	33.5	−0.1
BIII 82	intact cells	0.395	33.7	33.3	0.4	33.8	33.8	0
	cell extract	2.0 mg	33.3	32.6	0.7	33.2	33.3	−0.1
37	intact cells	0.510	32.9	32.3	0.6	32.9	32.9	0
	cell extract	2.1 mg	32.9	32.0	0.9	32.9	32.9	0

[a]The experimental procedure was as described by Ehrlich (1968) except that 0.1 g of reagent grade MnO_2 (Baker) were used instead of synthetic Mn-Fe oxide and the pH was maintained at 7.5 instead of 7.0.

[b]Strains BIII 39 and BIII 47 are *Vibrio;* strains BIII 45 and BIII 82 are *Oceanospirillum;* strain 37 is *Arthrobacter.*

[c]24-h-old cells were washed 3 times in sterile seawater. The optical density (OD) was measured at 450 nm, using a 10-fold diluted cell suspension.

[d]These values also correspond to the initial Mn^{2+} concentrations.

[e]The difference between the amounts of manganese in solution without and with cells or cell extract. Removal of manganese was due to biochemical action.

[f]The supernatant of a sonicate centrifuged at 12,000 *g* for 10 min at 4°C.

able to cells or cell extracts affects the rate of Mn^{2+} oxidation. This can be shown by determining the Mn^{2+} oxidation rate at different resting cell concentrations in the presence of an initial, fixed amount of Mn-Fe oxide. Such an experiment shows that there is an optimal cell concentration for fastest Mn^{2+} oxidation. Above or below this cell concentration, the oxidation rate is significantly slower (Ehrlich, 1966). Mn^{2+} oxidation in all these studies was measured in terms of Mn^{2+} removal with and without cells or cell extract.

The Mn^{2+} oxidase of Group II organisms is constitutive and oxygen-dependent. The activity is inhibited by a variety of compounds including rotenone, dicumarol, antimycin A, NOQNO, azide, and cyanide (Ehrlich, 1968, 1976, unpublished data). The enzyme activity requires the presence of cytoplasmic membrane and a heat-stable, soluble factor that is protein-associated (Arcuri and Ehrlich, 1979, 1980; Ehrlich, unpublished data). Mn^{2+} reduces a membrane-bound *c*-type cytochrome only in the presence of both MnO_2 and a heat-stable, soluble factor (Arcuri and Ehrlich, 1979).

These types of organisms have been isolated so far only from the Atlantic (Blake Plateau) and the eastern and central Pacific Oceans (Ehrlich et al., 1972). It is likely that the oxidizers of bound Mn^{2+} are favored where the dissolved Mn^{2+} concentration is low, whereas the oxidizers of the free Mn^{2+} are favored where the dissolved Mn^{2+} concentration is relatively high (Ehrlich, 1983).

Like the Group I organisms, this group can also derive energy from manganese oxidation (Ehrlich, 1976, 1978) by coupling it to ATP formation. Group II organisms have not been grown autotrophically.

The Group III organisms use H_2O_2 as an oxidant of Mn^{2+}. Hydrogen peroxide is a metabolic product formed in significant quantities by *Leptothrix pseudoochracea, Metallogenium personatum,* and *Arthrobacter siderocapsulatus* during glucose respiration (Dubinina, 1978a). The oxidation of Mn^{2+} to MnO_2 by H_2O_2 is catalase-dependent in these organisms. A steady supply of H_2O_2 from growing cultures and a catalase activity that is not so strong that it merely catalyzes the dismutation of H_2O_2 to water and oxygen, are important for this reaction. Mn^{2+} in these studies was measured by a formaldoxime reaction, Mn(IV) oxide by an *o*-tolidine reaction, and H_2O_2 by iodimetry (Dubinina, 1978a). This reaction was demonstrated *in vitro* by generating H_2O_2 by oxidation of glucose with glucose oxidase, thereby causing the oxidation of Mn^{2+} and also Fe^{2+} by H_2O_2 with catalase at neutral pH (Dubinina, 1978a). No useful energy is generated in these processes, and Dubinina (1978b) views them as means of detoxification of H_2O_2 in the presence of low catalase activity.

The coupling of ATP formation to Mn^{2+} oxidation was demonstrated with selected organisms from Groups I and II (Ehrlich, 1976, 1978, 1983). In these experiments, membrane vesicles were prepared by sonication and were tested in a reaction mixture containing adenosine 5′-diphosphate (ADP), inorganic phosphate, and $MnSO_4$ in appropriate salt solutions. The vesicles of Group II organisms were tested in the presence of MnO_2. ATP was formed outside the vesicles because, by inference, the vesicles must have had either a partial or a complete inside-out orientation.

Assuming that an electrochemical gradient across the plasma membrane of intact cells accounts for ATP synthesis coupled to Mn^{2+} oxidation, the following type of reaction may be viewed as occurring outside the plasma membrane (e.g., in the periplasmic space of gram-negative cells):

$$Mn^{2+} + 2\ H_2O \rightleftharpoons MnO_2 + 4\ H^+ + 2\ e^- . \qquad (5)$$

Whether the Mn(II) species is actually Mn^{2+} and the Mn(IV) is MnO_2 in the case of a particular organism will have to be determined. In any event, the electrons from reaction 5 will be transported to a cytochrome oxidase on the inner surface of the plasma membrane by way of the electron transport system of the membrane, to reduce oxygen to water according to the reaction:

$$\tfrac{1}{2}\ O_2 + 2\ H^+ + 2\ e^- \rightleftharpoons H_2O . \qquad (6)$$

The protons of reaction 5 will contribute to the pH gradient needed for energy conservation. This model, which conforms to the Mitchell theory (Mitchell, 1977), predicts that the oxygen in the MnO_2 formed in reaction 5 is derived from water and not from oxygen. This has yet to be tested.

None of the organisms that couple Mn^{2+} oxidation to ATP formation have been grown autotrophically. Mixotrophic growth of *Sphaerotilus discophorus (Leptothrix discophora)* has been claimed by Ali and Stokes (1971), but their claim has been disputed by Hajj and Makemson (1976). Mixotrophic growth by a marine pseudomonad has been reported by Kepkay and Nealson (1982). *Hyphomicrobium manganoxidans* has been reported by Eleftheriadis (1976) to oxidize Mn^{2+} for growth in an organic medium, and may be an obligate mixotroph.

Autotrophic growth by *Sphaerotilus discophorus (Leptothrix discophora)* in a medium in which $MnSO_4$ was the only energy source and CO_2 the only carbon source, with trace amounts of biotin, thiamin, and cyanocobalamin as added growth factors, has been claimed by Ali and Stokes (1971). This claim has been disputed on a quantitative basis (Van Veen, 1972), although on a qualitative basis it cannot be discounted without further investigation. Autotrophic growth by a marine pseudomonad that can grow mixotrophically on Mn^{2+} has been reported (Kepkay and Nealson, 1982).

At this point, it can be concluded that various unrelated bacteria have the capacity to oxidize Mn^{2+} enzymatically. The mechanism of oxidation is not the same in all cases, nor do all the oxidations yield useful energy to the organisms. Where the energy is metabolically available, it may be usable mixotrophically or autotrophically. Additional investigations are needed to determine the extent to which this occurs. Thermodynamically, reaction 1 is favorable for growth with a standard free-energy change at pH 7.0 of -68.18 kJ mol^{-1} (Ehrlich, 1978).

ACKNOWLEDGMENT

The experimental work in tables 1 and 2 was performed with the able assistance of Alice R. Ellett.

LITERATURE CITED

Ali, S. H., and J. L. Stokes. 1971. Stimulation of heterotrophic and autotrophic growth of *Sphaerotilus discophorus* by manganous ions. Antonie van Leeuwenhoek 37:519–28.

Arcuri, E. J., and H. L. Ehrlich. 1979. Cytochrome involvement in Mn(II) oxidation by two marine bacteria. Appl. Environ. Microbiol. 37:916–23.

Arcuri, E. J., and H. L. Ehrlich. 1980. Electron transfer coupled to Mn(II) oxidation in two deep-sea Pacific Ocean isolates. *In* P. A. Trudinger, M. R. Walter, and B. J. Ralph (eds.), Biogeochemistry of ancient and modern environments, pp. 339–44. Springer Verlag, New York.

Beijerinck, M. W. 1913. Oxydation des Mangancarbonates durch Bakterien und Schimmelpilze. Folia Microbiol. (Delft) 2:123–34.

Bromfield, S. M. 1956. Oxidation of manganese by soil microorganisms. Austr. J. Biol. Sci. 9:238–52.

Bromfield, S. M. 1978. The oxidation of manganous ions under acid conditions by an acidophilous actinomycete from acid soil. Aust. J. Soil Res. 16:91–100.

Bromfield, S. M. 1979. Manganous ion oxidation at pH values below 5.0 by cell-free substances from *Streptomyces* sp. cultures. Soil Biol. Biochem. 11:115–18.

Bromfield, S. M., and D. J. David. 1976. Sorption and oxidation of manganous ions and reduction of manganese oxide by cell suspensions of a manganese-oxidizing bacterium. Soil Biol. Biochem. 8:37–43.

Douka, C. E. 1977. Study of bacteria from manganese concretions. Precipitation of manganese by whole cells and cell-free extracts of isolated bacteria. Soil Biol. Biochem. 9:89–97.

Dubinina, G. A. 1978a. Mechanism of oxidation of divalent iron and manganese by iron bacteria growing in neutral medium. Mikrobiologiya 47:591–99.

Dubinina, G. A. 1978b. Functional role of bivalent iron and manganese oxidation in *Leptothrix pseudoochracea*. Mikrobiologiya 47:783–89.

Ehrlich, H. L. 1963. Bacteriology of manganese nodules. I. Bacterial action on manganese nodule enrichments. Appl. Microbiol. 11:15–19.

Ehrlich, H. L. 1966. Reactions with manganese by bacteria from ferromanganese nodules. Dev. Ind. Microbiol. 7:279–86.

Ehrlich, H. L. 1968. Bacteriology of manganese nodules. II. Manganese oxidation by cell-free extracts from a manganese nodule bacterium. Appl. Microbiol. 16:197–202.

Ehrlich, H. L. 1971. Bacteriology of manganese nodules. V. Effect of hydrostatic pressure on bacterial oxidation of Mn(II) and reduction of MnO_2. Appl. Microbiol. 21:306–10.

Ehrlich, H. L. 1974. Response of some activities of ferromanganese nodule bacteria to hydrostatic pressure. *In* R. R. Colwell and R. Y. Morita (ed.), Effect of the ocean environment on microbial activities, pp. 208–21. University Park Press, Baltimore.

Ehrlich, H. L. 1976. Manganese as energy source for bacteria. *In* J. O. Nriagu, ed., Environmental biogeochemistry, 2:633–44. Ann Arbor Science Publishers, Ann Arbor, Mich.

Ehrlich, H. L. 1978. Inorganic energy sources for chemolithotrophic and mixotrophic bacteria. Geomicrobiol. J. 1:65–83.

Ehrlich, H. L. 1981. Geomicrobiology. Marcel Dekker, New York.

Ehrlich, H. L. 1983. Manganese-oxidizing bacteria from a hydrothermally active area on the Galapagos Rift. Ecol. Bull. (Stockholm) 35:357–66.

Ehrlich, H. L., W. C. Ghiorse, and G. L. Johnson. 1972. Distribution of microbes in manganese nodules from the Atlantic and Pacific Oceans. Dev. Ind. Microbiol. 13:57–65.

Eleftheriadis, D. K. 1976. Mangan- und Eisenoxydation in Mineral- und Termal-Quellen–Mikrobiologie, Chemie und Geochemie. Ph.D. thesis. Universität des Saarlandes, Saarbrücken, Federal Republic of Germany.

Hajj, H., and J. Makemson. 1976. Determination of growth of *Sphaerotilus discophorus* in the presence of manganese. Appl. Environ. Microbiol. 32:699–702.

Jackson, D. D. 1901. A new species of *Crenothrix (C. manganifera)*. Trans. Am. Microsc. Soc. 23:31–39.

Johnson, A. H., and J. L. Stokes. 1966. Manganese oxidation by *Sphaerotilus discophorus*. J. Bacteriol. 91:1543–47.

Kepkay, P. E., and K. H. Nealson. 1982. Mixotrophic growth of a manganese oxidizing marine *Pseudomonad* in batch and chemostat culture. Ann. Meet. Am. Soc. Microbiol. Abstr. I 110, p. 112.

Marshall, K. C. 1979. Biogeochemistry of manganese minerals. *In* P. A. Trudinger and D. J. Swaine (eds.), Biogeochemical cycling of mineral-forming elements, pp. 253-92. Elsevier, Amsterdam.

Mills, V. H., and C. I. Randles. 1979. Manganese oxidation in *Sphaerotilus discophorus* particles. J. Gen. Appl. Microbiol. 25:205–7.

Mitchell, P. 1977. Vectorial chemiosmotic processes. Ann. Rev. Biochem. 46:996–1005.

Rosson, R. A., and K. H. Nealson. 1982. Manganese binding and oxidation by spores of a marine bacillus. J. Bacteriol. 151:1027–34.

Schuett, C. 1983. Some physiological properties of Mn(II)-oxidizing bacterial isolates. *In* Bacteriologie marine. Colloque International. C.N.R.S. 17–19 May 1982 (in press).

Schweisfurth, R. 1971. Manganoxidierende Pilze. I. Vorkommen, Isolierung, und mikrobiologische Untersuchungen. Z. Allg. Mikrobiol. 11:415–30.

Van Veen, W. L. 1972. Factors affecting the oxidation of manganese by *Sphaerotilus discophorus*. Antonie van Leeuwenhoek 38:623–26.

ABDUL MATIN

Mixotrophy in Facultative Thiobacilli

5

INTRODUCTION

Facultative thiobacilli, like other facultative chemolithotrophs, possess the unique capacity for autotrophic as well as heterotrophic growth. Their potential ability to combine the use of autotrophic and heterotrophic growth substrates is of considerable interest. In their natural habitat, these bacteria are likely to have both types of substrates simultaneously available; whether or not concomitant use of these is possible and of selective advantage to the organism are ecologically important questions. Moreover, the utilizable inorganic and organic energy substrates are dissimilated by pathways that do not share common catabolites. This situation is different from that encountered in heterotrophs supplied with multiple organic substrates, where shared catabolites modulate the activities of the pertinent metabolic pathways by the well-known phenomenon of catabolite repression (Magasanik, 1965). The question thus arises as to whether organic and inorganic energy substrates can modulate the activity of pathways of each other's metabolism and, if so, by what molecular mechanism.

Mixotrophy has been studied in both of the well-characterized groups of facultative chemolithotrophs, i.e., the hydrogen bacteria and the facultative thiobacilli, but the recent work has been concerned with the latter group, and is the main focus of this review. The early studies were conducted entirely using the batch culture technique (Rittenberg, 1969; Schlegel, 1975; Matin, 1978); more recently, following the realization that high nutrient concentration of batch culture is an exception rather than the rule in nature, mixotrophy has been examined in the thiobacilli under nutrient limitation using the continuous culture technique.

BATCH CULTURE STUDIES

Thiobacillus intermedius

The first thiobacillus in which mixotrophy was examined is *T. intermedius*; the studies were carried out by Rittenberg and his collaborators.

The growth rate as well as the yield of *T. intermedius* were higher in a thiosulfate-yeast extract medium than in either the autotrophic or the heterotrophic environment at corresponding substrate concentrations (London and Rittenberg, 1966). Both thiosulfate and yeast extract contributed to energy generation under the mixotrophic conditions. This was indicated by yield studies at various concentrations of thiosulfate and yeast extract, by the additive rate at which cells grown under these conditions oxidized thiosulfate and yeast extract, and by the fact that extracts of these cells could couple the oxidation of thiosulfate as well as yeast extract with the reduction of endogenous cytochrome. Nearly all the cell carbon, however, was derived from yeast extract, since ribulose-1,5-bisphosphate (RuBP) carboxylase activity and $^{14}CO_2$ fixation were greatly reduced under these conditions (London and Rittenberg, 1966).

Mixotrophic growth also occurred in a thiosulfate-glucose medium, but the amount of glucose consumed under these conditions, as well as the fact that extracts of these cells failed to couple glucose oxidation with cytochrome reduction, indicated that glucose provided only cell carbon and little or no energy (London and Rittenberg, 1966; Matin and Rittenberg, 1970a). As in thiosulfate-yeast extract medium, little carbon was derived from CO_2 (London and Rittenberg, 1966).

The physiological basis of the limited use of glucose in the mixotrophic environment by *T. intermedius,* which can use glucose for carbon and energy (Matin and Rittenberg, 1970a; Smith and Rittenberg, 1974), is the repression and/or inhibition by thiosulfate of the glucose transport system (Romano et al., 1975) and the glucose pathway enzymes in this bacterium (Matin and Rittenberg, 1970b). The autotrophic energy-generating system was also affected under mixotrophic conditions, since thiosulfate-glucose grown cells oxidized thiosulfate at a lower rate compared with the autotrophically grown cells (London and Rittenberg, 1966). Thus, a mutual antagonism existed between glucose and thiosulfate metabolisms but not to the extent of preventing mixotrophic growth. It is clear, however, that the full mixotrophic potential is not expressed under these conditions, since thiosulfate metabolic activity is decreased and glucose performs only a limited metabolic function.

Thiobacillus novellus

Growth pattern. Matin and his co-workers have conducted a thorough investigation of mixotrophy in *T. novellus*. Figure 1 presents the comparative growth pattern of *T. novellus* in autotrophic, heterotrophic, and mixotrophic batch culture environments (Perez and Matin, 1980). In autotrophic medium, *T. novellus* exhibited a generation time of approximately 20 h (fig. 1A), and the growth roughly paralleled thiosulfate consumption (fig. 1B). In heterotrophic glucose medium, growth occurred at an 8 h generation time and ceased upon glucose exhaustion. In contrast, a triphasic growth pattern was observed in mixotrophic medium. For the first three generations, the growth rate was similar to that in the heterotrophic medium; a slower growth rate was then established followed by another period of faster growth (fig. 1A). The disappearance of both glucose and thiosulfate was coincident with growth, but the kinetics differed from those observed in heterotrophic and autotrophic media (fig. 1B). Glucose was consumed at some 75% of the rate of heterotrophic culture, and thiosulfate at approximately 45% of the autotrophic rate for a substantial part of the growth period.

Although concurrent and complete disappearance of thiosulfate and glucose occurred from the medium, the final biomass in mixotrophic medium was somewhat lower than that found in heterotrophic medium (fig. 1); thus, the

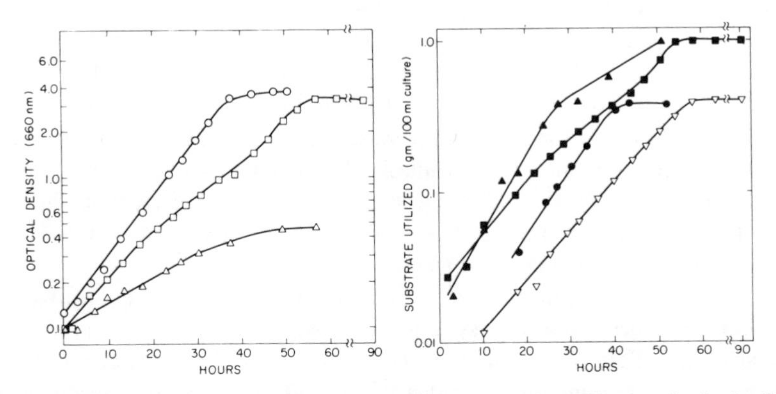

Fig. 1. Growth parameters of *T. novellus* in various media. The culture pH was maintained at 7.0 ± 0.1 by the use of a pH stat. (A) Kinetics of growth in 0.4% glucose medium (○), 1% $Na_2S_2O_3$ medium (△), and 1% $Na_2S_2O_3 - 0.4\%$ glucose medium (□). One optical density of 660 nm about equals 0.6 mg of cells (dry weight)/ml of culture. (B) Kinetics of substrate consumption: (●), glucose consumption in glucose medium; (▲), thiosulfate consumption in thiosulfate medium; glucose (▽) and thiosulfate (■) consumption in thiosulfate-glucose medium. From Perez and Matin (1980).

utilization of the additional substrate had no salutary effect on the amount of cell material synthesized. Studies involving the use of labeled glucose and titration of the amount of sulfate formed from thiosulfate oxidized showed that glucose and thiosulfate were both completely degraded in the mixotrophic medium. Thus, the lack of enhanced cell yield in the mixotrophic medium was not due to partial degradation of the growth substrates and consequent decrease in energy generation. It was concluded that the efficiency with which thiosulfate and/or glucose oxidation(s) were coupled to energy generation was decreased by some 25% in the mixotrophic environment.

Analysis showed accumulation of small amounts of sulfite in the mixotrophic media during growth, and since sulfite is a known growth inhibitor in this organism, it was inferred that it is this accumulation that accounted for the observed decreased growth rate in the mixotrophic medium as opposed to the heterotrophic medium.

Most of the cell carbon must have been derived from glucose in the mixotrophic medium, since Aleem and Huang (1965) have shown that organic compounds repress the RuBP carboxylase activity of *T. novellus* in batch culture.

Physiological basis of inhibition of glucose utilization in mixotrophic medium. Extensive studies were conducted to elucidate the physiological basis of the decreased rate of glucose utilization by *T. novellus* in mixotrophic media (fig. 1B). Using the glucose analogue 2-deoxy-D-glucose (2-DG), the transport of glucose in this bacterium was characterized, and it was concluded that glucose was probably transported as a symport with protons (Matin et al., 1980). The glucose transport system was constitutive in *T. novellus* because cells grown in thiosulfate medium possessed it (table 1). Growth in the presence of glucose stimulated the synthesis of this system, but to a lesser extent in the presence of thiosulfate than in its absence. This suggested that thiosulfate (or its metabolites) either repressed the glucose transport system or interfered with its induction by glucose.

Thiosulfate and its metabolite, sulfite, could also modulate the activity of the existing glucose transport system in glucose-grown cells. Kinetic studies, in which the effect of varying 2-DG concentration on the initial rate of 2-DG uptake was determined at individual fixed concentrations of thiosulfate or sulfite (figs. 2 and 3), showed that thiosulfate was an uncompetitive inhibitor of 2-DG uptake, with an apparent K_i of 3.1×10^{-2}M; and sulfite was a potent noncompetitive inhibitor, with an apparent K_i of 3.3×10^{-7}M.

Regulation of glucose metabolism also involved its metabolic enzymes. The activity of key pentose shunt and Krebs cycle enzymes exhibited a pattern

TABLE 1

INFLUENCE OF GROWTH ENVIRONMENT ON THE ACTIVITY OF 2-DEOXY-D-GLUCOSE TRANSPORT SYSTEM OF *T. NOVELLUS*

Concentration of 2-deoxy-D-glucose in reaction mixture (M)	TTRANSORT ACTIVITY[a] AFTER GROWTH IN SPECIFIED ENVIRONMENT		
	Heterotrophic	Mixotrophic	Autotrophic
10^{-5}	2.10	0.41	0.11
5×10^{-5}	5.80	1.24	0.22
10^{-4}	8.46	1.90	0.27

SOURCE: Matin et al. (1980).

[a]Nanomoles of 2-deoxy-D-glucose taken up per minute per milligram of protein. Rate was calculated from a time course of uptake at each concentration; the uptake was linear for 30 to 120 s.

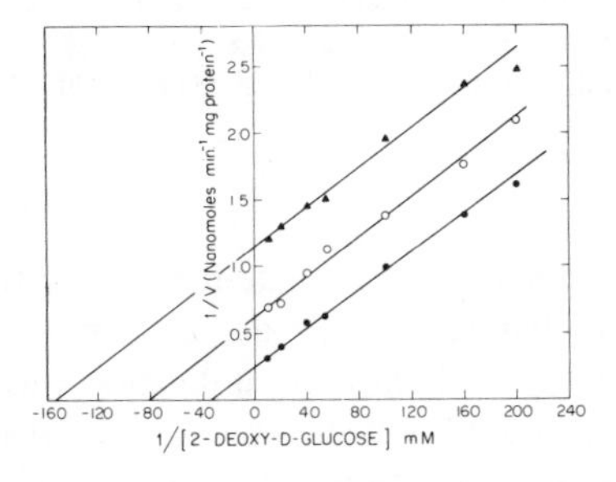

Fig. 2. Lineweaver-Burk plot of inhibition of 2-deoxy-D-glucose uptake by sodium thiosulfate in cell suspensions of *T. novellus*. Washed glucose-yeast extract grown cells were incubated with the specified concentrations of 2-deoxy-D-[^{14}C] glucose for 15 s in the absence (●), or presence of 0.005M (○) and 0.010 M (▲) thiosulfate. From Matin et al. (1980).

in cells grown in different environments that was very similar to that shown by the glucose transport activity (table 1), and led to the conclusion that these enzymes were induced by glucose but repressed by thiosulfate (Matin et al., 1980). Both thiosulfate and sulfite also markedly inhibited glucose-6-phosphate dehydrogenase activity in *T. novellus*.

Thiobacillus A2

In batch cultures containing 17 mM thiosulfate plus 12 mM acetate, simultaneous utilization of the two substrates by *Thiobacillus* A2 was dependent on inoculum history (Gottschal and Kuenen, 1980a). Acetate-grown cells grew

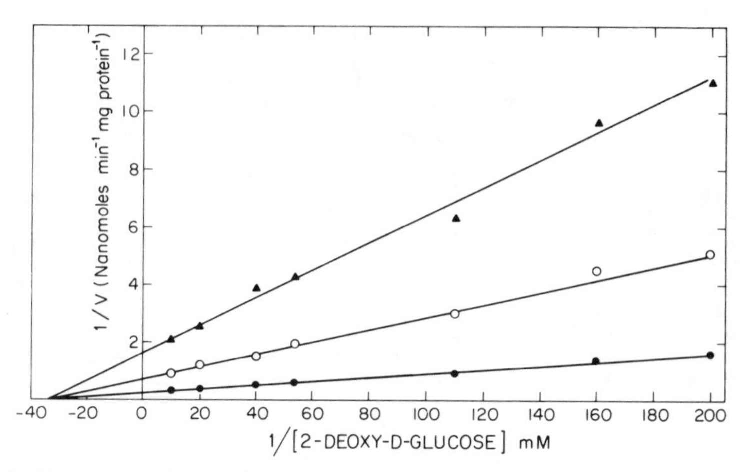

Fig. 3. Lineweaver-Burk plot of inhibition of 2-deoxy-D-glucose uptake by sodium sulfite. Symbols: (●): uptake kinetics in the absence of Na_2SO_3; (○,▲): Kinetics in the presence of 0.005 M and 0.010 M sodium sulfite. From Matin et al. (1980).

only on acetate and did not use thiosulfate for the first five hours of growth in mixotrophic medium, but later made use of both substrates (fig. 4). The transition from single to dual substrate utilization was not accompanied by any change in the growth rate, which remained poised at a generation time of about 3.6 h. In contrast, autotrophically grown cells utilized the two substrates simultaneously upon transfer to the mixotrophic medium but exhibited about the same growth rate as was observed with the heterotrophic inoculum. No information was presented on final biomass formation in autotrophic, heterotrophic, and mixotrophic media at corresponding substrate concentrations; and it is not known whether uncoupled substrate utilization, like that observed in *T. novellus* mixotrophic culture, also occurred in this bacterium.

With respect to the regulation of autotrophic and heterotrophic metabolic pathways in batch mixotrophic cultures, *T.* A2 behaved in a way similar to other facultative chemolithotrophs (Gottschal and Kuenen, 1980a). Thus, acetate antagonized thiosulfate metabolism, since in acetate-grown cells the rate of induction of thiosulfate-oxidizing activity in the presence of acetate was much lower than after acetate depletion. Similarly, thiosulfate-oxidizing capacity was reduced during mixotrophic growth on thiosulfate plus acetate, and the RuBP carboxylase activity was severely represessed in this growth environment (Taylor and Hoare, 1971).

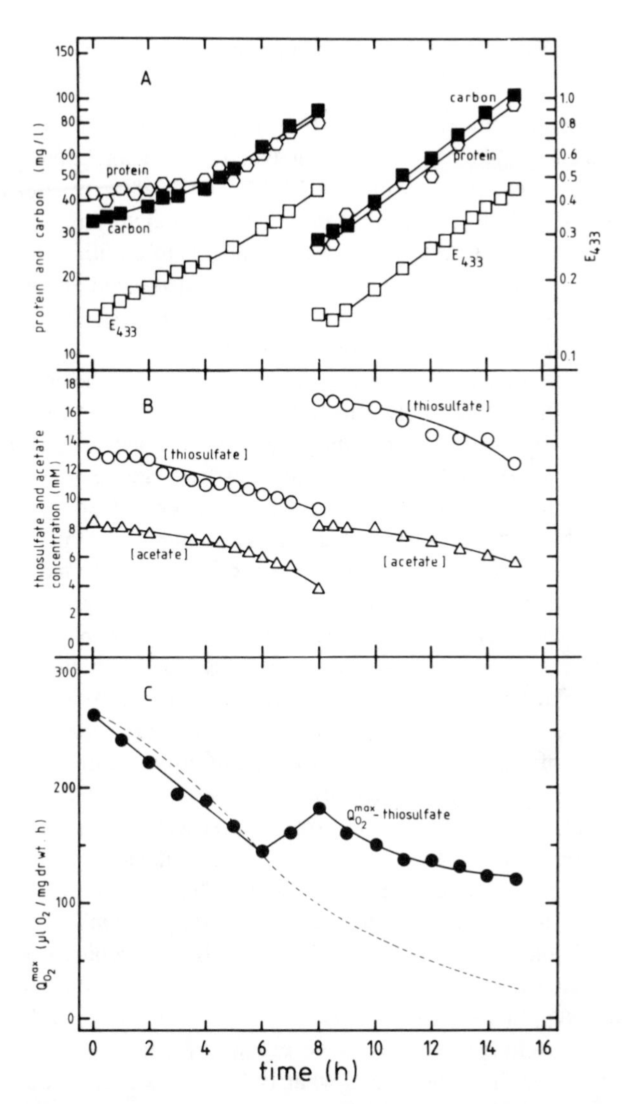

Fig. 4. Cell density and substrate concentrations during growth of *T.* A2 on a mixture of acetate and thiosulfate in batch culture (from Gottschal and Kuenen, 1980a). The inoculum consisted of autotrophically grown cells from a thiosulfate-limited chemostat culture at a dilution rate of 0.02 h^{-1}. Symbols: ⬡, protein; ■, organic cell carbon; □ absorbance; ○, thiosulfate concentration; △, acetate concentration. The dashed line in C represents the predicted decrease in the Q_{O2}^{max}-thiosulfate if the synthesis of this enzyme system had been repressed completely during the experiment.

RATIONALE OF REGULATORY STRATEGY UNDER HIGH NUTRIENT CONDITIONS

Among the three thiobacilli studied, only *T. intermedius* exhibits enhanced growth rate and yield under mixotrophic batch culture conditions; *T.* A2 does not appear to derive any benefit, and *T. novellus* is, in fact, hampered under these conditions. Indeed, the most pronounced regulatory response of these bacteria to mixotrophic batch culture conditions is to minimize the metabolic potential for the autotrophic and heterotrophic carbon and energy substrates and, in the case of *T. novellus,* to uncouple substrate oxidation from energy generation. Even *T. intermedius* did not express its full mixotrophic potential under these conditions: thiosulfate metabolic activity was decreased, glucose served only a limited metabolic function, and CO_2-fixing potential was probably not utilized. Very similar findings have been made with hydrogen bacteria where a pronounced antagonism between hydrogen-oxidizing potential, sugar metabolism, and CO_2-fixing activity has been well documented in mixotrophic batch cultures (Rittenberg, 1969; Schlegel, 1975; Matin, 1978).

Given the superabundance in which the various carbon and energy substrates are available to the organism in batch culture, it is logical that it should attempt to restrict their utilization. At such surfeit of nutrients, growth rate is unlikely to be limited by the rate of energy generation or the initial step in carbon assimilation, but rather by other factors imposed by the genetic constitution of the organism. Thus, rapid utilization of the various substrates is unlikely to confer advantage, and, in fact, can prove harmful by leading to overproduction of energy and excessive pool levels of various catabolites. For instance, excessive ATP production from unrestricted oxidation of inorganic and organic energy substrates could upset the regulatory mechanisms that respond to energy charge (Atkinson, 1969). Thus, the metabolic logic for restricting the use of autotrophic and heterotrophic growth substrates by a facultative chemolithotroph cultivated in a batch mixotrophic medium is the same that leads *Escherichia coli* to shut off the utilization of lactose in a batch culture also supplied with glucose.

It can be concluded that the above regulatory strategy is well suited to nutrient-excess environments. But what is its significance to the growth of facultative thiobacilli in their natural environment? In contrast to the conditions in batch culture, most natural environments are characterized by extremely low nutrient concentrations that permit growth only at submaximal rates (Duursma, 1963). Under such mixotrophic conditions, restriction of substrate utilization would not be a sensible strategy since the problem faced by the organism would be scarcity rather than excess of energy and carbon substrates. A different regulatory strategy is therefore called for under nutrient-limited conditions, as is illustrated by the case of *E. coli,* which, rather than exhibiting diauxie, makes concomitant use of lactose and glucose when they are supplied at limiting concentrations (Silver and Mateles, 1969).

If nutrient-excess conditions are an anomaly in nature, it might be asked why these organisms have taken the trouble to evolve the elaborate regulatory strategy appropriate to such environments. The answer must be that the vicissitudes of nature lead to the presence in the environment of excessive nutrients often enough to make the evolution of this strategy necessary.

CONTINUOUS CULTURE STUDIES

As discussed above, the early studies on mixotrophy were conducted using exclusively batch culture techniques, and the results were considered to be of direct ecological relevance. Based on the finding of mutual antagonism between the autotrophic and heterotrophic metabolisms, Whittenbury and Kelly (1977) suggested that the facultative chemolithotrophs existed as heterotrophs in nature, and not as mixotrophs. This view contradicted the hypothesis proposed earlier by Rittenberg (1969, 1972), and was challenged by Matin (1978) based on the points discussed above, namely, that conclusions of batch culture studies could not be directly applied to microbial growth in most natural environments. Studies have shown that the behavior of *T. novellus* and *T.* A2 under nutrient-limited mixotrophic environments is indeed quite different.

T. novellus

Growth pattern. Mixotrophy under nutrient-limited conditions was first studied in *T. novellus* (Matin, 1977, 1978; Leefeldt and Matin, 1980). In chemostat cultures, the steady-state cell biomass of *T. novellus* increased with increasing dilution rate (D) under autotrophic and heterotrophic conditions (fig. 5). However, under mixotrophic conditions, the biomass increased with D up to 0.05 h^{-1} and then declined. At all D values up to 0.05 h^{-1}, i.e., the highest D at which the organisms could be grown in all the three different nutritional environments, the mixotrophic biomass was additive of the autotrophic and heterotrophic biomass at corresponding D values. At D = 0.08 h^{-1}, however, the mixotrophic biomass was lower than the heterotrophic biomass.

The steady-state concentration of thiosulfate, glucose, or both in the culture fluid was very low at D = 0.06 h^{-1} or lower under all growth conditions (table 2), which is consistent with their being the growth-limiting nutrient. Furthermore, at D values of 0.05 h^{-1} or lower, the steady-state con-

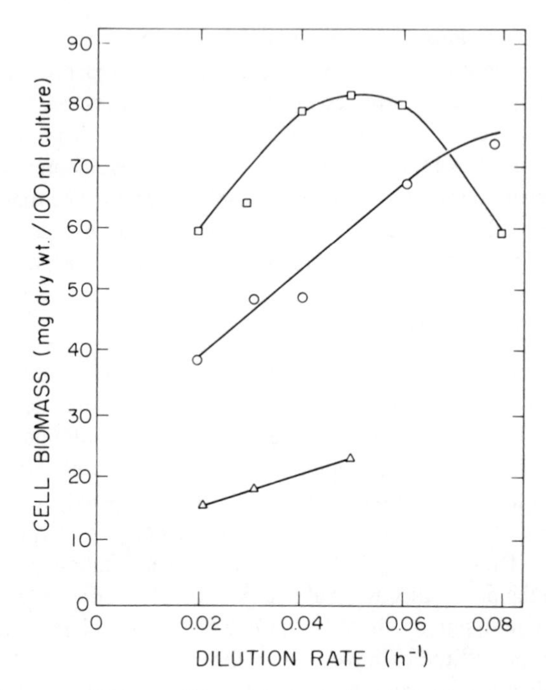

Fig. 5. Steady-state cell biomass of *T. novellus* in nutrient-limited heterotrophic (○), autotrophic (△), and mixotrophic (□) environments at specified dilution rates. Substrate concentrations in inflow media: glucose, 8 mM and $Na_2S_2O_3$, 40 mM. From Leefeldt and Matin (1980).

TABLE 2

STEADY-STATE RESIDUAL SUBSTRATE CONCENTRATIONS IN CULTURES *T. NOVELLUS* AT VARIOUS DILUTION RATES

	Residual substrate concn in various cultures (μM)			
Dilution rate (h^{-1})	Heterotrophic glucose	Mixotrophic thiosulfate	Mixotrophic glucose	Autotrophic thiosulfate
0.02	11	60	11	56
0.03	9	78	19	76
0.04	44	—[a]	44	—
0.05	44	78	44	115
0.06	46	75	44	—
0.08	11	8,500	3,000	—

SOURCE: Leefeldt and Matin (1980).

[a]—, Not determined.

centrations of thiosulfate and glucose in mixotrophic culture were similar to their concentrations in autotrophic and heterotrophic cultures, respectively, at corresponding D values. Thus, at these D values, there was no inhibition of thiosulfate or glucose utilization in the mixotrophic medium; had the inhibition occurred, a higher steady-state level of the affected substrate would be found in the mixotrophic culture compared with the single substrate cultures at the corresponding D value.

In contrast, at D = $0.08\ h^{-1}$, the steady-state concentrations of glucose and thiosulfate in mixotrophic culture were much higher than in single substrate cultures at corresponding D value. It is therefore evident that in contrast to D values of $0.05\ h^{-1}$ and lower, at D = $0.08\ h^{-1}$, the utilization of both thiosulfate and glucose is partially inhibited in mixotrophic medium. In this respect, as well as in not exhibiting an increased yield, the D = $0.08\ h^{-1}$ mixotrophic culture resembles the batch mixotrophic cultures.

Role of glucose. Since both glucose and thiosulfate contributed to growth under nutrient-limited conditions, the question arose as to whether glucose provided only cell carbon or both carbon and energy. A comparison of the amount of glucose carbon used up with the amount of carbon in the cell material synthesized showed that at various D values in the mixotrophic environment, some 15–50% glucose was used in energy generation along with thiosulfate. However, at D = $0.08\ h^{-1}$, nearly all the glucose taken up from the medium could be accounted for by the cell material synthesized. Thus, under such conditions *T. novellus* resembled *T. intermedius* in that glucose served only as a source of carbon.

Enzyme regulation. As might be expected from the finding that in mixotrophic cultures at D = $0.05\ h^{-1}$ or lower there was no inhibition of thiosulfate or glucose utilization, it was found that such cells did not exhibit enzyme repression characteristic of cells grown under nutrient-excess mixotrophic conditions. Indeed, such cells possessed a higher respiratory capacity for thiosulfate and a higher sulfite oxidase activity compared with autotrophic cells. They also exhibited a higher glucose-oxidizing ability, and they contained higher levels of many glucose enzymes than heterotrophic cells. As noted above, the D = $0.08\ h^{-1}$ mixotrophic cultures exhibited partial inhibition of substrate utilization; measurements showed that these cultures also exhibited the enzyme repression characteristic of mixotrophic nutrient-excess conditions (table 3).

It is evident that the behavior of D = $0.08\ h^{-1}$ and batch mixotrophic cultures is very similar. This is not surprising since at this D value, which is very close to the μ_{max} of the organism under these conditions, the cells grew

TABLE 3

EFFECT OF NUTRIENT-LIMITED GROWTH ENVIRONMENT ON RESPIRATORY AND ENZYMATIC ACTIVITIES OF *T. NOVELLUS* GROWN AT DILUTION RATES OF 0.05 AND 0.08 H^{-1}

Activity[a]	D = 0.05 h^{-1}			D = 0.08 h^{-1}	
	Heterotrophic	Mixotrophic	Autotrophic	Heterotrophic	Mixotrophic
Thiosulfate oxidation	—[b]	135	60	—	—
Sulfite oxidase (EC 1.8.3.1)	—	10,930	5,430	—	—
Glucose oxidation	11	22	—	20	5
Glucokinase (EC 2.7.1.2)	170	490	—	250	270
Glucose 6-phosphate dehydrogenase (EC 1.1.1.49)	27	210	—	60	23
Isocitrate dehydrogenase (EC 1.1.1.42)	4,830	4,220	—	4,700	4,980
cis-Aconitase (EC 4.2.1.3)	1,950	2,020	—	2,000	2,220
NADH oxidase (EC 1.6.99.3)	70	72	—	270	290

SOURCE: Leefeldt and Matin (1980).

[a]Oxygen uptake, nanomoles per minute per milligram of protein. Enzyme activity, 10^{-4} enzyme unit per mg of protein.

[b]—, not determined.

at near-saturating nutrient concentration, and thus made use of the regulatory strategy suited to nutrient-excess conditions (Leefeldt and Matin, 1980).

Role of CO_2. As discussed above, in batch mixotrophic cultures, CO_2 probably did not serve as a significant source of cell carbon for *T. novellus;* but under certain nutrient-limited mixotrophic conditions, it did. This was demonstrated by Perez and Matin (1982) through direct studies on the assimilation of $^{14}CO_2$ by chemostat cultures. *T. novellus* was grown in different environments, and after the steady-state had been reached, the cultures were pulsed with $NaH^{14}CO_3$ and initial rates of $^{14}CO_2$ uptake were determined without disturbing the steady-state. These rates were translated into the contribution of CO_2 to cell material synthesis by comparison with the $^{14}CO_2$ uptake rate of steady-state autotrophic culture, in which 90% of the cell carbon was provided by CO_2. Under heterotrophic conditions, some 13% of cell carbon was derived from CO_2. This value is comparable to the 10% amount that is generally derived from CO_2 by anaplerotic reactions during heterotrophic growth in batch culture. The contribution of CO_2 to the synthesis of cell material under mixotrophic conditions depended on the ratio of [thiosulfate] to [glucose] in the inflow medium. When this ratio was 5, nearly 17% of cell carbon came from CO_2, i.e., somewhat higher than under heterotrophic conditions. Reducing this ratio to 2.5 decreased CO_2 assimilation to the range of the heterotrophic culture, but when the ratio was increased to 10, some 32% of the cell carbon was derived from CO_2. Under the latter conditions, the steady-state biomass was 23 mg of cell dry wt, i.e., about 11.5 mg of cell carbon/ml of culture, and the amount of glucose supplied in the inflow medium (1.67 mM) was equivalent to 12 mg of carbon/100 ml of medium. Thus, the available glucose was just enough to serve as sole carbon source under these conditions. The fact that 32% of the cellular carbon was derived from CO_2 indicates that an equivalent amount of the available glucose was respired. There must have been a net loss of energy in these conversions, since the amount of energy gained from glucose oxidation is likely to be considerably less than that expended in generating an equivalent amount of organic material from CO_2.

Thiobacillus A2

Mixotrophy in *T.* A2 has been examined under thiosulfate-acetate limitation by Gottschal and Kuenen (1980a), and under thiosulfate-glucose limitation by Smith and colleagues (1980).

Mixotrophic growth under acetate plus thiosulfate limitation occurred through a range of [acetate]/[thiosulfate] ratios in the inflow medium at a fixed D value, as well as through a range of D values at a fixed [acetate]/

[thiosulfate] ratio. At D = 0.05 h^{-1}, through a large range of [acetate]/[thiosulfate] ratios in the inflow medium, both acetate and thiosulfate were growth limiting, since steady-state cultures exhibited no detectable residual of these substrates, and increase in the concentration of either in the inflow medium caused a proportional increase in steady-state biomass. The biomass formation during mixotrophic growth was greater than the additive yield of heterotrophic and autotrophic cultures, the precise value of the greater-than-expected yield depending on the [acetate]/[thiosulfate] ratio in the inflow medium. The CO_2-fixing capacity, as determined using aliquots of steady-state mixotrophic cultures, and the RuBP carboxylase activity of the cells also depended on this ratio, and exhibited progressive decreases as this ratio was increased; at [10 mM acetate]/[20 mM thiosulfate], no CO_2-fixing capacity remained.

The authors attributed the greater-than-expected increase in biomass formation under mixotrophic conditions to the elimination of CO_2 as major carbon source and the resultant saving of energy in cell biosynthesis from acetate. Their hypothesis would predict that the greatest increase in biomass formation should occur at [acetate]/[thiosulfate] ratios at which CO_2-fixing ability was completely eliminated. But, in fact, the highest increase (30% greater than expected) occurred under conditions where a significant potential for CO_2 fixation was present in the cells (compare figure 4B with figure 3 in Gottschal and Kuenen [1980a]). Thus, other factors probably also played a role in this phenomenon.

Gottschal and Kuenen (1980a) used measurements of O_2 and CO_2 concentrations in the outflowing air, as well as the assimilation of ^{14}C-acetate to estimate that at 8 mM acetate plus 24 mM thiosulfate in the inflow medium, about 40% cell carbon was derived from CO_2, the rest being contributed by acetate; energy was derived from both thiosulfate and acetate oxidation. Thus, *T.* A2, like *T. novellus,* can attain a fully mixotrophic existence.

Similar findings were made concerning mixotrophic growth of *T.* A2 on a mixture of thiosulfate plus glucose under nutrient-limited conditions (Smith et al., 1980). However, as was the case with *T. novellus* in this environment (Leefeldt and Matin, 1980), the mixotrophic biomass was additive of that found under heterotrophic and autotrophic conditions. CO_2, glucose, and thiosulfate were all utilized concurrently under mixotrophic conditions.

RATIONALE OF REGULATORY STRATEGY UNDER NUTRIENT LIMITATION

It is clear that in contrast to the nutrient-excess mixotrophic conditions, the facultative thiobacilli tend to attain a full expression of their mixotrophic potential under nutrient-limited conditions. There is no or minimal restriction of the use of potential carbon and energy substrates and little repression of enzyme synthesis. Thus, under conditions where the supply of carbon and energy substrates is limited, the facultative chemolithotroph is able to mobilize fully its ability to use diverse carbon and energy substrates. Such a strategy may give these bacteria a selective advantage over heterotrophs and obligate chemolithotrophs, which are restricted primarily to either the inorganic or the organic substrates. There is now direct experimental evidence in support of this view that will be discussed below.

Under certain mixotrophic environments, however, these organisms seem to overexpress their mixotrophic potential. Such is the case under conditions where enough organic carbon is supplied in the medium to fully satisfy the demand for cell biosynthesis, and yet significant cell carbon is derived from CO_2. This leads to the curious situation that equivalent amounts of organic material are generated from CO_2 and respired back to CO_2. In energetic terms, and in terms of biomass synthesis, this is not a sensible strategy because more energy has to be consumed to generate organic material from CO_2 than can be obtained from the complete oxidation of an equivalent amount of the organic material. However, this strategy may be advantageous as regards the growth rate that the organism can attain at relatively high [thiosulfate]/[organic matter] ratios under nutrient-limited conditions. The very minimal and subsaturating concentrations of the organic compound that can be expected to be present under these conditions would probably permit growth at a very low rate. This could conceivably be augmented by relying on CO_2 as a major carbon source, especially if it is present at saturating levels. Since in nature selection appears to favor a faster-growing organism rather than a more efficient one, this strategy might be of survival value to a facultative chemolithotroph; the energy wasted in the process, especially when a larger flux of thiosulfate ensures its relative abundance, is probably well worth the price.

COMPETITION STUDIES

The findings discussed above that facultative thiobacilli attain full expression of their mixotrophic potential under nutrient-limited mixotrophic conditions suggested that these bacteria should possess a selective advantage over heterotrophs and obligate chemolithotrophs (Matin, 1978). This prediction was first put to experimental test by Gottschal and colleagues (1979), who examined competition between the mixotrophic *T.* A2, a heterotrophic bacterium *Spirillum* G7, and an obligate chemolithotroph *Thiobacillus neapolitanus* under various nutrient-limited growth conditions.

Under autotrophic conditions of thiosulfate limitation, *T. neapolitanus* outcompeted *T.* A2 at most of the D values examined; and under heterotrophic conditions of acetate limitation, *Spirillum* G7 outcompeted *T.* A2. Thus, under strict autotrophic and heterotrophic conditions, the specialized organisms were at an advantage. Two other results of these experiments are worth noting. In competition in the autotrophic environment, where *T. neapolitanus* became the dominant population, *T.* A2 was never completely eliminated, but constituted about 9% of the steady-state population. This was ascribed to the growth of *T.* A2 on small amounts of glycollate that are known to be excreted by *T. neapolitanus* during growth (Cohen et al., 1979). The other noteworthy result is that at very low dilution rates, *T.* A2 won the competition against the obligate autotroph as well as the heterotroph. As discussed by Matin and Veldkamp (1978), this result was probably related to the role of minimal growth rate (μ_{min}) in the outcome of competition. These low D values were considerably below μ_{min} (which is believed to be about 6% of the μ_{max} of an organism [Pirt, 1972]) of *T. neapolitanus* but not that of *T.* A2, thus enabling the latter to establish itself during competition.

In contrast, under the mixotrophic conditions of dual limitation of acetate and thiosulfate, *T.* A2 had a clear selective edge over *T. neapolitanus* as well as *Spirillum* G7. The degree of this advantage depended on the relative proportion of thiosulfate and acetate in the inflow medium, and indeed, at certain proportions of these substrates, all three bacteria coexisted under steady-state conditions (fig. 6). The three bacteria were allowed to compete simultaneously at $D = 0.05\ h^{-1}$ using different ratios of [acetate]/[thiosulfate] in the inflow medium, and the relative cell numbers of the different organisms at steady-state were plotted (fig. 6). At all ratios of the two substrates, *T.* A2 was able to maintain its presence in the culture. In the range of [acetate]/[thiosulfate] ratio of 0/40 to 5/30, the spirillum was eliminated from the culture and increasing ratios favored the relative abundance of *T.* A2. Thus, the ability of *T.* A2 to make concurrent use of the two substrates enabled it to deny both the specialist bacteria their share of the substrates. In the range of substrate ratios of 15/10 to 20/0, *T. neapolitanus* was eliminated and decreasing ratios favored the dominance of *T.* A2 over the spirillum. Thus, two concentration ranges allowed the dominance of the facultative chemolithotroph *T.* A2: those of high [thiosulfate]/[acetate] ratios permitting coexistence with the obligate chemolithotroph, and those of high [acetate]/[thiosulfate] ratios allowing the coexistence with the heterotroph. The concentration ranges between these extremes allowed the coexistence of all three organisms. The authors concluded that it is the relative turnover ratio of inorganic and organic compounds in the natural environment that would determine the outcome of

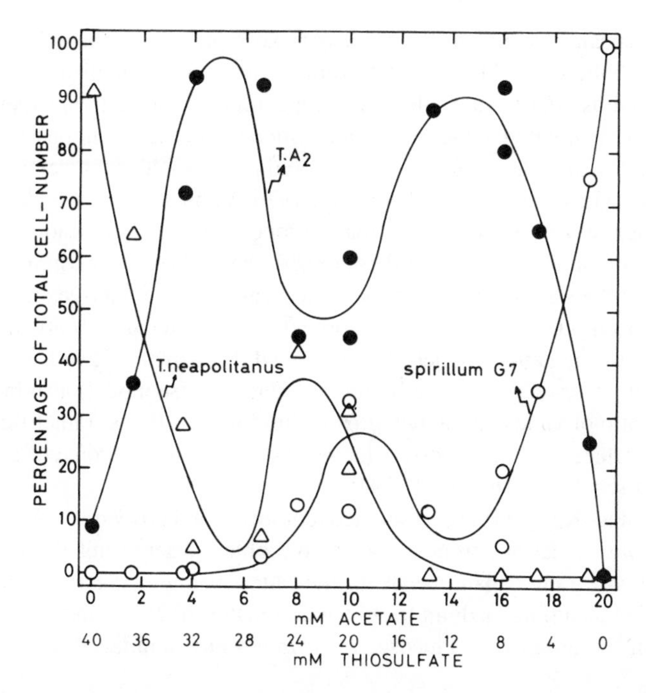

Fig. 6. Competition between *T.* A2, *T. neapolitanus,* and *Spirillum* G7 for thiosulfate and acetate as growth-limiting substrates in the chemostat at a dilution rate of 0.05 h^{-1}. Concentrations of these substrates in the inflowing medium ranged from 0-20 mM for acetate and 40–0 mM for thiosulfate. Relative steady-state cell numbers were determined at different ratios of [acetate]/[thiosulfate] in the feed. Symbols: ●, *T.* A2; △, *T. neapolitanus;* and ○, *Spirillum* G7. From Gottschal et al. (1979).

competition between obligate chemolithotrophs and chemoorganotrophs and the facultative chemolithotrophs.

In an extension of this work, Gottschal and colleagues (1981a) examined competition between the three organisms in cultures subjected to alternating heterotrophic and autotrophic conditions. The rationale for these experiments was provided by the finding (Gottschal et al., 1981b) that during an alternating regimen of 4 h acetate and 4 h thiosulfate feed, *T.* A2 exhibited continuous growth in the chemostat, and was able to utilize the two substrates without a lag. Nevertheless, under such alternating conditions *T.* A2 was outcompeted by *T. neapolitanus* and the spirillum. This finding emphasizes that the facultative chemolithotrophs like *T.* A2 owe their selective advantage over the specialized organisms to their mixotrophic potential rather than to their metabolic flexibility to switch from one type of metabolism to another.

The establishment of a stable mixed population in competition experiments is not consistent with the theory of competition under the conditions employed that predicts that the faster growing population would exclude the others completely (Jost et al., 1973; Veldkamp, 1976; Yoon et al., 1977). A possible explanation is that factors other than competition for a common substrate played a role in the outcome of these experiments. Another possibility is that because of the slow dynamics of the mixed-culture system a true steady-state had not been attained in these experiments. The latter possibility was supported by mathematical models that described the growth of two bacteria in a chemostat under dual substrate limitation. One of these models also indicated that the ratio of yields on the two growth-limiting substrates, rather than their absolute value, was the important factor in deciding the outcome of competition (Gottschal and Thingstad, 1981).

Smith and Kelly (1980) examined competition between *T.* A2 and *T. neapolitanus* under autotrophic and mixotrophic nutrient-limited conditions, using glucose rather than acetate in the latter medium. Again, addition of glucose to thiosulfate medium led to the dominance of *T.* A2 under conditions where, in the absence of glucose, *T. neapolitanus* dominated.

ENRICHMENT STUDIES

The finding that under nutrient-limited mixotrophic conditions, the facultative chemolithotrophs can benefit from their mixotrophic potential led to the suggestion (Matin, 1978) that mixotrophy may be the primary mode of existence of facultative chemolithotrophs in nature. Furthermore, this may be the reason why the enrichment, for instance, of facultative thiobacilli in classical batch cultures is often so difficult. In such cultures the high extant thiosulfate concentrations direct the selective pressure toward μ_{max} for autotrophic growth, which is almost invariably higher in obligate than in facultative thiobacilli. Matin (1978) suggested that the use of a nutrient-limited mixotrophic environment will prove to be a dependable and specific method for the enrichment of facultative chemolithotrophs from nature.

The results of the competition studies discussed above were consistent with this view, but the first experimental verification of it was provided by the chemostat enrichment studies of Gottschal and Kuenen (1980b). These workers conducted enrichment studies from a wide variety of aquatic environments using thiosulfate plus acetate-limited chemostat cultures. The [thiosulfate]/[acetate] ratio in the inflow medium was varied through a considerable range, and the D value employed was 0.05 h^{-1}. In most cases of enrichment from freshwater environments, the dominant population (up to 86%) at steady-state was that of facultative thiobacilli. One of these isolates that was examined further was similar to *T.* A2 but differed from the latter sufficiently to justify a new designation: *Thiobacillus* III. In many cases, the secondary population consisted of chemolithotrophic heterotrophs of *Thiobacillus perometabolis* type (London and Rittenberg, 1967), and in one enrichment, the dominant population consisted of such organisms. It is noteworthy that such organisms are also capable of mixotrophic growth (London and Rittenberg, 1967).

For unexplained reasons, enrichment from marine environments did not lead to the selection of facultative thiobacilli but to stable mixtures of a heterotroph and an obligate chemolithotroph (Gottshal and Kuenen, 1980b).

CONCLUSIONS

The studies with facultative thiobacilli have clearly demonstrated that the mixotrophic way of life is the special niche for these organisms in nature. The mixotrophic potential finds a full expression under the ecologically pertinent conditions of nutrient limitation. At least under certain conditions of relative turnover rates of inorganic and organic growth substrates, the mixotrophic potential confers on the facultative chemolithotroph a decisive advantage over strict heterotrophs and obligate chemolithotrophs. This point is clearly made by the competition studies as well as the enrichment studies discussed above. Why the enrichments from marine environments failed to select facultative chemolithotrophs is not known and deserves further examination.

The natural environment is, of course, rarely stable and uniform, and must present the resident microflora with considerable fluctuation in the concentration of available nutrients. Under conditions where utilizable carbon and energy substrates of a facultative chemolithotroph are available in high concentration, the mixotrophic potential may become a liability in leading to the generation of excessive energy and accumulation of high levels of carbon metabolites in the cell. Thus, a well-adapted facultative chemolithotroph has to possess sufficient regulatory resilience to make efficient and concurrent use of multiple carbon and energy substrates when they are in short supply, and yet to minimize their use and energy generation from this oxidation when their high concentration can become a threat to survival. As we have seen, the facultative thiobacilli do indeed possess this resilience to an exquisite degree. Nor are these bacteria alone in having a need for this resilience. Any organism possessing any degree of metabolic flexibility would require it, and it is now well established that many strict heterotrophs also exhibit a similar resilience. When supplied with mixtures of utilizable organic compounds, these heterotrophs can utilize them concurrently under nutrient limitation and yet are able to curtail their use in a high concentration environment (Harder and Hijkhuizen, 1976; Matin, 1978; Matin, 1981).

The molecular mechanism whereby facultative chemolithotrophs affect interaction between autotrophic and heterotrophic pathways that do not share common catabolites remains entirely unknown. For any progress in this area, genetic analysis of these organisms is a necessary prelude. The importance of genetic studies of chemolithotrophic bacteria for this and other compelling reasons is becoming realized. Recently a number of plasmids in hydrogen bacteria (Pootjes, 1977; Friedrich et al., 1981; Reh and Schlegel, 1981; Gerstenberg et al., 1982) and the thiobacilli (T. Nnalua, and A. Matin, unpublished) have been found, and a transformation system recently has been developed for *T. thioparus* by Stark and Yankofsky (1981).

In nature both nutrient-excess and nutrient-limited conditions must exist, with the latter being most probably the rule. However, most of our current understanding of the regulation of microbial metabolic function pertains to nutrient-excess conditions because of the nearly exclusive use of batch conditions in such studies. The studies on mixotrophy discussed above illustrate how different this regulation can be in environments of different nutritional status; similarly, the early controversy on the ecological relevance of the mixotrophic way of life demonstrates the perils of direct extrapolation of batch culture studies to nature.

LITERATURE CITED

Aleem, M. I. H., and E. Huang. 1965. Carbon dioxide fixation and carboxydismutase. Biochem. Biophys. Res. Commun. 20:515–20.

Atkinson, D. E. 1969. Regulation of enzyme function. Ann. Rev. Microbiol. 23:47–68.

Cohen, Y., I. deJorge, and J. G. Kuenen. 1979. Excretion of glycollate by *Thiobacillus neapolitanus* grown in continuous culture. Arch. Microbiol. 122:189–94.

Duursma, E. K. 1965. The dissolved organic constituents of sea water. *In* J. P. Riley and G. Skirrow (eds.), Chemical oceanography, 1:433–75. Academic Press, New York.

Friedrich, B., C. Hogrefer, and H. G. Schlegel. 1981. Naturally occurring genetic transfer of hydrogen-oxidizing ability between strains of *Alcaligenes eutrophus*. J. Bacteriol. 147:198–205.

Gerstenberg, C., B. Friedrich, and H. G. Schlegel. 1982. Physical evidence for plasmids in autotrophic, especially hydrogen-oxidizing bacteria. Arch. Microbiol. 133:90–96.

Gottschal, J. C., S. deVries, and J. G. Kuenen. 1979. Competition between the facultatively chemolithotrophic *Thiobacillus* A2, an obligately chemolithotrophic *Thiobacillus* and a heterotrophic spirillum for inorganic substrates. Arch. Microbiol. 121:241–49.

Gottschal, J. C., and J. G. Kuenen. 1980a. Mixotrophic growth of *Thiobacillus* A2 on acetate and thiosulfate as growth limiting substrates in the chemostat. Arch. Microbiol. 126:33–42.

Gottschal, J. C., and J. G. Kuenen. 1980b. Selective enrichment of facultatively chemolithotrophic thiobacilli and related organisms in continuous culture. FEMS Microbiol. Lett. 7:241–47.

Gottschal, J. C., H. J. Nanninga, and J. G. Kuenen. 1981a. Growth of *Thiobacillus* A2 under alternating growth conditions in the chemostat. J. Gen. Microbiol. 126:85–96.

Gottschal, J. C., A. Pol, and J. G. Kuenen. 1981b. Metabolic flexibility of *Thiobacillus* A2 during substrate transitions in the chemostat. Arch. Microbiol. 129:23–28.

Gottschal, J. C., and T. F. Thingstad. 1981. Mathematical description of competition between two and three bacterial species under dual substrate limitation in the chemostat: a comparison with experimental data. Biotechnol. Bioeng. 24:1403–18.

Harder, W., and L. Dijkhuizen. 1976. Mixed substrate utilization in microorganisms. *In* A. C. R. Dean, D. C. Elwood, C. G. T. Evans, and J. Melling (eds.), 6th international symposium of continuous culture, pp. 297–314. Ellis-Horwood Ltd., Chichester, England.

Jost, J. L., J. F. Drake, A. G. Fredrickson, and H. M. Tsuchiya. 1973. Interactions of *Tetrahymena pyriformis, Escherichia coli,* and *Azotobacter vinelandii* in a glucose minimal medium. J. Bacteriol. 113:834–40.

Leefeldt, R. H., and A. Matin. 1980. Growth and physiology of *Thiobacillus novellus* under nutrient-limited mixotrophic conditions. J. Bacteriol. 142:645–50.

London, J., and S. C. Rittenberg. 1966. Effects of organic matter on the growth of *Thiobacillus intermedius.* J. Bacteriol. 91:1062–69.

Magasanik, B. 1961. Catabolite repression. Cold Spring Harbor Symp. Quant. Biol. 26:249–56.

Matin, A., and S. C. Rittenberg. 1970a. Utilization of gluose in heterotrophic media by *Thiobacillus intermedius.* J. Bacteriol. 104:234–38.

Matin, A., and S. C. Rittenberg. 1970b. Regulation of glucose metabolism in *Thiobacillus intermedius.* J. Bacteriol. 104:239–47.

Matin, A. 1977. Mixotrophic growth of *Thiobacillus novellus* in batch and continuous cultures. Abstr. Ann. Meet. Amer. Soc. Microbiol. I 4, p. 155.

Matin, A. 1978a. Organic nutrition of chemolithotrophic bacteria. Ann. Rev. Microbiol. 32: 433–68.

Matin, A. 1978b. Microbial regulatory mechanisms at low nutrient concentrations as studied in chemostat. *In* M. Shilo (ed.), Strategies of microbial life in extreme environments, pp. 323–39. Life Science Research Report, Vol. 13. Verlag Chemie, Weinheim.

Matin, A., and H. Veldkamp. 1978. Physiological basis of the selective advantage of a *Spirillum* sp. in a carbon-limited environment. J. Gen. Microbiol. 105:187–97.

Matin, A., M. Schleiss, and R. C. Perez. 1980. Regulation of glucose transport and metabolism in *Thiobacillus novellus.* J. Bacteriol. 142:639–44.

Matin, A. 1981. Regulation of enzyme synthesis as studied in continuous culture. *In* P. Calcott (ed.), Continuous cultures of cells, 2:69–67. CRC Press, Boca Raton, Florida.

Perez, R., and A. Matin. 1980. Growth of *Thiobacillus novellus* on mixed substrates (mixotrophic growth). J. Bacteriol. 142:633–38.

Pirt, S. J. 1972. Prospects and problems in continuous flow culture of microorganisms. J. Appl. Chem. Biotechnol. 22:55–64.

Pootjes, C. F. 1977. Evidence for plasmid coding of the ability to utilize hydrogen gas by *Pseudomonas facilis*. Biochem. Biophys. Res. Commun. 76:1002–6.

Reh, M., and H. G. Schlegel. 1981. Hydrogen autotrophy as a transferable genetic character of *Nocardia opaca* lb. J. Gen. Microbiol. 126:327-336.

Rittenberg, S. C. 1969. The roles of exogenous organic matter in the physiology of chemolithotrophic bacteria. Adv. Microb. Physiol. 3:159–96.

Rittenberg, S. C. 1972. The obligate autotroph: the demise of a concept. Antonie van Leeuwenhoek 38:457–78.

Romano, A. H., N. J. van Vranken, P. Preisand, and M. Brustolon. 1975. Regulation of *Thiobacillus intermedius* glucose uptake system by thiosulfate. J. Bacteriol. 121:577–82.

Schlegel, H. G. 1975. Mechanisms of chemo-autotrophy. *In* O. Kline (ed.), Marine ecology, 2:9–59. John Wiley and Sons, New York.

Silver, R. S., and M. I. Mateles. 1969. Control of mixed-substrate utilization in continuous culture of *Escherichia coli*. J. Bacteriol. 97:535–43.

Smith, A. L., D. P. Kelly, and A. P. Wood. 1980. Metabolism of *Thiobacillus* A2 grown under autotrophic, mixotrophic, and heterotrophic conditions in chemostat culture. J. Gen. Microbiol. 121:127–38.

Smith, D. W., and S. C. Rittenberg. 1974. On the sulfur-source requirement for growth of *Thiobacillus intermedius*. Arch. Microbiol. 100:65–71.

Stark, A. A., and S. Yankofsky. 1981. Regulation of amino acid transport in *Thiobacillus thioparus*. J. Bacteriol. 148:966–72.

Taylor, B. F., and D. S. Hoare. 1971. *Thiobacillus denitrificans* as an obligate chemolithotroph. II. Cell suspension and enzymic studies. Arch. Mikrobiol. 80:262–76.

Veldkamp, H. 1976. Continuous culture in microbial physiology and ecology. Meadowfield Press, Durham, England.

Whittenbury, R., and D. P. Kelly. 1977. Autotrophy: a conceptual phoenix. *In* B. A. Haddock and W. A. Hamilton (eds.), Microbial energetics, pp. 121–49. Cambridge University Press, Cambridge.

Yoon, H., G. Klinzing, and H. W. Blanch. 1977. Competition for mixed substrates in microbial populations. Biotechnol. Bioeng. 19:1193–1211.

WILLIAM R. STROHL AND THOMAS M. SCHMIDT

Mixotrophy of the Colorless, Sulfide-Oxidizing, Gliding Bacteria *Beggiatoa* and *Thiothrix*

6

INTRODUCTION

The concept of bacterial chemolithoautotrophy was developed by Winogradsky (1887, 1888) as a result of his experiments on *Beggiatoa* nutrition.[1] Winogradsky observed that the addition of sulfide to slide cultures of *Beggiatoa* resulted in the deposition of sulfur inside the filamentous organisms. Upon removal of the sulfide, the sulfur deposits disappeared; Winogradsky assumed that the sulfur was oxidized to sulfate (Winogradsky, 1887, 1888). Winogradsky theorized that the oxidation of sulfide and, later, sulfur provided energy for the beggiatoas. Because he did not know of, or consider the possibility of, minute organic chemicals in his pond-water slide cultures, Winogradsky (1887, 1888) assumed that the beggiatoas obtained their carbon only from CO_2. Winogradsky (1888) did not purify *Beggiatoa*, however, and he later used axenic cultures of *Nitrobacter* to prove his chemoautotrophy theories (Winogradsky, 1890).

The term *mixotrophy* was first applied to bacterial nutrition by Pringsheim (1967) to describe the growth of certain *Beggiatoa* strains. Kowallik and Pringsheim (1966) had stated that several of their *Beggiatoa* strains grew autotrophically, but Pringsheim (1967, 1970a) retracted that claim due to his realization that the increase of cell mass by his beggiatoas was acetate-dependent (Pringsheim, 1967). He used the term *mixotrophy* to describe the acetate-stimulated growth of *Beggiatoa* with sulfide as an energy source and CO_2 (as well as the acetate) as carbon source (Pringsheim, 1967). Thus, even though the actual nutrition of various beggiatoas is still not clear today (Strohl and Larkin, 1978; Nelson and Castenholz, 1981; Kuenen and Beudeker,

1982), those organisms have been of primary importance in the development of new concepts of bacterial nutrition.

Rittenberg (1969, 1972) addressed the difficulties surrounding the concept of chemoautotrophy, and he further defined mixotrophy as the "concurrent use of inorganic and organic energy sources *or* the concurrent assimilation of carbon dioxide via the ribulose diphosphate cycle and organic carbon" (Rittenberg, 1969). Thus, the proof of either CO_2 fixation by ribulose bisphosphate carboxylase *or* the conservation of energy from sulfide (or thiosulfate) oxidation, along with the known *Beggiatoa* requirement for organic carbon (Strohl and Larkin, 1978), would be sufficient to characterize *Beggiatoa* as having mixotrophic nutrition. In this paper, we describe the facultative mixotrophy (of the *chemolithotrophic*-type) for *Beggiatoa alba* B18LD in terms of its potential for conservation of energy from sulfide oxidation. Furthermore, we propose mixotrophy as a mode of nutrition for *Thiothrix nivea* strains JP1 and JP3 because of their apparent requirement for both a reduced sulfur source and organic carbon.

The general physiological characteristics of *Beggiatoa* and *Thiothrix* are similar (table 1). They are both colorless, filamentous, and trichome-forming bacteria. They oxidize sulfide, deposit sulfur internally, and assimilate both acetate and CO_2 (although to varying degrees). *Beggiatoa* trichomes glide

TABLE 1

MORPHOLOGICAL AND PHYSIOLOGICAL COMPARISON OF *THIOTHRIX* AND *BEGGIATOA*

Characteristic	*Beggiatoa*	*Thiothrix*[a]
Trichome formation	+[b]	+
Sheath	−	+
Holdfast	−	+
Rosette formation	−	+
Hormogonium formation	+/−	+
Gliding of intact trichomes	+	−
Gliding of hormogonia	+	+
Sulfur inclusions from sulfide	+	+
Sulfur inclusions from thiosulfate	+/−	+
Cytochromes	*b,c,a,o*	*c*,(others?)
Quinones	Q_8	Q_8
Acetate assimilation	+	+
Acetate oxidation to CO_2	+	?
CO_2 assimilation	low	moderate
Reduced sulfur requirement	−	+
Sulfide-dependent oxygen consumption	+	+

[a]Data pertaining to *Thiothrix* strains used in this study.
[b]+, positive; −, negative; +/−, some strains positive; ?, unknown.

when in contact with solid surfaces, whereas only the hormogonial cells of *Thiothrix* glide (J. M. Larkin, personal communication). Although both organisms are found in lakes, streams, and ditches, their ecological niches appear to be quite different (Larkin and Strohl, 1983). *Beggiatoa* has been observed to form mats associated with sediment surfaces at the undisturbed interface between oxygen and sulfide concentration gradients (Jørgensen, 1982). Upon changes in the oxidized and reduced zones in the sediments, the beggiatoas move vertically to maintain their place in those gradients (Jørgensen, 1982). On the other hand, *Thiothrix* attaches by its holdfast to branches, rocks, seaweed, and other solid materials in flowing waters and where sulfide is detectable. *Thiothrix* often forms long streamers that apparently obtain sulfide and other nutrients from the flowing waters or from the interface at their holdfast. *Thiothrix* is often found in organic- and sulfide-rich environments such as effluents from sewage plants (Farquhar and Boyle, 1971; Merkel, 1975) and attached to rotting seaweed in marsh streams. *Beggiatoa* and *Thiothrix* are seldom, if ever, observed in the same locations; one organism or the other appears to be best suited for a particular environment.

Although the mechanisms for sulfide oxidation have been well established for organisms such as the thiobacilli and the photosynthetic bacteria, they are essentially uncharacterized for the long-recognized but poorly understood beggiatoas and thiothrixes. Hydrogen sulfide is a toxic gas that reacts with the hemes of *c*-type cytochromes and terminal cytochrome oxidases to form nonfunctional heme-sulfide complexes that inhibit respiration in some organisms (Nicholls and Kim, 1982). However, sulfide has also been shown to donate electrons to *c*-type flavocytochromes (Gray and Knaff, 1982), via FAD to *b*-type cytochromes (Peeters and Aleem, 1970), and to the photosynthetic reaction center I of some cyanobacteria (Garlick et al., 1977). In addition, sulfide oxidation has been linked to the reduction of $NADP^+$ via ATP-dependent reversed electron flow in thiobacilli (Roth et al., 1973).

One difficulty associated with studies of sulfide oxidation is that the biological oxidation of sulfide is in competition with chemical oxidation. Jørgensen (1982), however, has shown recently that the biological oxidation of sulfide by marine beggiatoas was about 10^5–10^6 fold greater than the chemical oxidation, demonstrating that those organisms easily outcompeted chemical oxidation for sulfide.

The oxidation of sulfide by both *Beggiatoa* and *Thiothrix* results in the formation of internal sulfur depositions that add to the weight of the organisms (Güde et al., 1981; Nelson and Castenholz, 1981). There is some question as to whether the apparent increase in growth is due only to the increment added by the weight (and light scattering) of the sulfur itself (Nelson and

Castenholz, 1981) or whether the increased growth yields in excess of the deposited sulfur are due to energy yields from sulfide oxidation (Güde et al., 1981).

The interpretation of results obtained by various investigators (Keil, 1912; Scotten and Stokes, 1962; Burton and Morita, 1964; Kowallik and Pringsheim, 1966; Pringsheim, 1967; Strohl and Larkin, 1978; Strohl et al., 1981a; Güde et al., 1981; Nelson and Castenholz, 1981) is difficult because several different *Beggiatoa* strains have been used. The mechanisms for, and the benefits derived from, sulfide oxidation may be different for each of those strains. The continuum observed among autotrophic, mixotrophic, and heterotrophic sulfur bacteria in general (Kuenen and Beudeker, 1982) may also occur within *Beggiatoa* itself (Larkin and Strohl, 1983).

It has been suggested that the oxidation of sulfide to sulfur at neutral pH is not thermodynamically beneficial to the beggiatoas (Hollis, 1979). However, the following half-reactions (CRC Chemistry Handbook, 3d ed.) can be used to calculate the thermodynamics of the biological oxidation of sulfide at pH 7:

	E_o'	
$\frac{1}{2}\,O_2 + 2\,H^+ + 2\,e^- \rightarrow H_2O$	$+0.816$	(1)
$H_2O_2 + 2\,H^+ + 2\,e^- \rightarrow 2\,H_2O$	$+1.357$	(2)
$S^o + 2\,H^+ + 2\,e^- \rightarrow H_2S$	-0.243	(3)

The above half-reactions can be coupled to yield the following summations:

	E_o'	
$H_2S + \frac{1}{2}\,O_2 \rightarrow S^o + H_2O$	$+1.059$	(4)
$H_2S + H_2O_2 \rightarrow S^o + 2\,H_2O$	$+1.60$	(5)

The potential energy of the above reactions can be described in terms of their thermodynamics ($\Delta G_o'$), where:

$$\Delta G_o' = -nF\,E_o' \,. \qquad (6)$$

Thus, for the oxygen-coupled biological oxidation of sulfide the $\Delta G_o'$ is -204.2 kJ mol^{-1}.

If the oxidation of sulfide were coupled to reduction of hydrogen peroxide, the $\Delta G_o'$ is -308.5 kJ mol^{-1}.

Thus, O_2- or H_2O_2-coupled sulfide oxidation is thermodynamically feasible as a means of energy conservation. Whether the beggiatoas and thiothrixes take advantage of the thermodynamics of this system is the subject of great controversy (Nelson and Castenholz, 1981; Güde et al., 1981; Kuenen and Beudeker, 1982; Jørgensen, 1982). It is also the subject of this paper.

THIOTHRIX

The major morphological and physiological characteristics of *Thiothrix* are shown in table 1. Several species of *Thiothrix* have been described, although their differentiation has been based on the trichome diameters of mixed cultures. The only currently accepted species of *Thiothrix* is *T. nivea* (Larkin and Strohl, 1983).

Keil (1912) obtained pure cultures of *Thiothrix* by allowing them to attach to the bottom of glass petri dishes. He then washed them with sterile sulfide-containing stream water until they were free from contaminants. Keil demonstrated the purity of his cultures by lack of microbial growth in nutrient-rich media such as beef extract or glucose and peptone media. Although Keil claimed his cultures were autotrophic, his culture conditions were poorly defined (Pringsheim, 1967, 1970a), so it is not known whether his *Thiothrix* strains were autotrophic or mixotrophic.

Clarification of the metabolic potential of *Thiothrix* has awaited re-isolation of pure cultures, and these were obtained only recently (Larkin, 1981). The appropriate balance between oxygen and sulfide, and carbon dioxide and organic carbon, is undoubtedly required for the isolation and growth of *Thiothrix* in axenic culture, and has probably hindered past attempts at its isolation. Larkin (1981) utilized a solid medium (MY) that was developed for maintenance of *Beggiatoa* cultures, and that contained a low organic carbon level and a high sulfide level, for the isolation of *Thiothrix*. Larkin has since isolated several strains of *Thiothrix* and *Thiothrix*-like organisms from many freshwater locations.

The two strains used in this study, JP1 and JP3, were isolated by Larkin from enrichments taken from John Pennycamp Park, Florida, and morphologically the isolates appear identical. These strains have been grown on modified mixotrophic media used by Güde et al. (1981) for *B. alba*. AC (heterotrophic) medium contained per liter: 200 mg NH_4Cl, 10 mg K_2HPO_4, 10 mg $MgSO_4 \cdot 7H_2O$, 75 mg $CaCl_2 \cdot 2H_2O$, 500 mg sodium acetate, and 5 ml of a microelement solution (Strohl and Larkin, 1978). MPO5 medium consisted of AC medium plus 480 mg sodium sulfide per liter, and TS medium consisted of AC medium plus 496 mg sodium thiosulfate per liter. All media were adjusted to pH 7.3 and were sterilized by autoclaving.

Thiothrix strains JP1 and JP3 did not grow on AC medium, or on any medium tested lacking a reduced inorganic sulfur source. Addition of sulfide or thiosulfate to AC medium promoted growth of *Thiothrix*, whereas yeast extract did not substitute for the reduced sulfur source requirement. Thus, the growth characteristics for these strains suggest a mixotrophic type of metabolism. Growth on MP05 (sulfide) or TS (thiosulfate) media resulted in excellent growth yielding an excess of 1 g l^{-1} wet weight of cells. Growth was better in media containing thiosulfate as a reduced sulfur source than in media containing sulfide. This may be related to the greater stability of thiosulfate in oxygenated media compared with sulfide. *Thiothrix* trichomes contained many sulfur inclusions when exposed to either sulfide or thiosulfate. When *Thiothrix* strain JP3 thrichomes were grown in MP05 medium, harvested and washed, and resuspended in AC medium containing $Na_2{}^{35}S$, the cells incorporated the $^{35}S^{2-}$ (fig. 1) and deposited sulfur in refractile inclusion bodies. This sulfur was readily extractable with benzene, acetone, ethanol, and carbon disulfide, but not with water or 5% TCA. Electron microscopy of the sulfur inclusions of *Thiothrix* has revealed that they are internal and are surrounded by a membrane (Bland and Stanley, 1978). Both sulfide and thiosulfate were oxidized by intact trichomes (fig. 2). Sulfide-dependent oxygen uptake was inhibited by about 50% by 100 μM KCN and 32% by 50 μM HOQNO (fig. 3). *Thiothrix* strain JP3 contained a *c*-type cytochrome (fig. 4(b)), ubiquinone 8 (table 1), and it was able to oxidize the basic dye TMPD (table 1). *Thiothrix* strain JP3 did not appear to contain a CO-binding *o*-type cytochrome (fig. 4(a)), as have all *Beggiatoa* strains thus far tested (fig. 4b; W. R. Strohl, T. M. Schmidt, and J. M. Larkin, unpublished).

Mixotrophically grown *Thiothrix* assimilated $^{14}CO_2$ at a rate of about 5 times that of *B. alba* (fig. 5). Whereas exogenous acetate stimulated the assimilation of $^{14}CO_2$ in *B. alba* (Strohl et al., 1981a), addition of acetate appeared to depress the assimilation of $^{14}CO_2$ by *Thiothrix* (fig. 5). Enzyme assays showed that phospho-enol-pyruvate carboxylase, an enzyme that is lacking in *B. alba* (Strohl et al., 1981a), was very active in crude extracts of *Thiothrix* JP1. Malic enzyme, an enzyme observed in *B. alba* extracts (Strohl et al., 1981a), was apparently at low levels in *Thiothrix* extracts, and pyruvate carboxylase has not been detected in either organism.

Both CO_2 and organic carbon appear to be used by *Thiothrix* for biosynthesis. It is apparent that the *Thiothrix* strains JP1 and JP3 required reduced forms of inorganic sulfur (e.g., sulfide, thiosulfate) for growth. It is probable that these *Thiothrix* strains are either obligate mixotrophs (because they do not grow heterotrophically as do *Beggiatoa*) or facultative mixotrophs that may have the (untested) capacity for autotrophic growth.

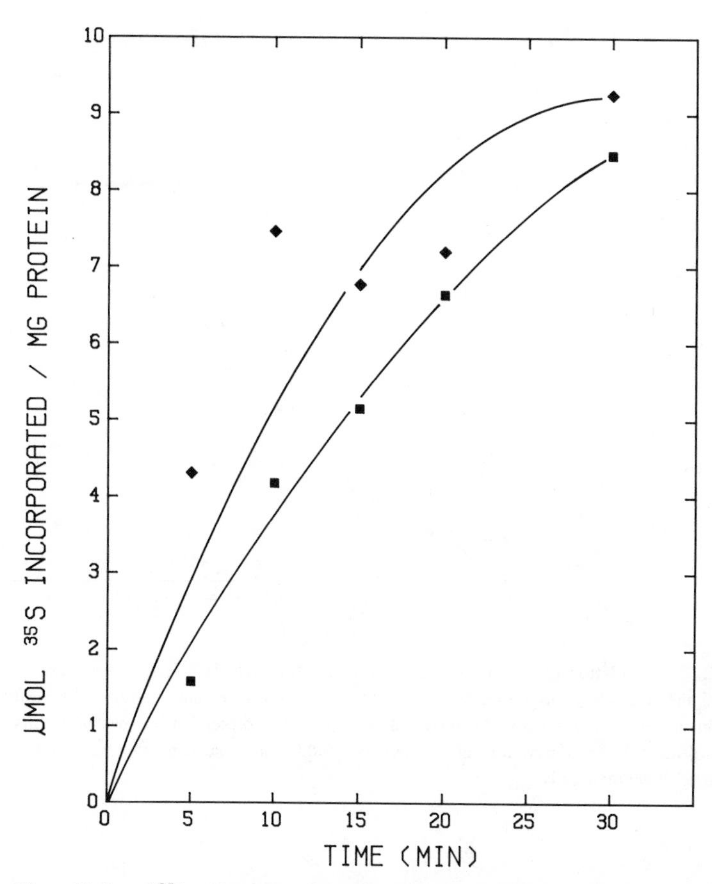

Fig. 1. The oxidation of ^{35}S-sulfide by *Beggiatoa alba* B18LD (■) and *Thiothrix* strain JP3 (◆). Washed log-phase trichomes from MPO5 media were suspended in AC media followed by the addition of 2 µCi (5 µmol) of ^{35}S-sodium sulfide (pH 7.5) at t_o. Whole cells were collected on glass fiber filters and were washed 3 times with the basal salts solution, pH 3 as described by Strohl and colleagues (1981a).

BEGGIATOA

There have been much controversy and disagreement surrounding the energy metabolism of *Beggiatoa*. Its modes of nutrition are still not fully understood (Pringsheim, 1967, 1970a, b; Strohl and Larkin, 1978; Nelson and Castenholz, 1981; Kuenen and Beudeker, 1982; Larkin and Strohl, 1983). Currently, there are three plausible explanations for the oxidation of sulfide

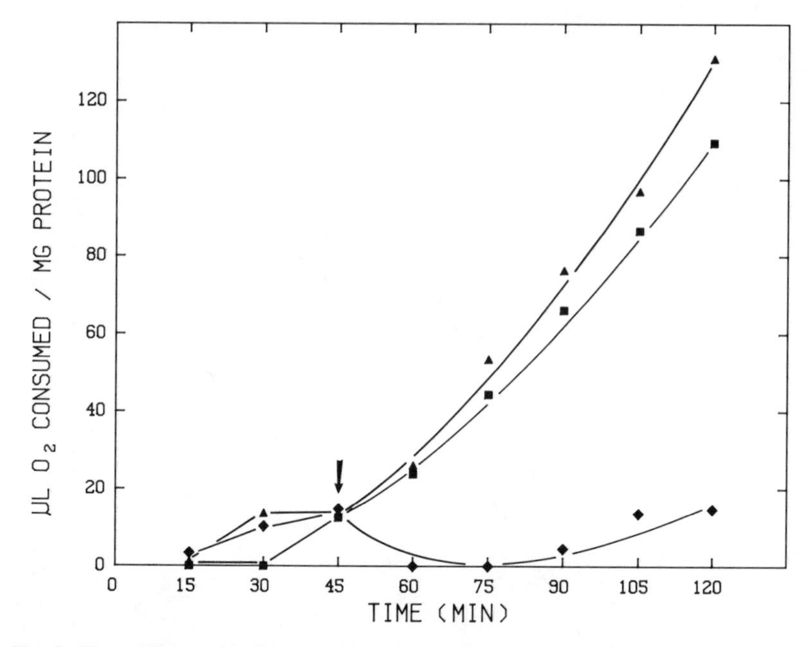

Fig. 2. The oxidation of sulfide and thiosulfate by *Thiothrix* JP3. Trichomes were grown in MPO5 medium and resuspended in AC medium for 12 h to exhaust intracellular sulfur. The trichomes were resuspended in AC media at *ca.* 1 mg ml^{-1} protein. At arrow, basal salts (◆), 1 mM sulfide (■), or 1 mM thiosulfate (▲) were added and the consumption of oxygen was measured manometrically.

by the beggiatoas. Scotten and Stokes (1962) isolated three strains of *Beggiatoa* on complex media; the isolates were only able to grow on defined media in the presence of sulfide. Although they ascribed the sulfide requirement to energy needs, it is possible that these strains required a reduced sulfur source for biosynthetic purposes, as has been observed with certain thiobacilli (Smith and Rittenberg, 1974). Secondly, because the beggiatoas lack catalase (Burton and Morita, 1964; Strohl and Larkin, 1978), sulfide may be involved in the detoxification of endogenously produced hydrogen peroxide (Burton and Morita, 1964; Nelson and Castenholz, 1981). In that way, sulfide (or thiosulfate) would replace the action of exogenously added catalase (Burton and Morita, 1964), which increases growth yields (Burton and Morita, 1964) and longevity (Strohl and Larkin, 1978), while it decreases the lag phase (Nelson and Castenholz, 1981) of *Beggiatoa* cultures. However, catalase is not required for growth of the beggiatoas (Strohl and Larkin, 1978), as has

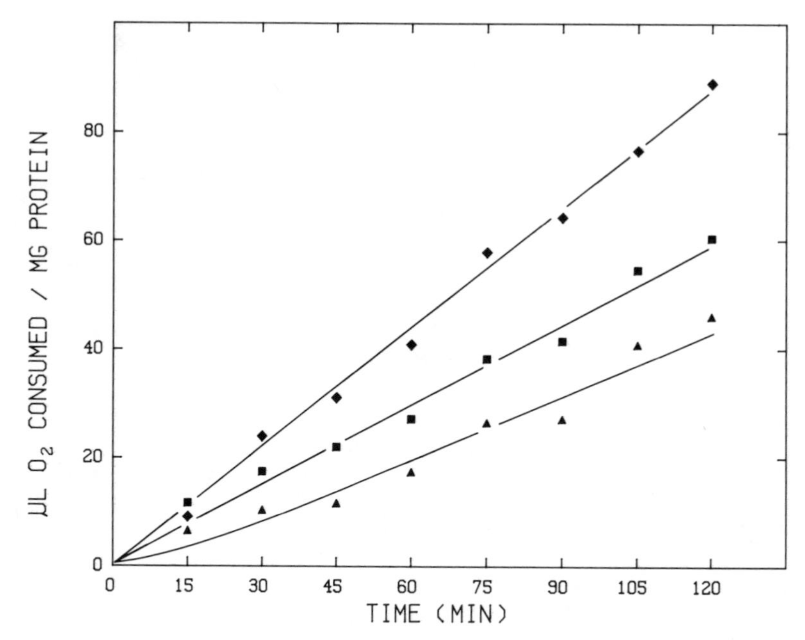

Fig. 3. The inhibition of sulfide oxidation in *Thiothrix* JP3 by electron transport inhibitors. *Thiothrix* trichomes were grown in MPO5 medium, harvested, washed, and standardized to approximately 1 mg ml^{-1} protein in basal salts with the following inhibitors present: control (◆); 100 μM KCN (▲); 50 μM HOQNO (■).

been sometimes implied (Hollis, 1979). Using strain 75-2a, Nelson and Castenholz (1981) showed that thiosulfate replaced the action of catalase and that the total yield of cellular material (as dry weight) grown with either sulfide or thiosulfate increased only by the increment of the deposited sulfur. Their data (Nelson and Castenholz, 1981) supported the concept that sulfide or thiosulfate oxidation was probably coupled with detoxification of hydrogen peroxide. The final possibility is that sulfide, and possibly thiosulfate, is used as an electron donor for energy conservation via an electron transport system (Cannon et al., 1979; Strohl and Larkin, 1980; Strohl et al., 1981a; Güde et al., 1981). Although oxygen would be the primary candidate as an electron acceptor for such a chemolithotrophic oxidative metabolism, it may also be possible that H_2O_2 acts as a (biological) terminal electron acceptor for sulfide oxidation (as shown in the Introduction). Thus, hydrogen peroxide may be detoxified *and* involved in energy metabolism.

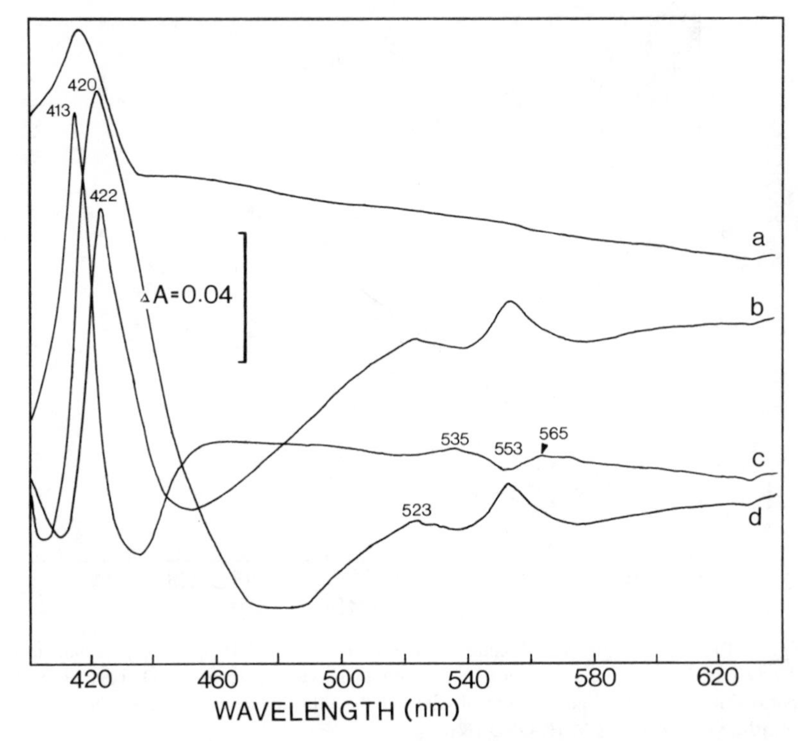

Fig. 4. Cytochrome spectra of the 20,000 × g (30 min) supernatant fractions of *Beggiatoa alba* B18LD and *Thiothrix* JP3. (a) Dithionite-reduced-CO (gassed for 2 min) minus dithionite-reduced spectrum of *Thiothrix* JP3. (b) Dithionite-reduced minus air-oxidized spectrum of *Thiothrix* JP3. (c) Dithionite-reduced-CO minus air-oxidized spectrum of *B. alba* B18LD.

Burton and Morita (1964) stated that *B. leptomitiformis* did not contain cytochromes. Instead, they proposed the existence of a flavin-linked electron transport system in *Beggiatoa*, similar to that observed in lactobacilli, for the production of peroxide (Burton et al., 1966). Later, Burton and Lee (1978) suggested that azide may be used to assist in the isolation of beggiatoas, based on their earlier data on one strain (Burton and Morita, 1964). However, several *Beggiatoa* strains were observed to be sensitive to azide and cyanide (Strohl and Larkin, 1978). This led to the eventual observation of *c*-type (Cannon et al., 1979), *b*-type, *a*-type, and CO-binding *o*-type cytochromes (Strohl and Larkin, 1980) in *B. alba* B18LD. The spectra in fig. 4(c) and (d) indicate the presence of the CO-binding *o*-type cytochrome and the *c*-type

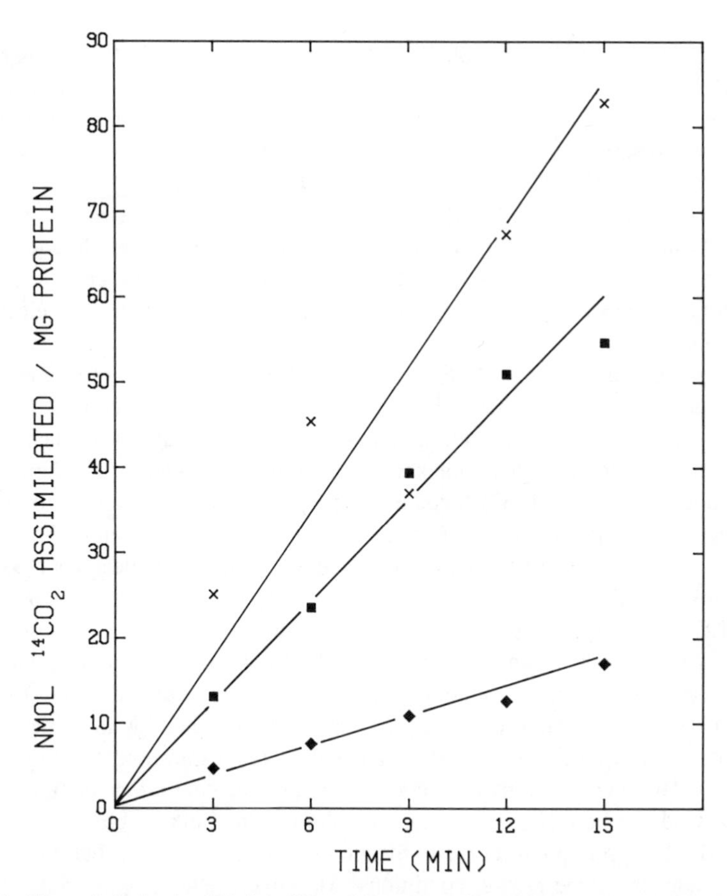

Fig. 5. CO_2 assimilation by *Thiothrix* JP3 and *Beggiatoa alba* B18LD. Both organisms were grown to mid-log phase in MPO5 medium, washed, and resuspended in basal salts plus 2 mM Na_2S (pH 7.5). Ten μCi (1 μmol) of $NaH^{14}CO_3$ were added and the assimilation of $^{14}CO_2$ by whole cells was measured as per fig. 1 legend. $^{14}CO_2$ assimilation by *Thiothrix* (×); $^{14}CO_2$ assimilation by *Thiothrix* in the presence of 5 mM acetate (■); $^{14}CO_2$ assimilation by *Beggiatoa alba* B18LD (◆).

cytochrome, respectively, in *B. alba* B18LD. These two cytochromes have been shown to be located in the soluble fraction (144,000 × g for 90 min) of the cells (T. M. Schmidt and W. R. Strohl, unpublished). Other electron transport components in *B. alba* B18LD include ubiquinone 8 (table 1), NADPH and NADH oxidoreductases (Vinci, 1982), and superoxide dismutase (W. R. Strohl, unpublished). Burton and colleagues (1966) observed NADH oxidoreductase activity in *B. leptomitiformis*, and Carr and his co-workers (1967) observed ubiquinone 8 and traces of a naphthaquinone in a Pringsheim strain (L-1401-5) of *B. leptomitiformis*. Furthermore, a survey of three *B. leptomitiformis* strains (L-1401-6, PD1, ATCC 15551) and two other *B. alba* strains (B15LD, B25RD) showed that they also contained *c*-, *b*-, *a*-, and CO-binding *o*-type cytochromes (W. R. Strohl, T. M. Schmidt, and J. M. Larkin, unpublished). Thus, the beggiatoas contain relatively complete respiratory systems that potentially could be used for chemolithotrophic electron transport.

Although Burton and Morita (1964) stated that CO_2 was not evolved by *Beggiatoa* during heterotrophic metabolism, Scotten and Stokes (1962) demonstrated acetate, lactate, glutamate, α-ketoglutarate, succinate, and pyruvate-dependent oxygen uptake with a strain of *Beggiatoa*. Furthermore, Carr and co-workers (1967) stated that the endogenous respiratory activity by *B. leptomitiformis* strain L-1401-5 (Pringsheim's isolate) was quite high and was stimulated further by acetate and glucose. Strohl and colleagues (1981a) demonstrated the oxidation of both carbons of ^{14}C-acetate to $^{14}CO_2$ by *B. alba* B18LD. They observed that the K_m for mixotrophic acetate oxidation was about 2 times higher than that of heterotrophic cells. It was postulated that the decrease in the acetate oxidation rate by mixotrophic cells was due to the oxidation of sulfide that putatively provided energy, thereby allowing more acetate to be used for biosynthesis rather than for respiration (Strohl et al., 1981a). These conclusions were corroborated by the continuous culture studies of Güde and colleagues (1981), who showed that mixotrophic cells of *B. alba* B18LD grew about twice as efficiently on acetate as did heterotrophic cells under the same growth conditions. Moreover, heterotrophic cells of *B. alba* B18LD grown in continuous culture demonstrated several possible aspects of energy limitation, including incomplete utilization of organic carbon, massive synthesis of poly-β-hydroxybutyrate (as a replacement for energy-expensive biosynthesis), and a lack of steady-state growth kinetics (Güde et al., 1981). Thus, it appears that under mixotrophic growth conditions, *B. alba* B18LD was much more efficient in its utilization of organic nutrients than under heterotrophic growth conditions.

The oxidation of Na_2S to S^o by *B. alba* B18LD is coupled with oxygen uptake (fig. 2). Furthermore, the incorporation of ^{35}S from $Na_2{}^{35}S$ into an acetone-, ethanol-, and benzene-soluble fraction (ostensibly, S^o inclusions) has been used as a method to verify the oxidation of sulfide to sulfur by *B. alba* B18LD (fig. 1). The electron transport inhibitors 8-HQ, 1,10-*o*-phenanthroline, HOQNO, NaN_3, and KCN inhibited the oxidation of $Na_2{}^{35}S$

to $^{35}S^o$ by *B. alba* B18LD (table 2). The protonophores CCCP and FCCP and the uncoupler DNP did not inhibit sulfide oxidation (table 2). The oxidation of [2-^{14}C]-acetate to $^{14}CO_2$ was tested in the same manner, using techniques described by Strohl and colleagues (1981a). Amytal, TTFA, 8-HQ, KCN, CCCP, FCCP, HOQNO, NaN_3, and DNP each inhibited acetate oxidation (table 2). These data suggest that the heterotrophic oxidation mechanisms are separate from those used for the oxidation of sulfide as summarized in figure 6.

Electron microscopic observations have shown that the sulfur inclusions in *B. alba* B18LD are bound by a single electron-dense envelope that is about 4–6 nm in width (Strohl et al., 1981b). The sulfur inclusions lie external and in close physical proximity to the cytoplasmic membrane in *B. alba* B18LD (Strohl et al., 1981b; fig. 7). In heterotrophic cells, the sulfur inclusions, per se, are absent, although rudimentary envelope material may be present, as was observed with *Beggiatoa* sp. strain B15LD (Strohl et al., 1982). Thus, when *Beggiatoa* comes in contact with sulfide, it can oxidize the sulfide immediately to sulfur, which it deposits in the ''external'' (e.g., periplasmic)

TABLE 2

EFFECTS OF ELECTRON TRANSPORT INHIBITORS AND ENERGY CONSERVATION UNCOUPLERS ON $NA_2{}^{35}S$ ASSIMILATION AND [1 – ^{14}C]-ACETATE OXIDATION TO $^{14}CO_2$ IN *BEGGIATOA ALBA* B18LD

Inhibitor	Concentration	Site of Action	% OF CONTROL	
			$Na_2{}^{35}S$ assimilation	^{14}C-acetate oxidation
None	NA[a]	NA	100[b]	100
Ethanol	NA	NA	116	100
Amytal	1 mM	NADH → FP	106	45
TTFA	1 mM	Succ → FP	65	1
8-HQ	1 mM	FAD → cyt *b*	69	20
Phen[c]	2 mM	FAD → cyt *b*	39	ND
Urethane	1 mM	cyt *b*	111	100
HOQNO	50 μM	cyt *b* → cyt *c*	49	100
NaN_3	1 mM	cyt oxidase	35	106
KCN	1 mM	cyt oxidase	5	3
CO gassing	2 min	cyt oxidase	100	ND
DNP	500 μM	uncoupler	108	103
CCCP	10 μM	protonophore	101	61
FCCP	10 μM	protonophore	95	17

[a]NA, not applicable; ND, not done.

[b]Control rates: $Na_2{}^{35}S$ (4 μCi; 0.8 μmol) assimilation; 50,000 cpm min^{-1} mg $protein^{-1}$; [1-^{14}C]-acetate oxidation to $^{14}CO_2$ (1 μCi; 10 μmol), 10,500 cpm min^{-1} mg $protein^{-1}$.

[c]Phen, 1,10-*o*-phenanthroline; for other inhibitors, see abbreviations on page 93.

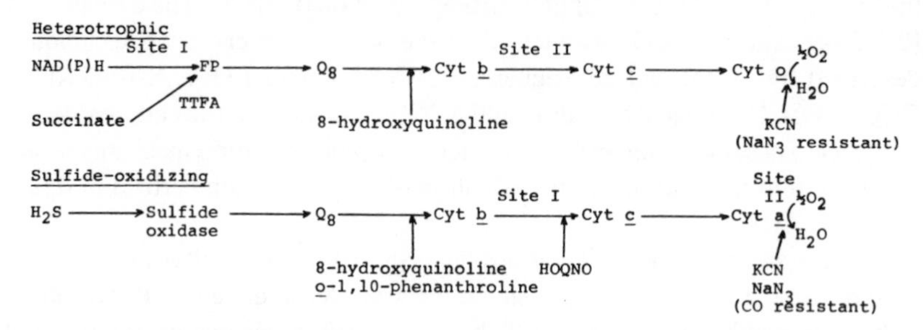

Fig. 6. Possible electron transport systems of *Beggiatoa alba* B18LD based on data presented in this paper.

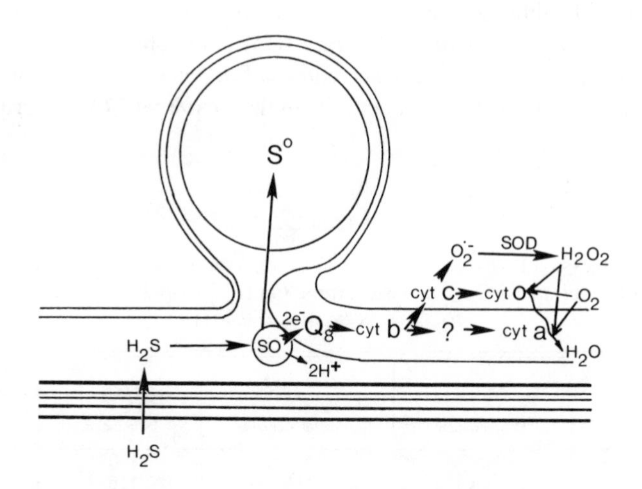

Fig. 7. Ultrastructural model for the coupling of sulfide oxidation by *Beggiatoa alba* B18LD with the electron transport system. SO, sulfide oxidase system; SOD, superoxide dismutase; Q_8, ubiquinone 8; S°, sulfur deposited within sulfur inclusion envelope, but external to the cytoplasmic membrane.

envelopes. Furthermore, the external oxidation of sulfide would protect enzymes and other intracellular components from sulfide toxicity. For these reasons and because the protonophores CCCP and FCCP do not inhibit sulfide oxidation, the first step of sulfide oxidation may occur on the outer surface of the cytoplasmic membrane (fig. 7). The results of electron transport inhibition studies suggest that the electrons from sulfide oxidation by *B. alba* B18LD are probably passed on to the electron transport system at the level of ubiquinone or cytochrome *b* (figs. 6, 7). Although the entire scheme for electron transport-mediated oxidation of sulfide has not been completely elucidated for *B. alba* B18LD, potentially there are one or two coupling sites for ATP synthesis (figs. 6, 7). Enough energy could be conserved from sulfide oxidation to account for the observed alleviation of acetate oxidation under mixotrophic conditions (Strohl et al., 1981a; Güde et al., 1981).

The recent description of *in situ* sulfide and oxygen gradients formed by marine *Beggiatoa* mats (Jørgensen and Revsbech, 1983) and the demonstration of chemoautotrophic growth in sulfide-gradient cultures of a marine *Beggiatoa* (Nelson and Jannasch, 1983) lend further credence to the sulfide-linked chemolithotrophic mechanisms of energy metabolism by beggiatoas.

ACKNOWLEDGEMENTS

We thank Dr. John Larkin for supplying us with the *Thiothrix* strains, for giving us unpublished information concerning them, and for stimulating our interest in these organisms. We also thank Drs. Leonard Hook and Olli Tuovinen for their critical reading of this manuscript. This work was supported by the OSU Graduate School Small Grants Program and partially by NSF Grant PCM-8204778.

1. Abbreviations: AC, acetate heterotrophic medium; MPO5, acetate plus sulfide mixotrophic medium; TS, acetate plus thiosulfate mixotrophic medium; MY, sulfide medium with 0.001% each of yeast extract, sodium acetate, and nutrient broth; HOQNO, 2-n-heptyl-4-hydroxyquinoline N-oxide; CCCP, carbonyl cyanide m-chlorophenylhydrazone; FCCP, carbonylcyanide-p-trifluoromethoxyphenylhydrazone; TTFA, thenoyltrifluoroacetone; 8-HQ, 8-hydroxyquinoline; DNP, 2,4-dinitrophenol; FAD, flavine adenine dinucleotide; TMPD, N,N,N′,N′-tetramethylphenylenediamine; Q8, ubiquinone 8.

LITERATURE CITED

Bland, J. A., and J. T. Staley. 1978. Observations on the biology of *Thiothrix*. Arch. Microbiol. 117:79–87.

Burton, S. D., and R. Y. Morita. 1964. Effect of catalase and cultural conditions on growth of *Beggiatoa*. J. Bacteriol. 88:1755–61.

Burton, S. D., R. Y. Morita, and W. Miller. 1966. Utilization of acetate by *Beggiatoa*. J. Bacteriol. 91:1192–1200.

Burton, S. D., and J. D. Lee. 1978. Improved enrichment and isolation procedures for obtaining pure cultures of *Beggiatoa*. Appl. Environ. Microbiol. 35:614–17.

Cannon, G. C., W. R. Strohl, J. M. Larkin, and J. M. Shively. 1979. Cytochromes in *Beggiatoa alba*. Curr. Microbiol. 2:263–66.

Carr, N. G., G. Exell, V. Flynn, M. Hallaway, and S. Talukdar. 1967. Minor quinones of some *Myxophyceae*. Arch. Biochem. Biophys. 120:503–7.

Farquhar, G. J., and W. C. Boyle. 1971. Occurrence of filamentous microorganisms in activated sludge. J. Water Pollut. Control. Fed. 43:779–98.

Garlick, S., A. Oren, and E. Padan. 1977. Occurrence of facultative anoxygenic photosynthesis among filamentous and unicellular cyanobacteria. J. Bacteriol. 129:623–29.

Gray, G. O., and D. B. Knaff. 1982. The role of a cytochrome *c*-552-cytochrome *c* complex in the oxidation of sulfide in *Chromatium vinosum*. Biochim. Biophys. Acta 680:290–96.

Güde, H., W. R. Strohl, and J. M. Larkin, 1981. Mixotrophic and heterotrophic growth of *Beggiatoa alba* in continuous culture. Arch. Microbiol. 129:357–60.

Hollis, J. P. 1979. Ecology of *Beggiatoa*. Acta Phytopath. Acad. Sci. Hung. 14:419–39.

Jørgensen, B. B. 1982. Ecology of the bacteria of the sulphur cycle with special reference to anoxic-oxic interface environments. Phil. Trans. R. Soc. Lond. B 298:543–61.

Jørgensen, B. B., and N. P. Revsbech. 1983. Colorless sulfur bacteria, *Beggiatoa* spp. and *Thiovulum* spp., in O_2 and H_2S microgradients. Appl. Environ. Microbiol. 45:1261–70.

Keil, F. 1912. Beiträge zur Physiologie der farblosen Schwefelbakterien. Beitr. Biol. Pflanz. 11:335–72.

Kowallik, U., and E. G. Pringsheim. 1966. The oxidation of hydrogen sulfide by *Beggiatoa*. Am. J. Bot. 53:801–6.

Kuenen, J. G., and R. F. Beudeker. 1982. Microbiology of thiobacilli and other sulphur-oxidizing autotrophs, mixotrophs, and heterotrophs. Phil. Trans. R. Soc. Lond. B 298:473–97.

Larkin, J. M. 1981. Isolation of *Thiothrix* in pure culture and observation of a filamentous epiphyte on *Thiothrix*. Curr. Microbiol. 4:155–58.

Larkin, J. M., and W. R. Strohl. 1983. *Beggiatoa*, *Thiothrix*, and *Thioploca*. Ann. Rev. Microbiol. 37:341–67.

Merkel, G. J. 1975. Observations on the attachment of *Thiothrix* to biological surfaces in activated sludge. Water Res. 9:881–85.

Nelson, D. C., and R. W. Castenholz. 1981. Use of reduced sulfur compounds by *Beggiatoa* sp. J. Bacteriol. 147:140–54.

Nelson, D. C., and H. W. Jannasch. 1983. Chemoautotrophic growth of a marine *Beggiatoa* in sulfide-gradient cultures. Arch. Microbiol. 136:262–69.

Nicholls, P., and J.-K. Kim. 1982. Sulphide as an inhibitor and electron donor for the cytochrome *c* oxidase system. Can. J. Biochem. 60:613–23.

Peeters, T., and M. I. H. Aleem. 1970. Oxidation of sulfur compounds and electron transport in *Thiobacillus denitrificans*. Arch. Mikrobiol. 71:319–30.

Pringsheim, E. G. 1967. Die Mixotrophie von *Beggiatoa*. Arch. Mikrobiol. 59:247–54.

Pringsheim, E. G. 1970a. Prefatory chapter: contributions toward the development of general microbiology. Ann. Rev. Microbiol. 24:1–16.

Pringsheim, E. G. 1970b. Die Lebensbedingungen des farblosen Schwefelorganismus *Beggiatoa*. Beitr. Biol. Pflanz. 46:323–36.

Rittenberg, S. C. 1969. The roles of exogenous organic matter in the physiology of chemolithotrophic bacteria. Adv. Microb. Physiol. 3:159–95.

Rittenberg, S. C. 1972. The obligate autotroph: the demise of a concept. Antonie van Leeuwenhoek 38:457–78.

Roth, C. W., W. P. Hempfling, J. N. Conners, and W. V. Vischniac. 1973. Thiosulfate- and sulfide-dependent pyridine nucleotide reduction and gluconeogenesis in intact *Thiobacillus neapolitanus*. J. Bacteriol. 114:592–99.

Scotten, H. L., and J. L. Stokes. 1962. Isolation and properties of *Beggiatoa*. Arch. Mikrobiol. 42:353–68.

Smith, D. W., and S. C. Rittenberg. 1974. On the sulfur-source requirement for growth of *Thiobacillus intermedius*. Arch. Microbiol. 100:65–71.

Strohl, W. R., and J. M. Larkin. 1978. Enumeration, isolation, and characterization of *Beggiatoa* from freshwater sediments. Appl. Environ. Microbiol. 36:755–70.

Strohl, W. R., and J. M. Larkin. 1980. Sulfide oxidation and metabolism by *Beggiatoa alba*. Abstr. Ann. Meet. Am. Soc. Microbiol. K 11, p.128.

Strohl, W. R., G. C. Cannon, J. M. Shively, H. Güde, L. A. Hook, C. M. Lane, and J. M. Larkin. 1981a. Heterotrophic carbon metabolism by *Beggiatoa alba*. J. Bacteriol. 148:572–83.

Strohl, W. R., I. Geffers, and J. M. Larkin. 1981b. Structure of the sulfur inclusion envelopes from four beggiatoas. Curr. Microbiol. 6:75–79.

Strohl, W. R., K. S. Howard, and J. M. Larkin. 1982. Ultrastructure of *Beggiatoa alba* strain B15LD. J. Gen. Microbiol. 128:73–84.

Vinci, V. A. 1982. Sulfide oxidation by *Beggiatoa*. M.S. thesis, Ohio State University, Columbus, Ohio.

Winogradsky, S. 1887. Über Schwefelbakterien. Bot. Z. 45:489–610.

Winogradsky, S. 1888. Zur Morphologie und Physiologie der Schwefelbakterien. *In* Beiträge zur Morphologie und Physiologie der Bakterien 1:1–107.

Winogradsky, S. 1890. Recherches sur les organismes de la nitrification. Ann. Inst. Pasteur 4:213–31.

JANE GIBSON

Intermediary Carbon Metabolism in Autotrophic and Mixotrophic Microorganisms

7

INTRODUCTION

The macromolecular composition of growing bacteria of many very different physiological types appears remarkably similar. Protein is likely to account for 50–60% of the total dry weight, with nucleic acids usually conributing an additional 15–20%. Lipid and carbohydrate components of membranes and envelopes add usually about 10%. Small molecule pools, including inorganic as well as organic components, round out the whole. This distribution, of course, may be very substantially modified by exposure to unbalanced proportions of nutrients; and most, if not all, microorganisms tend to accumulate osmotically inert polymers, which may subsequently be mobilized and redistributed to what can be called the essential working machinery—informational, biosynthetic, enzymatic, and structural elements essential to cell growth.

To generalize a bit further, one might suggest that there is also a high degree of uniformity in the pathways used in the synthesis of biosynthetic intermediates. So far as is known, there is indeed a broad tendency to use the same synthetic reactions to form the nucleotides, amino acids, carbohydrate units, and fatty acids that are ultimately assembled into the macromolecules in organisms of widely different metabolic types.

If a generalized similarity in the pathways leading to the production of biosynthetic monomers is allowed, then it is possible to draw up a kind of biosynthetic flow chart for cell carbon, branching from a common core of metabolism in which C_3 components can be considered central (fig. 1). Using the generalized values for cell composition, it becomes possible to calculate approximately how much of the cell's carbon is derived from each major

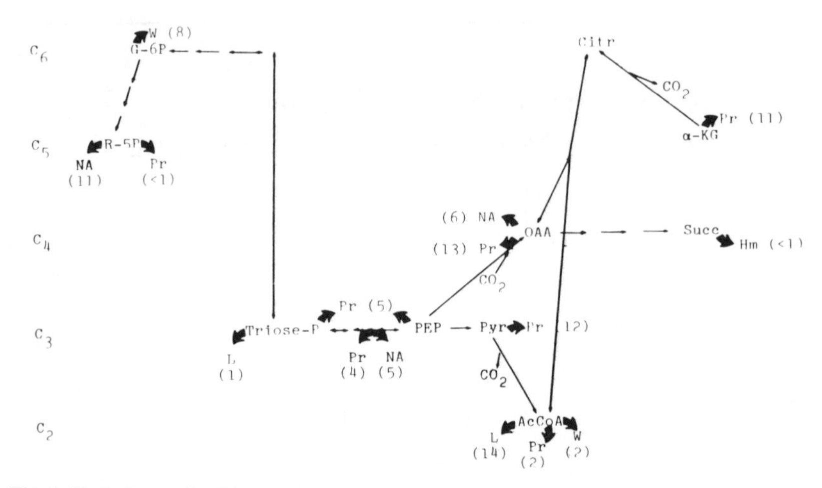

Fig. 1. Central core of carbon metabolism drawn to emphasize biosynthetic branch points (curved arrows) and the magnitude of carbon flow to macromolecular synthesis in exponentially growing cells. (Pr, protein; NA, nucleic acids; L, lipid; W, walls; Hm, pyrrols and heme.) Numbers in parentheses indicate the approximate percentage of carbon removed at each branch point.

branch. Note that from a biosynthetic point of view, α-ketoglutarate serves as a branch point in a long and quantitatively major chain of reactions, while succinate serves as an intermediate in a series of reactions leading to pyrrols that make up only a very small bulk of cell material. No connection between the two reaction sequences is required, and Gest (1981) has argued most persuasively that the connection allowing cyclic flow of carbon was a major innovation required to exploit the potentials opened up by the appearance of oxygen in the atmosphere. Whether as a primitive remnant or through secondary loss, many of the organisms that are the subject of this volume do not, in practice, connect these biosynthetic pathways, but continue to use them as linear reaction sequences. Even that most typical of heterotrophs, *Escherichia coli,* fails to produce α-ketoglutarate dehydrogenase, and thus a cyclic sequence, when grown anaerobically (Amarasingham and Davis, 1965).

Any microorganism, then, no matter what substrates are used, faces the same basic problems in assuring an adequate and balanced flow of carbon through this central core of reactions. Supplied with a single substrate, glucose, an organism must ensure adequate production of α-ketoglutarate, while one given succinate must ensure a considerable flow through gluconeogenesis channels. When C_1 compounds are the sole carbon source, then the products of the major C-C bond synthesizing reaction or reactions likewise have to be distributed appropriately. The outstanding difference between the two heterotrophic examples and the autotrophic would be, of course, that major portions of the core distribution reactions serve that purpose alone in autotrophs, whereas in organotrophs some or all of these reactions may be serving dual functions in biosynthesis and in energy generation. It is also true that incorporation of some CO_2 is probable in any organism that provides a balanced mixture of intermediates to its macromolecular assembly machinery, a point made by Woods and Lascelles (1953) in their introduction to an earlier symposium on autotrophic microorganisms. However, the autotrophs are able to form all, rather than only a part, of the carbon-carbon bonds that appear in their macromolecules. With the possible exception of the methanogens, this ability depends on the existence of quite elaborate cyclic processes whereby a C_1 acceptor molecule is regenerated; all the schemes overlap with what has been referred to as the central core of reactions, providing for distribution. Because the nature of these pathways is comparatively well known, and has been ably discussed many times (e.g., Quayle and Ferenci, 1980), they will be reviewed only briefly. The main emphasis in what follows will lie on CO_2-fixation sequences, and hence on autotrophy, in attempting to bridge the gap between biochemistry in the test tube (or spectrophotometer cell) and in the cellular environment, and in making some rather tentative suggestions about how assimilatory carbon metabolism is governed in cells living in the real world. Obviously, only a small number of critical reactions can be treated in any detail, and limited examples given. Among these, however, will be some drawn from photoautotrophic systems, since carbon assimilation in chemoautotrophic and photoautotrophic organisms can be very similar in spite of different energy-generating mechanism. The recent demonstrations that many organisms previously considered primarily, or even essentially, phototrophic may in fact be more flexible, and include chemoautotrophy and mixotrophy among possible life styles (Kondratieva et al., 1976; Kämpf and Pfennig, 1980) also justified their inclusion.

MAJOR FIXATION REACTIONS AND REGENERATIVE CYCLES

Although CO_2 fixation occurs to a minor extent in virtually all cells, there is very little diversity in the reactions that have become dominant in autotrophic organisms. Formation of the new C-C bond and reduction may be sequential or simultaneous, and some physiologically irreversible steps are found in all sequences, providing specially suitable points for metabolic regulation.

Elucidation of the pathways involved has grown out of several types of experimentation. The analytical procedures developed by Calvin and his group (Bassham and Calvin, 1957) and applied to products formed over very short periods of isotopic CO_2 assimilation have, of course, been crucial. It has to be remembered, however, that the results obtained in such experiments can be greatly influenced by experimental conditions. For example, the main pathway of CO_2 fixation in cyanobacteria is unquestionably the reductive pentose phosphate cycle (Pelroy and Bassham, 1972; Ihlenfeldt and Gibson, 1975); but if cells previously starved for CO_2 are used, malate and aspartate dominate in early fixation products (Döhler, 1974). The elegant methods for analyzing pools and their redistribution kinetics developed by Bassham (e.g., Bassham and Kirk, 1968; Pelroy and Bassham, 1972) provide for very desirable adjunct experiments confirming pathways *in vivo* as well as exploring regulation. As has been emphasized by the recent work on methanogen carbon assimilation, to be discussed later, long-term isotope incorporation during exponential growth can also bring major clarification. The importance of experiments with cell extracts and isolated enzymes is obvious, and has allowed detailed model experiments on regulation to be carried out. Instances where reaction sequences clearly occur *in vivo*, and yet some crucial enzymes are not measurable in extracts, are also known. As in so many types of investigations, the use of several experimental approaches is essential to the understanding of carbon assimilation.

THE REDUCTIVE PENTOSE PHOSPHATE CYCLE

The reduction of CO_2 following its incorporation into organic linkage by the action of ribulose bisphosphate carboxylase (fig. 2) is quantitatively by far the dominant fixation reaction in the modern world. Not only is this the sequence adopted by the eukaryotic plant world but it is also the commonest among autotrophic prokaryotes, being used by chemoautotrophic and photoautotrophic sulfur and hydrogen bacteria as well as the nitrifiers. Exceptions occur only among some strictly anaerobic organisms whose metabolism is geared to very low redox potentials. Two unique and irreversible reactions are involved:

$$\text{ribulose 5-phosphate} + \text{ATP} \xrightarrow{\text{ribulose 5-phosphate kinase}} \text{ribulose 1,5-bisphosphate} \qquad (1)$$

$$\text{ribulose 1,5-bisphosphate} + CO_2 \xrightarrow{\text{ribulose bisphosphate carboxylase}} \text{2 3-phosphoglycerate .} \qquad (2)$$

Regeneration of the acceptor molecule involves, in addition to the main re-

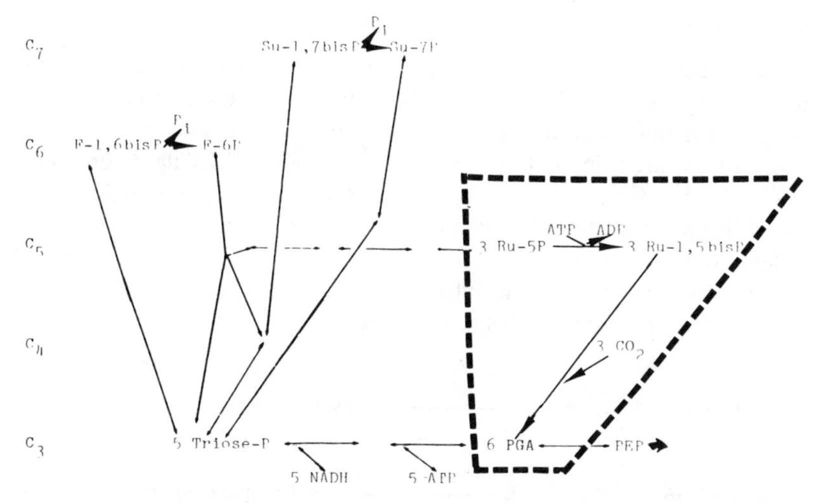

Fig. 2. Outline of the reductive pentose phosphate cycle. Unique reactions are enclosed in dashed lines, and physiologically irreversible steps are indicated by heavy arrows. Curved arrows indicated carbon flow to biosynthesis.

ductive sequence, a complex series of sugar rearrangements that include two more irreversible steps (fructose 1,6-bisphosphatase and sedoheptulose 1,7-bisphosphatase), which are important control points in intact *Chlorella* and spinach chloroplasts (Bassham and Kirk, 1968). It is worth pointing out that though a great deal is known about the activity of the unique reactions, particularly RuBP carboxylases, in different microorganisms grown under varied conditions, the acceptor-regenerating steps have been generally little studied beyond the early work in which the reaction sequence was established (Bassham and Calvin, 1957) and the evidence for regulation at the carboxylase and phosphatase steps accumulated (e.g., Bassham and Kirk, 1968). As a result of the continued operation of the cyclic pathway, the accumulated products could be drawn off at the level of hexose, or pentose, or, and perhaps more usefully from the point of view of this discussion of the provision of biosynthetic intermediates, by the generation of an additional C-3 unit after three turns of the whole. Organisms using this mechanism will need supplementary carboxylation of a C_3 compound to fix one-fourth of all carbon in biosynthetic monomers derived from oxalacetate or succinate, and one-fifth of all that in compounds derived from glutamate.

ANAEROBIC ORGANISMS

The reductive pentose phosphate pathway for CO_2 fixation very probably arose during the anaerobic phase of the earth's evolution, and is still found today in a major group of the anoxygenic prokaryotes, the purple bacteria, whether these are using reduced sulfur compounds or molecular hydrogen as reductant. However, the existence of a different pathway, resembling a reversed dicarboxylic and tricarboxylic acid pathway (fig. 3), in green sulfur bacteria was first proposed by Evans and colleagues (1966). The critical fixation reactions involve the inherently reversible, ferredoxin-dependent, carboxylations of acetyl and succinyl CoA:

$$\text{acetyl CoA} + Fd_{red} + CO_2 \xrightarrow{\text{pyruvate synthase}} \text{pyruvate} + Fd_{ox} \quad (3)$$

$$\text{succinyl CoA} + Fd_{red} + CO_2 \xrightarrow{\alpha\text{-ketoglutarate synthase}} \alpha\text{-ketoglutarate} + Fd_{ox} \quad (4)$$

and in addition, the already familiar carboxylations of phosphoenolpyruvate and of α-ketoglutarate. The formation of phosphoenolpyruvate from pyruvate

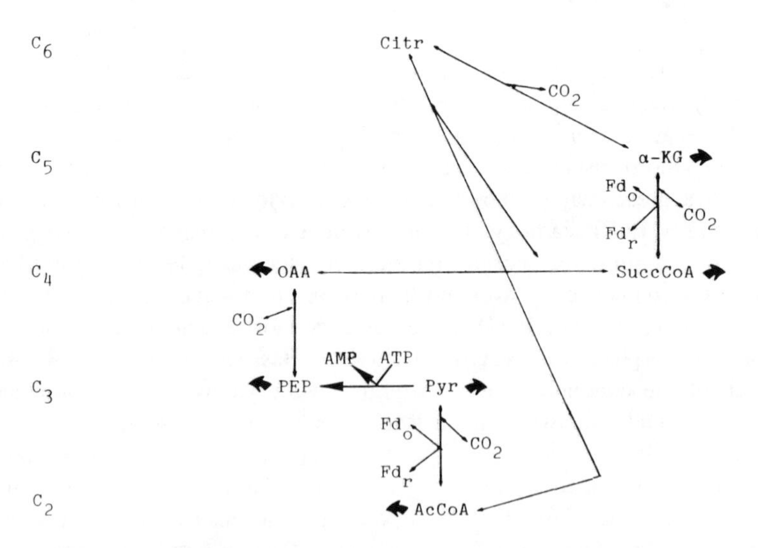

Fig. 3. Outline of the reductive carboxylate cycle. Heavy arrow indicates physiologically irreversible reaction, and curved arrows are biosynthetic branch points.

+ ATP is physiologically irreversible, in contrast to the other reactions involved. In spite of controversy resulting largely from enzyme studies in extracts (Tabita et al., 1974; Buchanan and Sirevåg, 1976; Beuscher and Gottschalk, 1972), the functional dominance of this cycle in green sulfur bacteria has recently been vindicated. Carbon isotope fractionation experiments demonstrated clear differences between green bacteria and reductive pentose phosphate cycle using organisms (Sirevåg et al., 1977). The careful and elegant experiments of Fuchs and colleagues (Fuchs et al., 1980a,b) determined the specific activities and labeling patterns of cell consituents formed during growth in the presence of labeled pyruvate or propionate. The results demonstrated both that the reductive carboxylate cycle was, and that the reductive pentose phosphate cycle was not, the major route of CO_2 fixation under what are, of course, the ultimate physiological conditions. Interestingly, *Chloroflexus aurantiacus,* a green bacterium that is preferentially photoheterotrophic, appears to use the reductive pentose phosphate cycle when made to grow autotrophically (Madigan and Brock, 1977; Sirevåg and Castenholz, 1979). If reduced ferredoxin is required to drive acetyl and succinyl CoA carboxylation, major CO_2 incorporation by the reductive carboxylate cycle would indeed be expected to occur only in organisms functional at very low redox potentials. The suggestion of Strohl and colleagues (1981) that such carboxylations may occur in *Beggiatoa alba* deserves further study.

METHANOGENS

Perhaps the most tantalizing of the remaining problems in C-C bond formation are to be found in the mechanisms employed by the autotrophic methanogens. The earlier work of Thauer and Zeikus and their colleagues (Zeikus et al., 1977; Daniels and Zeikus, 1978) demonstrated that alanine, aspartate, and glutamate were early products of CO_2 metabolism presumably unrelated to methane formation. However, they were unable to detect citrate lyase or isocitric dehydrogenase, so that a cyclic system analogous to the reductive carboxylate cycle of green bacteria did not appear to exist (fig. 4). More recent work (Fuchs and Stupperich, 1980, 1982; Stupperich and Fuchs, 1981) has confirmed the central importance of acetyl and succinyl CoA carboxylations in *Methanobacterium thermoautotrophicum*. They also showed that the labeling patterns obtained excluded the formation of acetate from 2 CO_2 by the kind of cyclic mechanism that has been clarified for *Clostridium thermoaceticum*, in which pyruvate is an intermediate (Hu et al., 1982; Drake et al., 1982). How acetate is made, therefore, is still unknown.

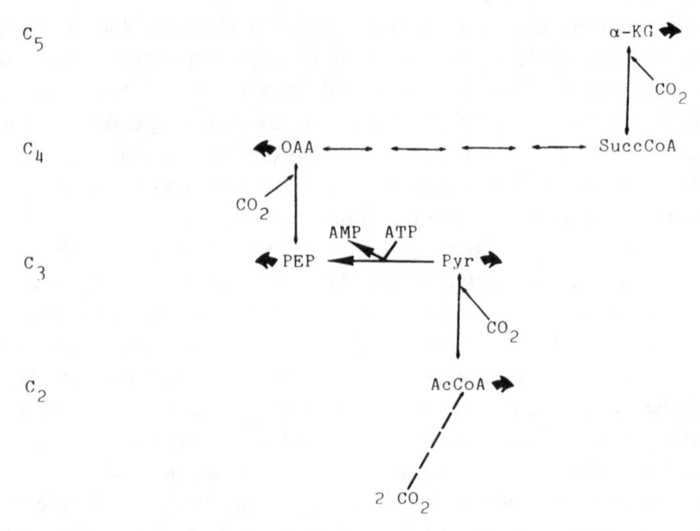

Fig. 4. Tentative scheme for carbon assimilation in *Methanobacterium thermoautotrophicum*, after Fuchs and Stupperich (1982). Symbols as in fig. 3. The mechanism of acetyl CoA formation from 2 CO_2 is not known.

METHANOTROPHS AND METHYLOTROPHS

Although all C-C bonds have to be formed by organisms belonging to these groups, they have the great advantage of starting with fully or partially reduced compounds. The main pathways whereby biosynthetic intermediates are generated are therefore energetically less costly than is the case for CO_2-using organisms. Two pathways have been elucidated. The hexulose 6-phosphate system (Quayle, 1972; Quayle and Ferenci, 1978) involves addition of formaldehyde to ribulose 5-phosphate, followed by a series of reorganization reactions reminiscent of those involved in the reductive pentose phosphate cycle. These may in fact be species-dependent, so that the energetics of the process can be quite varied. Redistribution of fixed carbon from the level of glyceraldehyde 3-phosphate would lead to the same extent of involvement of secondary carboxylation reactions as is needed for the reductive pentose phosphate cycle.

The serine pathway (Anthony, 1975; Higgins et al., 1981) has greater complexity than the hexulose phosphate pathway. Again, however, the C_1 unit is derived from formaldehyde, in this case by chemical reaction with tetrahydrofolate, from which it is transferred to glycine to form serine. In this cycle the reactions involved in regeneration of the acceptor glycine molecule are further removed from the common core of reactions involved in carbon distribution to biosynthetic branchpoints, and indeed the existence of this pathway is usually based on measurements of glycerate dehydrogenase activity. The further steps in glycine regeneration involve carboxylations leading to C_4 dicarboxylic acids or to isocitrate, so that, since two of the three carbons incorporated by three turns of the cycle are derived from formaldehyde and one from CO_2, the secondary carboxylation reactions are quantitatively more important than they are in other cycles, and contribute to the total cell carbon derived from C_3 and hexose intermediates as well as to those derived from C_4 and C_5 dicarboxylates.

ARE THE KNOWN MAJOR CARBOXYLATION PATHWAYS MUTUALLY EXCLUSIVE?

It has generally been assumed that a single pathway is to be found in any one organism, and that short-term experiments showing, for example, ribulose 1,5-bisphosphate stimulation of $^{14}CO_2$ fixation in crude cell extracts, is sufficient to prove that the reductive pentose phosphate cycle is the characteristic CO_2-fixation mechanism in any organism. It may be well to keep an open mind, however, since a range of growth conditions has rarely been investigated. Differences in results of short-term fixation experiments depending on experimental conditions have already been referred to. Striking examples of metabolic versatility are provided by *Methylococcus capsulatus* (Bath), which appears to use the reductive pentose phosphate cycle, the hexulose phosphate and the serine pathway for carbon assimilation (Taylor et al., 1981), and the radiorespirometric demonstration of simultaneous use of different pathways of glucose breakdown in *Thiobacillus* A2 (Wood et al., 1977). If there can be some questions about the main fixation reactions, there are certainly many more about the acceptor-regenerating mechanisms. As already mentioned, these have been very little studied, and may change with growth conditions as well as vary with species.

REGULATORY ASPECTS

Regulation of autotrophic metabolism should be inherently simpler than that of heterotrophs, since the pathways concerned serve assimilatory needs only. It is well known, however, that all potentially autotrophic organisms will, under appropriate conditions, incorporate some exogenous organic material, and changes both in enzyme content and enzyme activity may be expected when this occurs. There are difficulties in applying the results of experiments in which the specific activities of extracts prepared from cells grown under different conditions are compared, to real activities within whole cells. The same is inevitably true of model *in vitro* experiments showing stimulation or inhibition of an enzyme reaction by metabolite effectors. Attractive models can be based on the experiments of Taylor (1970) and Weitzman (1981) that demonstrate that the citrate synthase of specialist organisms is regulated by α-ketoglutarate. In organisms that use this sequence for energy generation, regulation by ATP and NADH is also common, but there are as yet too few complete investigations that involve measurements of intracellular pools and transients when changed conditions are imposed to assess the relevance of these effects in the functional cell. Nonetheless, since the complex regulatory behavior of one of the key enzymes of the reductive pentose phosphate cycle, ribulose bisphosphate carboxylase, has been greatly clarified by work over the last years, an attempt to consider how these findings may apply *in vivo* is made in what follows.

RUBP CARBOXYLASE AND ITS REGULATION

The properties of the RuBP carboxylase purified from a number of different sources has been discussed in recent reviews (McFadden, 1980; Lorimer, 1981), and some of these are summarized in table 1. The isolated enzymes are always polymeric, and commonly very large molecules. Large subunits of M_r 55,000 are invariably found and bear the catalytic sites. In most, 8 large subunits are associated with an equal number of small subunits, having M_r of 11,000–15,000, whose function is thought to be regulatory, but is still not understood. The enzyme from *Rhodospirillum rubrum* is exceptional in consisting of a dimer of large subunits only (Tabita and McFadden, 1974). *Rhodopseudomonas sphaeroides* can contain both a complex enzyme with large and small subunits and a form consisting of large subunits only (Gibson and Tabita, 1977).

The kinetic parameters of the isolated enzyme were uncertain until the careful work of Lorimer and colleagues (1976), and Laing and Christeller (1976) clarified the regulatory behavior of the spinach enzyme. To reach maximum specific activity, a molecule of CO_2 and one Mg^{2+} must be bound to the large subunit of the enzyme, at sites distinct from the catalytic. Activation is reversible, has a half time often of several minutes, and requires quite high concentrations of CO_2; a K_m of about 0.1 mM has been measured (Laing and Christeller, 1976). Ribulose bisphosphate slows activation *in vitro*, whereas 6-phosphogluconate accelerates it. As a result of activation, the specific activity of the enzyme is increased one hundred-fold. The reaction K_m for CO_2 of about 20 μM is in the range of the concentration calculated to exist in chloroplasts, and the observed rate of fixation corresponds to the

TABLE 1

PROPERTIES OF SOME RIBULOSE 1,5-BISPHOSPHATE CARBOXYLASES

Source	M_r	Subunit Composition[a]	Specific Activity[b]	Reference
R. rubrum	112 000	2 L	9.1	Schloss et al., 1978
R. sphaeroides	360 000	6 L	3.3	Gibson and Tabita, 1977
	550 000	8 L, 8 S	2.98	Gibson and Tabita, 1977
T. intermedius	550 000	8 L, 8 S	2.76	Purohit et al., 1976
T. neapolitanus		8 L, 8 S		McFadden, 1980
CO_2 limited:				
soluble			0.26[c]	Beudeker et al., 1981b
particulate			1.18[c]	Beudeker et al., 1981b
$S_2O_3^{2-}$-limited:				
soluble			0.78[c]	Beudeker et al., 1981b
particulate			0.72[c]	Beudeker et al., 1981b
C. fritschii				
Log phase:				
soluble			6.5[c]	Lanaras and Codd, 1982
particulate			3.8[c]	Lanaras and Codd, 1982
Early stationary:				
soluble			0.6[c]	Lanaras and Codd, 1982
particulate			0.4[c]	Lanaras and Codd, 1982
Spinach	560 000	8 L, 8 S	1.8	Lorimer et al., 1976

[a]Subunit composition refers to number of large (L) and small (S) subunits in the isolated enzyme.
[b]Specific activity is given in μmol CO_2 fixed min^{-1} mg $protein^{-1}$ at 30°C and saturating concentrations of substrates; isolated enzyme, preactivated, unless indicated.
[c]Specific activity in μmol CO_2 fixed min^{-1} mg RuBP $carboxylase^{-1}$, determined immunochemically in crude extracts.

RuBP carboxylase content. The light-dependent activation of RuBP carboxylase in *Chlorella* already mentioned (Bassham and Kirk, 1968) appears to be explained by pH and Mg^{2+} concentration changes accompanying proton movement into the thylakoids (Heldt et al. 1973). The enzymes from all sources are activated in a comparable way, and, although recent investigations of RuBP carboxylase in autotrophic microorganisms (e.g., Beudeker et al. 1980; 1981a) have been carried out with appropriate preincubation to ensure full activation, much of the earlier work is hard to interpret.

Even when fully activated, the specific activity of isolated RuBP carboxylases is unimpressive, and is commonly about 2–5 μmol CO_2 or RuBP min^{-1} mg $protein^{-1}$. This presumably accounts for the very high levels that this enzyme can reach in chloroplasts or in microorganisms; up to 15% in *Hydrogenomonas facilis* (Bowien and Schlegel, 1981) and an astonishing 45% of soluble protein in *R. rubrum* after serial subculture in H_2/CO_2 (Schloss et al., 1979).

A further important property of RuBP carboxylase was discovered by Bowes and colleagues (1971), who showed that it catalyzed the reaction of O_2 as well as CO_2 with RuBP, yielding phosphoglycolate as well as phosphoglycerate. The same catalytic site has been shown to be involved, so that oxygenase is activated in the same way as carboxylase. This potentially wasteful reaction forms the basis of photorespiration in plants.

IS THE OXYGENASE ACTIVITY OF RUBP CARBOXYLASE OF ANY RELEVANCE IN GROWING MICROBIAL AUTOTROPHS?

The purple phototrophs using the reductive pentose phosphate cycle for CO_2 fixation anaerobically should not encounter any problems due to the dual substrate specificities of the carboxylase; indeed, it can be suggested that the evolution of this fixation mechanism occurred at a period when atmospheric O_2 was low, and so not competitive. But it has been demonstrated repeatedly that glycolate can be produced by many aerobic autotrophs. For example, the purple sulfur bacterium *Chromatium* released significant amounts of glycolate when exposed to $^{14}CO_2$ in oxygen (Asami and Akazawa, 1975), and this release was reduced by at least 50% in the presence of 5 mM $NaHCO_3$. Oxygen isotope incorporation experiments demonstrated conclusively that the oxygenase activity of RuBP carboxylase was responsible for the glycolate production (Lorimer et al., 1978). The specialist *Thiobacillus neapolitanus* maintained in continuous culture with limiting electron donor, but with excess CO_2 and oxygen supply, excreted about 15% of the carbon fixed during a short pulse, and one-fourth of this material was shown to be glycolate (Cohen et al., 1979). If CO_2 was made limiting, as much as 27% of the carbon fixed appeared outside the cells as glycolate (Beudeker et al., 1981a). This quantity was decreased if aeration of the culture was reduced. Such extreme losses of fixed carbon may, therefore, be uncommon in natural populations, which presumably make greater use of S^{2-} than of $S_2O_3{}^{2-}$ as electron donor, and are therefore likely to accumulate in regions of low oxygen tension. However, these experiments demonstrate both that RuBP carboxylase remains active in cells growing in low CO_2 tensions, and that the oxygenase activity of the enzyme is significant *in vivo*.

ARE THERE SPECIAL MECHANISMS TO MAINTAIN HIGH RATES OF CO_2 FIXATION?

There is good evidence that active transport of CO_2 occurs in unicellular green algae (Raven, 1980). Recently, investigations with *Anabaena variabilis* (Badger et al., 1978; Kaplan et al., 1980) and *Coccochloris peniocystis* (Miller and Colman, 1980) have demonstrated that HCO_3^- can be transported by these cyanobacteria, and that a concentration gradient of up to 10^3 can be maintained in light. The activity of these systems is greater in cells grown with limiting CO_2 (Marcus et al., 1982), and the internal concentrations achieved should, at the intracellular pH found in these and related cyanobacteria (Kaplan et al., 1980; Miller and Colman, 1980; Gibson, 1981), provide CO_2 levels of the order needed for *in vitro* activation, as well as for saturating activated enzyme.

At present, clear evidence for prokaryotic bicarbonate transport is limited to the oxygenic cyanobacteria, in which competition between O_2 and CO_2 would be expected to present the most acute problem. It would be of considerable interest to find out whether the potentially wasteful effects of low CO_2/O_2 ratios are mitigated by active transport in other aerobic autotrophs using this fixation mechanism. As already mentioned, sulfur-oxidizing organisms may be restricted to environments of low oxygen tension by their energy-generating mechanisms, but this is not obviously necessary for nitrifiers.

An as yet highly tentative alternative method for ensuring effective CO_2 fixation is suggested from recent work on the polyhedral bodies seen in electron micrographs of thiobacilli, nitrifiers, and many cyanobacteria, though not, apparently, in hydrogen bacteria or the anaerobic autotrophs. These particles were shown by Shively and colleagues (1973) to have RuBP carboxylase activity and so were named carboxysomes. Ribulose bisphosphate carboxylase exists in both soluble and particulate forms in thiobacilli and cyanobacteria, and the ratio of the activities in the two locations can be varied systematically. In *Thiobacillus neapolitanus,* both the total enzyme content, as measured by quantitative immunochemical methods, and the proportion in carboxysomes are increased by growth with limiting CO_2 (Beudeker et al., 1980). Although the maximal rate of CO_2 fixation by whole cells appeared to correspond with the ratio of soluble to particulate enzyme, carboxysomal RuBP carboxylase was found to be potentially as active, and in some instances more so, as the soluble enzyme. A number of attempts to demonstrate other enzyme activities associated with carboxysomes have been made in the past (e.g., Shively et al., 1973). Beudeker and Kuenen (1981) partially purified *Thiobacillus neapolitanus* carboxysomes through two sucrose gradients, and demonstrated RuBP carboxylase, fructose 1,6-bisphosphate aldolase, and sedoheptulose 1,7-bisphosphate aldolase activity in the intact particles. After sonication, however, a complete battery of reductive pentose phosphate cycle activities could be detected, and the specific activities of some, such as sedoheptulose 1,7-bisphosphatase, ribose isomerase, and ribulose 5-phosphate kinase, were significantly higher than those found in crude cell extracts. The limiting glycoprotein layer of the carboxysomes thus appears to be selectively permeable, and this raises some interesting possibilities. Beudeker and Kuenen (1981) suggest that carboxysome location for the reductive pentose phosphate cycle enzymes may facilitate the localized production of NADH, since the particles also contained high malic dehydrogenase activity. Other possibilities could also be considered, such as maintenance of locally high CO_2 tensions through a malate shuttle reminiscent of that found in C_4 plants. If pools were small and rapidly turned over, they might not be detected in short-term CO_2-fixation experiments such as were carried out. The studies on *Thiobacillus* and on *Chlorogloeopsis* (Lanaras and Codd, 1981) carboxysomal and soluble RuBP carboxylases also seem to suggest that irreversible inactivation of the enzyme may occur *in vivo,* since true specific activities in extracts are not constant with growth conditions (Beudeker et al., 1981b) or growth phase (Lanaras and Codd, 1982). The possibility that carboxysomes facilitate continued CO_2 reduction at low external CO_2 concentrations in these and other autotrophs deserves further study, as do the indications of differences in the apparent specific activity of "fully" activated enzyme.

OTHER ASSIMILATORY INPUTS

All the organisms that are able to use CO_2 as sole carbon source are nonetheless willing to take some contribution from organic compounds; the main difference between the specialist and the versatile types can thus appear to be quantitative rather than qualitative. There appear to be a number of underlying reasons for continued dependence on CO_2 for the bulk of cellular carbon needs. Organisms that cannot oxidize acetate will be unable to use this molecule for growth in the absence of exogenous CO_2, since it must be carboxylated before incorporation into any component except lipids. Another limitation is found in *Thiocapsa pfennigii,* in which isotope-labeling patterns are consistent with a complete tricarboxylate cycle and glyoxylate bypass, but which appears to lack means for forming phosphoenolpyruvate from TCA intermediates. All cellular components derived from sugars, therefore, are necessarily derived from the reductive pentose phosphate cycle (D. G. Taff and J. Gibson, unpublished).

A commonly cited explanation for failure to grow on organic components is that autotrophic organisms possess no means for concentrating potentially useful metabolites within the cell. There is indeed an encouraging correlation between the ability to make use of glucose as a carbon source for mixotrophic or heterotrophic growth and the possession of an active transport system. Unicellular strains of cyanobacteria that can or cannot grow on glucose in the dark have similar activities of the key enzymes involved in metabolizing this compound by the oxidative pentose phosphate cycle (Pelroy et al., 1972), but only those that can grow on glucose possess an active transport system (Beauclerk and Smith 1978; Raboy and Padan, 1978). This moves the molecule across the cell membrane without concomitant phosphorylation. In mixotrophic thiobacilli, similar systems have been described (Romano et al., 1975; Matin et al., 1980). Transport activity is related to growth conditions (Wood and Kelly, 1982), and it appears that restriction of rates of glucose entry may play some part in the regulaton of glucose catabolism under mixotrophic conditions. Matin and colleagues (1980) have also suggested that $S_2O_3^{2-}$ may influence glucose entry, since this compound was shown to be a noncompetitive inhibitor of glucose analogue uptake; however, very high $S_2O_3^{2-}$ concentrations would be required for this to be effective, since a K_i of 30 mM was found, and it is questionable whether this effect would be significant in nature.

THE SPECIAL FEATURES OF ACETATE

It has been found repeatedly that acetate is the most generally used of organic compounds added to cultures of autotrophic microorganisms (Smith et al., 1967; Smith and Hoare, 1977), although the extent to which carbon from this compound can be incorporated would, naturally, depend on the enzymatic capability of the organism to which it is fed. Moreover, acetate can be taken up from the medium until the remaining concentration is of the order of micromolar or less (Ihlenfeldt and Gibson, 1977; Strohl et al., 1981). There is as yet no instance known in which the entry of acetate is governed by any transport system. As has been shown for mutant unicellular cyanobacteria, which cannot incorporate acetate, the distribution of the molecule appears to depend on a high degree of differential permeability of acetic acid and acetate ion. The internal concentration will thus depend both on the external total concentration and the pH difference between the cell and its environment (Gibson, 1981; fig. 5). This failure to retain the molecule presumably accounts for the ready production of acetate as a fermentation end product. Its incorporation into the general carbon economy of the cell will depend, however, on its activation to acetyl CoA, which may then be carboxylated in low redox systems, or combined with a C_2 or C_4 acceptor molecule.

STORAGE

An obvious advantageous response to the intermittent and unbalanced nutrient supply that must characterize most natural habitats is the sequestering

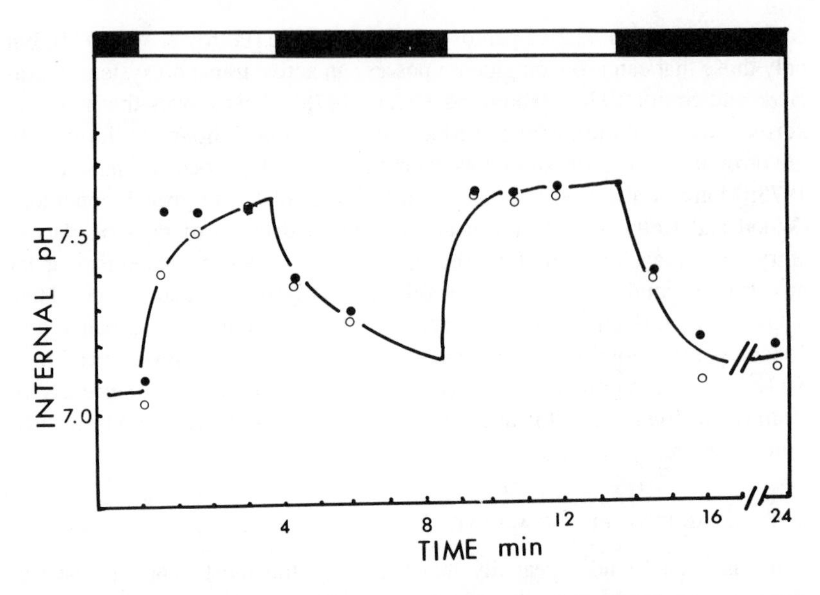

Fig. 5. Internal pH in a mutant of cyanobacterium *Synechococcus* that cannot incorporate acetate calculated from distribution of the weak acids dimethyloxazolidenedione (○) or acetic acid (●). The bars at the top of the figure indicate dark (filled) and light (open) periods.

of high molecular weight, and thus osmotically inactive, carbon-rich storage products; this ability does indeed seem to be well developed among autotrophic microorganisms. Accumulation of large quantities of glucose polymers under nitrogen source limitation has been shown in many types; for example, cyanobacteria (Allen and Smith, 1969; Lehmann and Wöber, 1976; Grillo and Gibson, 1979), green sulfur bacteria (Sirevåg and Ormerod, 1970), thiobacilli (Beudeker et al., 1981c), and a methanotroph (Linton and Cripps, 1978). As Wilkinson (1959) pointed out, it is important to demonstrate that any suggested storage product can be mobilized and converted to components of active cellular machinery. This has been shown to occur in each of these instances, although the methanotroph appeared to use stored polyglucose primarily in energy generation, rather than for biosynthetic purposes. It is interesting that polyglucose can also be fermented slowly, at least in a cyanobacterium (Oren and Shilo, 1979) and in *Thiobacillus neapolitanus* (Beudeker et al., 1981d). Since both of these types of organisms are most numerous in regions where marked shifts in the position of the aerobic/anaerobic interface occur diurnally, this fermentative ability, although slow, may be very important as a means of providing maintenance energy during darkness. It may indeed be that the almost universal ability to activate acetate, noted earlier, is of greater significance when working in reverse for energy-scavenging during anaerobic periods provided exogenous reductant is available, than for carbon gain during normal, more active metabolism.

In addition to polyglucose accumulation, many reports of poly β-hydroxybutyrate accumulation in autotrophs have appeared (Dawes and Senior, 1973), although the number of instances in which its later mobilization has been followed is more limited. Appropriate conditions of energy and CO_2 supply may be needed to permit incorporation of this polymer's carbon into other cellular macromolecules. Storage of nitrogen compounds appears to be a more-specialized trait, but is certainly a characteristic of cyanobacteria, in which two candidates for this function have been found. An unusual polymer of arginine and aspartate accumulates to a minor extent under normal culture conditions, but may also be found at least transiently in great excess (Simon, 1971; Allen et al., 1980). The phycobiliproteins, a characteristic and photochemically active group of proteins found in cells grown with adequate supplies of nitrogen, are quite rapidly degraded when nitrogen is withdrawn (Allen and Smith, 1969; Boussiba and Richmond, 1980), and the products appear to be reused for other protein synthesis. This interesting example of a metabolically functional "storage" product may perhaps have less colorful, and so less obvious, parallels in other organisms, for there are many instances in which changed growth conditions lead to decreased activities of specific enzymes. It is becoming apparent that many bacteria produce a range of proteases, which have been particularly well explored in the bacilli (Maurizi and Switzer, 1980). Breakdown and reutilization of enzyme components not required in new conditions seem particularly desirable abilities for autotrophic microorganisms, but have yet to be demonstrated.

EXPORT OF FIXED CARBON

Loss of carbon compounds from the cell may serve as a last regulatory resort when homeostatic mechanisms fail. In addition to those already mentioned, many investigations have shown that autotrophic microorganisms may release significant quantities of fixed carbon into their environment, particularly when nutrient supply is unbalanced. *Thiobacillus neapolitanus* released pyruvate, succinate, and α-ketoglutarate in NH_4^+-limited continuous culture to almost 25% of the total carbon fixed (Beudeker et al., 1981e), and green sulfur bacteria fixing CO_2 in N-free medium released almost half the total gained as α-ketoglutarate and α-ketomethylvalerate (Sirevåg and Ormerod, 1970). In both organisms export was reduced if NH_4^+ was supplied. Thus, it appears that the regulation of citrate synthase by α-ketoglutarate, which can be demonstrated *in vitro* (Taylor, 1970), may not be adequate to control carbon flow in intact cells. Whether the spectrum of metabolites lost under stress conditions such as NH_4^+ shortage reflects simply those whose internal concentrations rise to the highest levels, and results from nonspecific leakage, or whether there is some selectivity to the process, is not known. Studies of prokaryote metabolite export are few, but it is intriguing that lactate export, which occurs through facilitated diffusion (Harold and Levin, 1974), can energize membrane vesicles of *Streptococcus cremoris* (Otto et al., 1982) or of *Escherichia coli* (Ten Brink and Konings, 1980). It is questionable whether export has any energetic significance in autotrophs, but the possibility is worth exploring further.

There is, however, one recently discovered situation in which export of fixed carbon by autotrophs may be the basis of a thriving multiorganismal symbiosis. The deep-sea hydrothermal vents that occur at several sites north of the Galàpagos Islands are surrounded by a remarkably rich and apparently fast-growing fauna. The primary producers in this region are believed to be the sulfide-oxidizing bacteria, which are numerous and active in the vent water (Karl et al., 1980). The large vestamentiferous worms that surround the vents have no filtering apparatus, and also no gut; but the trophosome, which occupies a large part of the body, appear to be a packed mass of bacteria, including sulfide oxidizers (Cavanaugh et al., 1981; H. W. Jannasch, personal communication). High activities of RuBP carboxylase and Ru 5-P kinase have been found in this tissue (Felbeck, 1981; Felbeck and Somero, 1982), and the basis for the high productivity of the areas appears to be at least in part based on the export of autotrophically fixed carbon in return, presumably, for relatively constant sulfide, carbon dioxide, and oxygen supply.

CONCLUSIONS

Recent work has recognized, and indeed emphasized, the complexities of the interactions involved in the organization of carbon assimilation in autotrophic, as well as in more versatile, microorganisms. Studies have progressed beyond the performance of model experiments *in vitro* to observations of the behavior of growing cultures. The potential of continuous culture methods is being exploited increasingly as a means of measuring actual, as opposed to potential, rates of assimilation. It is becoming obvious that cells whose growth is limited by the availability of any of a number of nutrients will tend to carry excess assimilatory capacity (e.g. Kuenen and Veldkamp, 1973; Beeftinck and van Gemerden, 1979; Beudeker et al., 1980) allowing for rapid storage during any fluctuations in conditions. The importance of such phenomena in the ecology of autotrophs is as yet poorly recognized, but beautifully illustrated in competition between two *Chromatium* species (van Gemerden, 1974); although the smaller invariably outgrew the larger in steady-state conditions, stable co-culture could be established if alternating light and dark conditions, and fluctuating sulfide concentrations, were imposed. New methods, notably quantitative immunochemistry and ^{31}P NMR analysis (Naron et al., 1977; Kallas and Dahlquist, 1980) offer exciting possibilities for detailed analysis of metabolites within the intact cell and the study of integrated systems.

LITERATURE CITED

Allen, M. M., and A. J. Smith. 1969. Nitrogen chlorosis in blue-green algae. Arch. Mikrobiol. 69:114–20.

Allen, M. M., F. Hutchison, and P. M. Weathers. 1980. Cyanophycin granule polypeptide formation and degradation in the cyanobacterium *Aphanocapsa* 6308. J. Bacteriol. 141:687–93.

Amarasingham, C. R., and B. D. Davis. 1965. Regulation of α-ketoglutarate dehydrogenase formation in *Escherichia coli*. J. Biol. Chem. 240:3664–68.

Anthony, C. 1975. The biochemistry of methylotrophic bacteria. Sci. Progr. (London) 62:167–206.

Asami, S., and T. Akazawa. 1975. Oxidative formation of glycolic acid in photosynthesizing cells of *Chromatium*. Plant Cell. Physiol. 15:571–76.

Badger, M. R., A. Kaplan, and J. A. Berry. 1978. A mechanism for concentrating CO_2 in *Chlamydomonas reinhardtii* and *Anabaena variabilis* and its role in photosynthetic CO_2 fixation. Carnegie Inst. Wash. Yearbook 77:251–61.

Bassham, J. A., and M. Calvin. 1957. The path of carbon in photosynthesis, 107 p. Prentice Hall, New York.

Bassham, J. A., and M. Kirk. 1968. Dynamic metabolic regulation of the photosynthetic carbon reduction cycle. *In* K. Shibata, A. Takamiya, A. T. Jagendorf, and R. C. Fuller (eds.), Comparative biochemistry and biophysics of photosynthesis, pp. 365–78. University of Tokyo Press, Tokyo.

Beauclerk, A. A. D., and A. J. Smith. 1978. Transport of D-glucose and 3-0-methyl glucose in the cyanobacteria *Aphanocapsa* 6714 and *Nostoc* strain Mac. Eur. J. Biochem. 82:187–97.

Beeftinck, H. H., and H. van Gemerden. 1979. Actual and potential rates of substrate oxidation and product formation in continuous cultures of *Chromatium*. Arch. Microbiol. 121:161–67.

Beudeker, R. F., G. C. Cannon, J. G. Kuenen, and J. M. Shively. 1980. Relations between D-ribulose 1,5-bisphosphate carboxylase, carboxysomes and CO_2-fixing capacity in the obligate chemolithotroph *Thiobacillus neapolitanus* grown under different limitations in the chemostat. Arch. Microbiol. 125:185–89.

Beudeker, R. F., and J. G. Kuenen. 1981. Carboxysomes: "Calvinosomes"? FEBS Lett. 131:269–74.

Beudeker, R. F., J. G. Kuenen, and G. A. Codd. 1981a. Glycollate metabolism in the obligate chemolithotroph *Thiobacillus neapolitanus* grown in continuous culture. J. Gen. Microbiol. 126:337–46.

Beudeker, R. F., G. A. Codd, and J. G. Kuenen. 1981b. Quantification and intracellular distribution of ribulose 1,5-bisphosphate carboxylase in *Thiobacillus neapolitanus* as related to possible functions of carboxysomes. Arch. Microbiol. 129:361–67.

Beudeker, R. F., J. W. M. Kerver, and J. G. Kuenen. 1981c. Occurrence, structure, and function of intracellular polyglucose in the obligate chemolithotroph *Thiobacillus neapolitanus*. Arch. Microbiol. 129:221–26.

Beudeker, R. F., W. deBoer, and J. G. Kuenen. 1981d. Heterolactic fermentation of intracellular polyglucose by the obligate chemoautotroph *Thiobacillus neapolitanus* under anaerobic conditions. FEMS Microbiol. Lett. 10:337–42.

Beudeker, R. F., R. Riegman, and J. G. Kuenen. 1981e. Regulation of nitrogen assimilation by the obligate chemolithotroph *Thiobacillus neapolitanus*. J. Gen. Microbiol. 128:39–47.

Beuscher, N., and G. Gottschalk. 1972. Lack of citrate lyase—the key enzyme of the reductive carboxylic acid cycle—in *Clorobium thiosulfatophilum* and *Rhodospirillum rubrum*. Z. Naturforsch. 276:967–73.

Boussiba, S., and A. E. Richmond. 1980. C-phycocyanin as a storage protein in the blue-green alga *Spirulina platensis*. Arch. Microbiol. 125:143–47.

Bowes, G., W. L. Orgen, and R. H. Hageman. 1971. Phosphoglycolate production catalyzed by ribulose 1,5-diphosphate carboxylase. Biochem. Biophys. Res. Commun. 45:716–22.

Bowien, B., and H. G. Schlegel. 1981. Physiology and biochemistry of aerobic hydrogen-oxidizing bacteria. Ann. Rev. Microbiol. 35:405–52.

Buchanan, B. B., and R. Sirevåg. 1976. Ribulose 1,5-diphosphate carboxylase and *Chlorobium thiosulfatophilum*. Arch. Microbiol. 109:15–19.

Cavanaugh, C. M., S. L. Gardiner, M. L. Jones, H. W. Jannasch, and J. B. Waterbury. 1981. Prokaryotic cells in the hydrothermal vent tube worm *Riftia pachyptela* Jones: possible chemoautotrophic symbionts. Science 213:338–40.

Cohen, Y., I. de Jonge, and J. G. Kuenen. 1979. Excretion of glycollate by *Thiobacillus neapolitanus* in continuous culture. Arch. Microbiol. 122:189–94.

Daniels, L., and J. G. Zeikus, 1978. One carbon metabolism in methanogenic bacteria: analysis of short-term fixation products of $^{14}CO_2$ and $^{14}CH_3OH$ incorporated into whole cells. J. Bacteriol. 136:75–84.

Dawes, E. A., and P. J. Senior. 1973. The role and regulation of energy reserve polymers in microorganisms. Adv. Microb. Physiol. 10:136–266.

Döhler, G. 1974. C_4 Weg der Photosynthese in der Blaualge *Anacystis nidulans*. Planta 111: 259–69.

Drake, H. L., S.-I. Hu, and H. G. Woods. 1981. Purification of five components from *Clostridium thermoaceticum* which catalyzed synthesis of acetate from pyruvate and methyltetrahydrofolate: properties of the phosphotransacetylase. J. Biol. Chem. 256:11137–44.

Evans, M. C. W., B. B. Buchanan, and D. I. Arnon. 1966. A new ferredoxin dependent carbon reduction cycle in a photosynthetic bacterium. Proc. Natl. Acad. Aci. USA 55:928–34.

Felbeck, H. 1981. Chemoautotrophic potential of the hydrothermal vent tube worm *Riftia pachyptela* Jones (Vestamentifera). Science 213:336–38.

Felbeck, H., and G. N. Somero. 1982. Primary production in deep sea hydrothermal vent organisms: roles of sulfide-oxidizing organisms. Trends Biochem. Sci. 7:201–5.

Fuchs, G., and E. Stupperich. 1980. Acetyl CoA, a central intermediate in autotrophic CO_2 fixation in *Methanobacterium thermoautotrophicum*. Arch. Microbiol. 127:267–72.

Fuchs, G., and E. Stupperich. 1982. Autotrophic CO_2 fixation in *Methanobacterium thermoautotrophicum*. Zbl. Bakt. Hyg. I Abt. Orig. C3:277–88.

Fuchs, G., E. Stupperich, and R. Jaenchen. 1980a. Autotrophic CO_2 fixation in *Chlorobium limicola:* evidence against the operation of the Calvin cycle in growing cells. Arch. Microbiol. 128:56–63.

Fuchs, G., E. Stupperich, and G. Eden. 1980b. Autotrophic CO_2 fixation in *Chlorobium limicola:* evidence for the operation of a reductive tricarboxylic acid cycle in growing cells. Arch. Microbiol. 128:64–71.

Gest, H. 1981. Evolution of the citric acid cycle and respiratory energy conversion in prokaryotes. FEMS Microbiol. Lett. 12:209–15.

Gibson, J. L., and F. R. Tabita. 1977. Different molecular forms of D-ribulose 1,5-diphosphate carboxylase from *Rhodopseudomonas sphaeroides*. J. Biol. Chem. 252:943–49.

Gibson, J. 1981. Movement of acetate across the cytoplasmic membrane of the unicellular cyanobacteria *Synechococcus* and *Aphanocapsa*. Arch. Microbiol. 130:175–79.

Grillo, J. F., and J. Gibson. 1979. Regulation of phosphate accumulation in the unicellular cyanobacterium *Synechococcus*. J. Bacteriol. 140:508–17.

Harold, F. M., and E. Levin, 1974. Lactic acid translocation: terminal step in glycolysis by *Streptococcus faecalis*. J. Bacteriol. 117:1141–48.

Heldt, H. W., K. Werdan, M. Milovancev, and G. Geller. 1973. Alkalinization of the chloroplast stroma caused by light-dependent proton flux into the thylakoid space. Biochim. Biophys. Acta 314:224–41.

Higgins, I. J., D. J. Best, R. C. Hammond, and D. Scott. 1981. Methane-oxidizing microorganisms. Microbiol. Rev. 45:556–90.

Hu, S.-I., H. L. Drake, and H. G. Wood. 1982. Synthesis of acetyl coenzyme A from carbon monoxide, methyltetrahydrofolate, and coenzyme A by enzymes from *Clostridium thermoaceticum*. J. Bacteriol. 149:440–48.

Ihlenfeldt, M. J. A., and J. Gibson. 1975. CO_2 fixation and its regulation in *Anacystis nidulans (Synechococcus)*. Arch. Microbiol. 102:13–21.

Ihlenfeldt, M. J. A., and J. Gibson. 1977. Acetate uptake by the unicellular cyanobacteria *Synechococcus* and *Aphanocapsa*. Arch. Microbiol. 113:231–44.

Kallas, T., and F. W. Dahlquist. 1981. Phosphorus-31 NMR analysis of internal pH during photosynthesis in the cyanobacterium *Synechococcus*. Biochemistry 20:5900–5907.

Kämpf, C., and N. Pfennig. 1980. Capacity of *Chromatiaceae* for chemotrophic growth: specific respiration rates of *Thiocystis violaceae* and *Chromatium vinosum*. Arch. Microbiol. 127:125–235.

Kaplan, A., M. R. Badger, and J. A. Berry. 1980. Photosynthesis and the inorganic carbon pool in the blue-green alga *Anabaena variabilis*: response to external CO_2 concentration. Planta 149:219–21.

Karl, D. M., C. O. Wirsen, and H. W. Jannasch. 1980. Deep sea primary production at the Galápagos hydrothermal vents. Science 207:1345–47.

Kondratieva, E. N., V. G. Zhukov, R. N. Ivanosky, Yu. P. Petuschkova, and E. Z. Monosov. 1976. The capacity of phototrophic bacterium *Thiocapsa roseopersicina* for chemosynthesis. Arch. Microbiol. 108:287–92.

Kuenen, J. G., and H. Veldkamp. 1973. Effects of organic compounds on the growth of chemostat cultures of *Thiomicrospira pelophila, Thiobacillus thioparus,* and *Thiobacillus neapolitanus.* Arch. Mikrobiol. 94:173–90.

Laing, W. A., and J. T. Christeller. 1976. A model for the kinetics of activation and catalysis of ribulose 1,5-bisphosphate carboxylase. Biochem. J. 159:563–70.

Lanaras, T., and G. A. Codd. 1981. Structural and immunoelectrophoretic comparison of soluble and particulate ribulose bisphosphate carboxylase from the cyanobacterium *Chlorogloeopsis fritschii.* Arch. Microbiol. 130:213–17.

Lanaras, T., and G. A. Codd. 1982. Variations in D-ribulose 1,5-bisphosphate carboxylase protein levels, activity, and subcellular distribution during photoautorophic batch culture of *Chlorogloeopsis fritschii.* Planta 154:284–88.

Lehmann, M., and G. Wöber. 1979. Accumulation and mobilization of glycogen in the blue-green bacterium *Anacystis nidulans.* Arch. Microbiol. 111:93–97.

Linton, J. D., and R. E. Cripps. 1978. The occurrence and identification of intracellular polyglucose storage granules in *Methylococcus* NCIB 11083 grown in chemostat culture on methane. J. Gen. Microbiol. 117:41–48.

Lorimer, G. H. 1981. The carboxylation and oxygenation of ribulose 1,5-bisphosphate: the primary events in photosynthesis and photorespiration. Ann. Rev. Plant Physiol. 32:349–83.

Lorimer, G. H., M. R. Badger, and J. Andrews. 1976. The activation of ribulose bisphosphate carboxylase by CO_2 and Mg ions: equilibria, kinetics, a suggested mechanism, and physiological implication. Biochemistry 15:529–36.

Lorimer, G. H., C. B. Osmond, T. Akazawa, and S. Asami. 1978. On the mechanism of glycolate synthesis by *Chromatium.* Arch. Biochem. Biophys. 185:49–56.

Madigan, M. T., and T. Brock. 1977. CO_2 fixation in photosynthetically grown *Chloroflexus aurantiacus.* FEMS Microbiol. Lett. 1:301–4.

Marcus, Y., D. Zenwirth, E. Harel, and A. Kaplan. 1982. Induction of bicarbonate transporting capability and high photosynthetic affinity to inorganic carbon by low concentrations of CO_2 in *Anabaena variabilis.* Plant Physiol. 69:1008–12.

Matin, A., M. Schleiss, and R. C. Perez. 1980. Regulation of glucose transport and metabolism in *Thiobacillus novellus.* J. Bacteriol. 142:639–44.

Maurizi, M. R., and R. L. Switzer. 1980. Proteolysis in bacterial sporulation. Curr. Topics Cell Reg. 16:164–225.

McFadden, B. A. 1980. A perspective of ribulose bisphosphate carboxylase/oxygenase, the key catalyst in photosynthesis and photorespiration. Acc. Chem. Res. 13:394–99.

Miller, A. G., and B. Colman. 1980. Active transport and accumulation of bicarbonate by a unicellular cyanobacterium. J. Bacteriol. 143:1253–59.

Naron, G., S. Ogawa, R. G. Shulman, and T. Yamane. 1977. High-resolution ^{31}P nuclear magnetic resonance studies of metabolism in aerobic *Escherichia coli.* Proc. Natl. Acad. Sci. USA 74:888–91.

Oren, A., and M. Shilo. 1979. Anaerobic heterotrophic dark metabolism in the cyanobacterium *Oscillatoria limnetica:* sulfur respiration and lactate fermentation. Arch. Microbiol. 122: 77–84.

Otto, R., R. G. Lageveen, H. Veldkamp, and W. N. Konings. 1982. Lactate efflux-induced electrical potential in membrane vesicles of *Streptococcus cremoris*. J. Bacteriol. 149: 733–38.

Pelroy, R. A., and J. A. Bassham. 1972. Photosynthetic and dark metabolism in unicellular blue-green algae. Arch. Mikrobiol. 86:25–38.

Pelroy, R. A., R. Rippka, and R. Y. Stanier. 1972. Metabolism of glucose by unicellular blue-green algae. Arch. Mikrobiol. 87:303–22.

Purohit, K., B. A. McFadden, and A. L. Cohen. 1976. Purification, quaternary structure, composition, and properties of D-ribulose 1,5-bisphosphate carboxylase from *Thiobacillus intermedius*. J. Bacteriol. 127:505–15.

Quayle, J. R. 1972. The metabolism of one carbon compound by microorganisms. Adv. Microb. Physiol. 7:119–203.

Quayle, J. R., and T. Ferenci. 1978. Evolutionary aspects of autotrophy. Bacteriol. Rev. 42:251–73.

Raboy, B., and E. Padan. 1978. Active transport of glucose and α-methyl glucoside in the cyanobacterium *Plectonema boryanum*. J. Biol. Chem. 253:3287–91.

Raven, J. A. 1980. Nutrient transport on microalgae. Adv. Microb. Physiol. 21:47–226.

Romano, A. H., N. J. van Vranken, P. Preisand, and M. Brustolon. 1975. Regulation of the *Thiobacillus intermedius* glucose uptake systems by thiosulfate. J. Bacteriol. 121:856–71.

Schloss, J. V., E. F. Pares, M. V. Long, I. L. Norton, C. D. Stringer, and F. C. Hartman. 1979. Isolation, characterization, and crystallization of ribulose 1,5-bisphosphate carboxylase from autotrophically-grown *Rhodospirillum rubrum*. J. Bacteriol. 137:490–501.

Shively, J. M., F. Ball, D. H. Brown, and R. E. Saunders. 1973. Functional organelles in prokaryotes: polyhedral inclusions (carboxysomes) of *Thiobacillus neapolitanus*. Science 182:584–86.

Simon, R. D. 1971. Cyanophycin granules from the blue-green alga *Anabaena cylindrica:* a reserve material consisting of copolymers of aspartic acid and arginine. Proc. Natl. Acad. Sci. USA 68:265–67.

Sirevåg, R., and J. G. Ormerod. 1970. Carbon dioxide fixation in green sulfur bacteria. Biochem. J. 120:399–408.

Sirevåg, R., and R. W. Castenholz. 1979. Aspects of carbon metabolism in *Chloroflexus*. Arch. Microbiol. 120:151–53.

Sirevåg, R., B. B. Buchanan, J. A. Berry, and J. H. Troughton. 1977. Mechanism of CO_2 fixation in bacterial photosynthesis studies by the carbon isotope fractionation technique. Arch. Microbiol. 112:35–38.

Smith, A. J., J. London, and R. Y. Stanier. 1967. Biochemical basis of obligate autotrophy in blue-green algae and thiobacilli. J. Bacteriol. 94:972–83.

Smith, A. J., and D. S. Hoare. 1977. Specialist phototrophs, lithotrophs, and methylotrophs: a unity among a diversity of procaryotes? Bacteriol. Rev. 41:419–48.

Strohl, W. R., G. C. Cannon, J. M. Shively, H. Güde, L. A. Hook, C. M. Lane, and J. M. Larkin. 1981. Heterotrophic carbon metabolism in *Beggiatoa alba*. J. Bacteriol. 148:572–83.

Stupperich, E., and G. Fuchs. 1980. Acetyl CoA, a central intermediate of autotrophic CO_2 fixation in *Methanobacterium thermoautotrophicum*. Arch. Microbiol. 127:267–72.

Tabita, F. R., and B. A. McFadden. 1974. D-ribulose 1,5-diphosphate carboxylase from *Rhodospirillum rubrum* II: quaternary structure, composition, catalytic and immunological properties. J. Biol. Chem. 249:3459–64.

Tabita, F. R., B. A. McFadden, and N. Pfennig. 1974. D-ribulose 1,5-diphosphate carboxylase in *Chlorobium thiosulfatophilum* Tassajara. Biochim. Biophys. Acta 321:187–94.

Taylor, B. F. 1970. Regulation of citrate synthase activity in strict and facultatively autotrophic thiobacilli. Biochem. Biophys. Res. Commun. 40:957–73.

Taylor, S. C., H. Dalton, and C. S. Dow. 1981. Ribulose 1,5-bisphosphate carboxylase/oxygenase and carbon assimilation in *Methylococcus capsulatus* (Bath). J. Gen. Microbiol. 122:89–94.

Ten Brink, B., and W. N. Konings. 1980. Generation of an electrochemical proton gradient by lactate efflux in membrane vesicles of *Escherichia coli*. Eur. J. Biochem. 111:59–66.

van Gemerden, H. 1974. Coexistence of organisms competing for the same substrate: an example among the purple sulfur bacteria. Microb. Ecol. 1:104–19.

Weitzman, P. D. J. 1981. Unity and diversity in some bacterial citric acid cycle enzymes. Adv. Microb. Physiol. 22:185–244.

Wilkinson, J. F. 1959. The problem of energy storage compounds in bacteria. Exptl. Cell Res. Suppl. 7:111–30.

Wood, A. P., and D. P. Kelly. 1982. Kinetics of sugar transport by *Thiobacillus* A2. Arch. Microbiol. 131:156–59.

Wood, A. P., D. P. Kelly, and C. F. Thurston. 1977. Simultaneous operation of three catabolic pathways in the metabolism of glucose by *Thiobacillus* A2. Arch. Microbiol. 113:265–74.

Woods, D. D., and J. Lascelles. 1953. The no-man's land between the autotrophic and heterotrophic ways of life. In B. A. Fry and J. L. Peel (eds.), Autotrophic micro-organisms, pp. 1–23. Cambridge University Press, Cambridge.

Zeikus, J. G., G. Fuchs, W. Kenealy, and R. K. Thauer. 1977. Oxidoreductases involved in cell carbon synthesis in *Methanobacterium thermoautotrophicum*. J. Bacteriol. 132:604–13.

JOHN G. COBLEY

The Maintenance of pH Gradients in Acidophilic and Alkalophilic Bacteria: Gibbs-Donnan Equilibrium Calculations

8

Most prokaryotes have a pH optimum for growth close to neutrality (pH 5.0–8.5). In contrast, acidophilic bacteria are characterized by optimal growth rates in media of pH 4.0 or lower. Alkalophiles grow optimally in media of pH 9.0 or above. In spite of the requirements for acidic or alkaline growth media, organisms of the latter two groups maintain an intracellular pH value of close to neutrality. This is inferred from (1) the pH profiles of activity and stability for isolated enzymes; and (2) the steady-state distribution of weak acids or bases between the intracellular space and the suspending medium.

With the advent of the chemiosmotic hypothesis of oxidative phosphorylation (Mitchell, 1966), in which a transmembrane gradient of proton electrochemical activity is considered as the driving force for ADP phosphorylation, much attention has been paid to the magnitude and maintenance of transmembrane pH gradients. In acidophiles pH gradients of up to 4.5 units have been recorded (Cox et al., 1979; Matin et al., 1982). In some acidophiles (Cox et al., 1979), the pH gradient (ΔpH) is maintained in the presence of protonophores such as 2,4-dinitrophenol that are known to render cellular membranes permeable to protons. In other acidophiles and in alkalophiles, the ΔpH is at least partially collapsed by protonophores (Matin et al., 1982; Krulwich et al., 1978). The ΔpH that is refractory to protonophores is considered by most authors to be "passively" maintained. To account for the "passive" maintenance, it has been proposed (Hsung and Haug, 1977) that "a Donnan potential exists, possibly generated by charged macromolecules impermeable to the cell membrane." It has also been stated (Matin et al.,

1982) that "an unusual resistance to ionic movements, presumably mainly to the efflux of cellular cations, holds the key to acidophilism." However, nowhere in the literature has any attempt been made to calculate the concentration of nondialyzable charge required to balance a given ΔpH in Gibbs-Donnan equilibrium (Bolam, 1932). The purpose of this paper is to make such calculation. The physiology of acidophiles has recently been reviewed (Cobley and Cox, 1983) and the energy conservation in *Thiobacillus ferroxidans* has been discussed (Cox and Brand, 1984).

ACIDOPHILES

Let us consider a two-phase system separated by a membrane that is impermeable to R^+ and permeable to the species H^+ and A^-:

OUT	M	IN
A^-	|	R^+ A^-
H^+ OH^-	|	H^+ OH^-

Here the OUT phase represents the growth medium and IN the intracellular space. M denotes the cell membrane, and HA represents a strong monoprotic acid. Let us assume that pressure is applied to the IN phase to prevent its increase in volume. In acidophilic eubacteria, this is achieved by the rigid cell wall, which prevents increases in cell volume. At equilibrium and if the principle of electrical neutrality is closely adhered to:

$$[H^+]_i + [R^+]_i = [A^-]_i + [OH^-]_i \tag{1}$$

and

$$[H^+]_o = [A^-]_o + [OH^-]_o \,. \tag{2}$$

At equilibrium the electrochemical potential difference for H^+ between the two phases, $\Delta\mu_{H^+}$, is zero by definition and

$$\Delta\bar{\mu}_{H^+} = 0 = \frac{RT}{F} \ln \frac{[H^+]_o}{[H^+]_i} + \Delta\Psi \tag{3}$$

where $\Delta\Psi$ is the electrical potential difference between the two phases. Similarly for $\Delta\bar{\mu}_{A^-}$

$$\Delta\bar{\mu}_{A^-} = 0 = \frac{RT}{F} \ln \frac{[A^-]_i}{[A^-]_o} + \Delta\Psi \,. \tag{4}$$

Therefore, since the value of $\Delta\Psi$ will be the same in equations (3) and (4),

$$\frac{[H^+]_o}{[H^+]_i} = \frac{[A^-]_i}{[A^-]_o} \,. \tag{5}$$

Substituting equations (1) and (2) into equation (5)

$$\frac{[H^+]_o}{[H^+]_i} = \frac{[H^+]_i + [R^+]_i - [OH^-]_i}{[H^+]_o - [OH^-]_o}$$

and solving for $[R^+]_i$

$$[R^+]_i = \frac{[H^+]_o^2 - [H^+]_o[OH^-]_o - [H^+]_i^2 + [H^+]_i[OH^-]_i}{[H^+]_i} \,. \tag{6}$$

Equation (6) simplifies to

$$[R^+]_i = \frac{[H^+]_o^2 - [H^+]_i^2}{[H^+]_i} \,. \tag{7}$$

when $\Delta\mu_{H_2O}$ is zero. Equation (7) is a very good approximation to equation (6) even if $[H_2O]_i$ is substantially decreased so long as $[H^+]_i \geq 10^{-7}$M and $[H^+]_o \geq 10^{-6}$M. Equation (7) is also applicable if we consider the membrane impermeable to H^+ and A^-, and if we include in the membrane a symporter (Mitchell, 1966) that translocates one H^+ with every A^-.

Having derived an equation (7) that relates $[R^+]_i$ to ΔpH for the case of a strong monoprotic acid, an analogous equation can be presented for the case of a strong diprotic acid, H_2A:

OUT	M	IN
A^{2-}	|	R^+ A^{2-}
H^+ OH^-	|	H^+ OH^-

Here the membrane is considered permeable to the species H^+ and A^{2-}. At equilibrium and with the same assumptions as before:

$$[H^+]_i + [R^+]_i = 2\,[A^{2-}]_i + [OH^-]_i \tag{8}$$

$$[H^+]_o = 2\,[A^{2-}]_o + [OH^-]_o \tag{9}$$

since at equilibrium,

$$\Delta\bar{\mu}_{H^+} = 0 = \frac{RT}{F} \ln \frac{[H^+]_o}{[H^+]_i} + \Delta\Psi \tag{10}$$

and

$$\Delta\bar{\mu}_{A^{2-}} = 0 = \frac{RT}{2F} \ln \frac{[A^{2-}]_i}{[A^{2-}]_o} + \Delta\Psi$$

then

$$\left(\frac{[H^+]_o}{[H^+]_i}\right)^2 = \frac{[A^{2-}]_i}{[A^{2-}]_o}. \tag{11}$$

Substituting equations (8) and (9) into equation (11)

$$\left(\frac{[H^+]_o}{[H^+]_i}\right)^2 = \frac{\tfrac{1}{2}([R^+]_i + [H^+]_i - [OH^-]_i)}{\tfrac{1}{2}([H^+]_o - [OH^-]_o)}$$

and solving for $[R^+]$

$$[R^+]_i = ([H^+]_o - [OH^-]_o) \left(\frac{[H^+]_o}{[H^+]_i}\right)^2 - [H^+]_i + [OH^-]_i\,. \tag{12}$$

This equation (12) is very closely approximated by

$$[R^+]_i = [H^+]_o \left(\frac{[H^+]_o}{[H^+]_i}\right)^2 \tag{13}$$

when $[H^+]_i \geq 10^{-7}$M and $[H^+]_o \geq 10^{-6}$M. Equation (13) can also be derived for a system in which the membrane is not freely permeated by H^+ or A^{2-} but contains a symporter that translocates $2H^+$ for every A^{2-}.

Mitchell (1966) defines the proton motive force ($\Delta\bar{\mu}_{H^+}$) by the following equation

$$\Delta\bar{\mu}_{H^+} = \Delta\Psi - \frac{2.303RT}{F}\Delta pH \ . \tag{14}$$

In this equation $\Delta\Psi$ is positive when the OUT phase in the above diagrams is electropositive and for ΔpH

$$\Delta pH = pH_o - pH_i \ . \tag{15}$$

When this definition of ΔpH is used, ΔpH values for acidophiles are negative. This has created confusion in the literature where ΔpH is often described as "increasing" when pH_i remains constant and pH_o is decreased. Strictly speaking, under these circumstances ΔpH decreases in value, i.e., it becomes a larger negative number. Therefore, in this paper, when discussing the acidophiles (but not the alkalophiles), the pH gradient is presented as $-\Delta pH$ rather than ΔpH.

In figure 1 the relationship between ΔpH and $[R^+]_i$ for three values of $[H^+]_i$ (10^{-5}M, 10^{-6}M, and 10^{-7}M) is plotted. The cases of both mono-

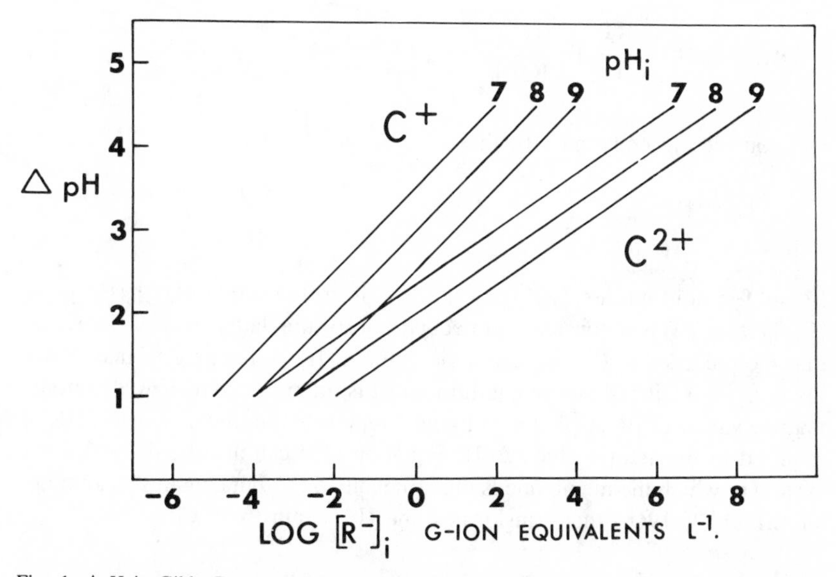

Fig. 1. ΔpH in Gibbs-Donnan equilibrium: ΔpH values and the minimum concentrations of nondialyzable ion, $[R^+]$, required for maintenance. The values of $[R^+]$ for the case of a strong monoprotic acid, HA, have been calculated from equation (7). The values of $[R^+]$ for the case of a diprotic acid, H_2A, have been calculated from equation (13).

protic and diprotic strong acids are included. From this graph two significant features emerge. First, in the case of the strong monoprotic acid, HA, an increase by one unit of the $-\Delta pH$ maintained in Gibbs-Donnan equilibrium requires an increase in $[R^+]_i$ by 10^2. In the case of the strong diprotic acid, H_2A, to increase $-\Delta pH$ by one unit requires an increase in $[R^+]_i$ by 10^3. The second important feature is that in order to maintain a given value of ΔpH if $[H^+]_i$ is increased tenfold, then $[R^+]_i$ must also be increased tenfold. This second feature can be seen in the plots for both monoprotic and diprotic acids.

In the preceding discussion, the only salts considered were RA and R_2A, which were confirmed to the IN phase. What would be the effect on ΔpH if the salt CA were present where C^+ permeates the membrane? For example, in the case of the strong monoprotic acid, HA, by employing the same logical procedures as before the following equation can be derived:

$$[R^+]_i = \frac{[H^+]_o^2 + [c^+]_o[H^+]_o - [H^+]_i^2 - [C^+]_i[H^+]_i}{[H^+]_i} . \qquad (16)$$

Since at equilibrium

$$\Delta\bar{\mu}_{C^+} = 0 = \frac{RT}{F} \ln \frac{[C^+]_o}{[C^+]_i} + \Delta\Psi \qquad (17)$$

and considering equation (10) then:

$$\frac{[C^+]_o}{[C^+]_i} = \frac{[H^+]_o}{[H^+]_i} .$$

Since for acidophiles $[H^+]_o > [H^+]_i$, it follows that $[C^+]_o[H^+]_o > [C^+]_i[H^+]_i$. When considered in equation (16) this latter inequality reveals that the presence of C^+ increases the value of $[R^+]_i$ required to maintain a given ΔpH in Gibbs-Donnan equilibrium. It is important to realize, therefore, that the values of $[R^+]_i$ plotted in figure 1 represent minimum values of $[R^+]_i$ required to maintain a given ΔpH. Equation (16) can also be derived for a system in which the membrane is impermeable to C^+ but contains an antiporter (Mitchell, 1966) that translocates one H^+ against one C^+.

ALKALOPHILES

The maintenance of ΔpH in alkalophilic bacteria can be considered as follows:

$$\begin{array}{cc|cc} & \text{OUT} & \text{M} & \text{IN} \\ C^+ & & C^+ & R^- \\ H^+ & OH^- & H^+ & OH^- \end{array}$$

where C^+ is a monovalent cation that permeates the membrane and R^- is confined to the inner phase. The logical approach is similar to that applied to the acidophiles. For the case of the strong base, COH, an equation analogous to that for the strong monoprotic acid, HA, can be derived

$$[R^-]_i = \frac{[H^+]_i[OH^-]_o - [H^+]_o[OH^-]_i}{[H^+]_o} . \tag{18}$$

In the case of the strong base, $C(OH)_2$, where the divalent cation, C^{2+}, permeates the membrane, the relationship between $[R^-]_i$ and ΔpH can be derived:

$$[R^-]_i = [OH^-]_o \left(\frac{[H^+]_i}{[H^+]_o}\right)^2 \tag{19}$$

and is analogous to equation (13) derived for the strong diprotic acid situation. Both equations (18) and (19) are very good approximations when $[H^+]_i \leq 10^{-7}M$ and $[H^+]_o \leq 10^{-8}M$. Equations (18) and (19) also apply to systems in which the membrane is impermeable to cations and protons but which contain antiporters that respectively translocate either one C^+ against one H^+ or one C^{2+} against two H^+.

In figure 2 the relationships between $[R^-]$ and ΔpH for the two strong bases, COH and $C(OH)_2$, at three values of $[H^+]_i$ ($10^{-9}M$, $10^{-8}M$, and $10^{-7}M$) are plotted. It is clear that these relationships are quantitatively similar to those illustrated for acidophiles in figure 1. The values of $[R^-]_i$ calculated for a given ΔpH are again minimal values since the inclusion of salts such as CA or CA_2 increase the value of $[R^-]_i$ required to maintain a given ΔpH in Gibbs-Donnan equilibrium.

THE NATURE OF R^+ AND R^-

Though no attempt will be made to establish the specific chemical nature of R^+ or R^-, it is important to note that the valencies of R^+ and R^- are irrelevant to the calculations when the concentrations of these hypothetical substances have been expressed in units of charge concentration. The unit chosen to express $[R^+]_i$ and $[R^-]_i$ is the gram-ion equivalent per liter, which denotes the idea of moles of singly charged ions or groups per liter. In charge terms this is equivalent to Faradays per liter. It is also important to note that neither R^+ nor R^- need stand for a single chemical species. In the above calculations, square brackets have been used to denote concentration; activity coefficients for the various ions have not been included. However, since activity coefficients for ions decrease with increasing ionic strength, I can reaffirm that the values of $[R^+]_i$ and $[R^-]_i$ illustrated in figures 1 and 2 are the minimum concentrations required to generate the specified ΔpH values.

The values of $[R^+]_i$ or $[R^-]_i$ that might be realistically expected within a cell are presently unknown. The neutrophilic prokaryote *Klebsiella aerogenes* is osmotically stressed when grown in media containing 1 M NaCl. Under these conditions, elevated levels of K^+ (625 mg-ions l^{-1}) and glutamate (750 mg-ions l^{-1}) have been reported (Measures, 1975). It seems that for $[R^+]_i$ and $[R^-]_i$, values of the order of 1 g-ion equivalent l^{-1} are possible.

APPLICATION OF THE EQUATIONS

The above calculations were made on the assumption that the cell wall exerts a pressure on the intracellular space to prevent water entry. Although

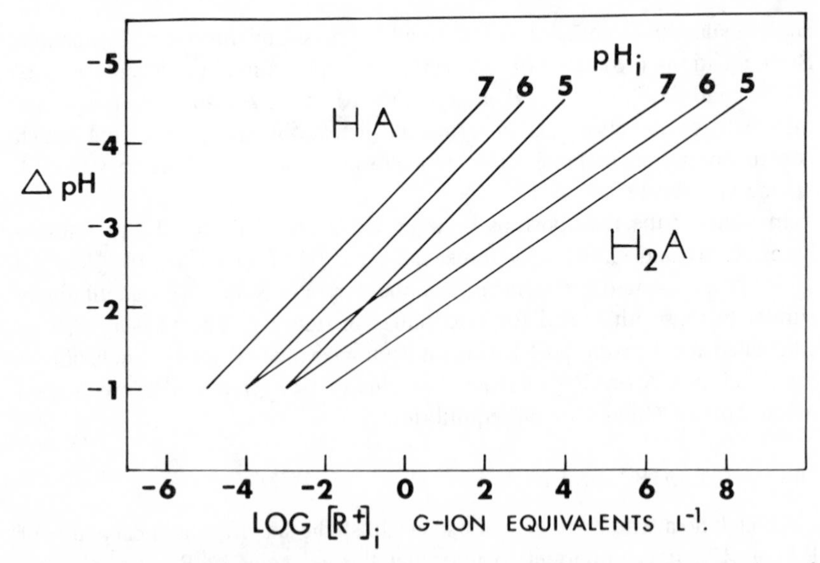

Fig. 2. ΔpH in Gibbs-Donnan equilibrium: ΔpH values and the minimum concentrations of nondialyzable ion, $[R^-]$, required for maintenance. The values of $[R^-]$ for the case of a strong base, COH, have been calculated from equation (18). The values of $[R^-]$ for the case of a strong base, $C(OH)_2$, have been calculated from equation (19).

this is true for acidophilic eubacteria, it is not true for *Thermoplasma acidophilum,* an acidophilic archaebacterium that lacks a cell wall. This organism, which maintains a ΔpH of at least -3.5 units and which grows optimally and in osmotic balance with a medium of 0.2 osmolal (Searcy, 1976), would require an $[R^+]_i$ of at least 32 g-ion equivalents l^{-1}. Such a concentration is unrealistically high, and would clearly generate an osmotic pressure adequate to cause cell lysis. The proposal (Hsung and Haug, 1977) that ΔpH is maintained in *Thermoplasma acidophilum* by a Gibbs-Donnan equilibrium is therefore untenable.

Table 1 lists published values for pH_i and ΔpH for three acidophilic eubacteria. Also presented are the minimum theoretical values for $[R^+]_i$ required to balance ΔpH in Gibbs-Donnan equilibrium. In each case the value for $[R^+]_i$ is inordinately high. In these organisms and under these conditions, the maintenance of ΔpH by a Gibbs-Donnan equilibrium is therefore impossible.

There is a second and equally compelling reason to dismiss the notion that in acidophiles ΔpH is maintained in Gibbs-Donnan equilibrium. Since ΔpH, $[H^+]_i$, and $[R^+]_i$ are not independent variables, it follows that if the value of $[R^+]_i$ is constant then when $[H^+]_o$ is varied, $[H^+]_i$ must also vary. Empirically this is not the case. In both *Thiobacillus ferrooxidans* and in *T. acidophilus,* $[H^+]_i$ remains almost constant as $[H^+]_o$ is varied from 10^{-1}M to 10^{-5}M (Cox et al., 1979; Matin et al., 1982).

The empirical values in table 1 were obtained in experiments with a duration of minutes (approximately). Matin and colleagues (1982) have shown that in *T. acidophilus* with a pH_i value of 5.4 a ΔpH of -2.4 units could be maintained for 48 hours. The maintenance of this ΔpH is compatible with a Gibbs-Donnan equilibrium, since an $[R^+]_i$ value of only 0.46 g-ion equivalents l^{-1} would be required. It could be proposed that in *T. acidophilus* the ΔpH that persists on the order of days is maintained in Gibbs-Donnan equilibrium. Such a proposal could be tested by measuring the dependence of pH_i on pH_o in experiments lasting 48 hours.

TABLE 1

ΔpH VALUES IN ACIDOPHILIC EUBACTERIA

Organism	Observed Values		Minimum Theoretical Values	Reference
	pH_i	ΔpH	$[R^+]$ g-ion Equivs l^{-1}	
Thiobacillus ferrooxidans	6.5	-5.5	3.2×10^4	Cox et al., 1979
Thiobacillus acidophilus	5.5	-4.5	3.2×10^3	Matin et al., 1982
Bacillus acidocaldarius	6.15	-4.15	1.6×10^2	Krulwich et al., 1978

Less data are available on the parameters of ΔpH in alkalophilic bacteria. In a study of *Bacillus alkalophilus* (Guffanti et al., 1978), a maximum ΔpH of 2.5 units was reported when the pH_i value was 9.0. In this situation ΔpH could be in Gibbs-Donnan equilibrium if 1 g-ion equivalent l^{-1} of R^- were found inside the cell. However, pH_i remained almost constant at 9.0–9.5 when pH_o was varied from 9.0–11.5. It therefore seems unlikely that in this alkalophile the Gibbs-Donnan equilibrium plays a significant role in the maintenance of ΔpH.

ALTERNATE HYPOTHESIS AND UNANSWERED QUESTIONS

If a Gibbs-Donnan equilibrium cannot be evoked to explain ΔpH maintenance in acidophiles, how might this maintenance be achieved? The membranes defined in the above model systems were either freely permeable to cations and anions (except R^+) or impermeable to all ions but containing exchange-diffusion porters. Let us consider a third type of membrane freely permeable to A^- or A^{2-} but impermeable to H^+ and R^+. A membrane with these properties would not permit the generation of ΔpH in response to $[R^+]_i$ because the movement of anions would be severely restricted by the generation of $\Delta\Psi$. However, if ΔpH were *generated* by an alternative mechanism, such a membrane would be capable of *maintaining* ΔpH. If acidophiles possess such membranes, protonophores would be predicted to totally collapse ΔpH because H^+ entry would not be limited by the generation of an opposing $\Delta\Psi$. Collapse of ΔpH by protophores occurs in *B. acidocaldarius* (Krulwich et al., 1978) but not in *T. ferrooxidans* or *T. acidophilus* (Cox et al., 1979; Matin et al., 1982). The cell membranes in the acidophilic thiobacilli, therefore, appear highly impermeable to protons *and* anions. This conclusion is supported by the finding (Matin et al., 1982) that in *T. acidophilus* ΔpH can be totally collapsed by the action of a protonophore if a sufficient concentration of the membrane-permeating anion SCN^- is included in the reaction medium. A membrane with a low permeability to ions is, of course, one of the basic requirements for chemiosmotic coupling (Mitchell, 1966).

Although membrane impermeabilities may account for ΔpH maintenance, the generation of ΔpH remains obscure. Even more of a mystery is the mechanism by which acidophiles might actively accumulate phosphate. In neutrophilic bacteria, this is usually achieved by means of specific proton/anion symporters (Harold, 1972). As we have seen above, if such mechanisms were present in acidophiles, it would be expected that pH_i would vary with pH_o and that ΔpH would be easily dissipated unless maintained by an adequate $[R^+]_i$.

CONCLUSIONS

It has been proposed in the literature that acidophilic and alkalophilic bacteria maintain ΔpH by means of a Gibbs-Donnan equilibrium. Equations have been derived that relate ΔpH and pH_i to the concentration of intracellular nondialyzable ions required to maintain ΔpH. The published empirical values of ΔpH and pH_i for acidophiles and alkalophiles are quantitatively incompatible with Gibbs-Donnan equilibria. As an alternative hypothesis to explain ΔpH maintenance, cell membranes are considered to exhibit extremely low nonfacilitated permeability to ions, especially protons.

ACKNOWLEDGMENTS

I would like to acknowledge financial support from the National Science Foundation (Grant no. PCM 7816096) and from the Research Corporation (a Cottrell College Science Grant).

LITERATURE CITED

Bolam, T. R. 1932. The Donnan Equilibria. G. Bell and Sons, London.

Cobley, J. G., and J. C. Cox. 1983. Energy conservation in acidophilic bacteria. Microbiol. Rev. 47: 579–95.

Cox, J. C., and M. D. Brand. 1984. Iron oxidation and energy conservation in the chemoautotroph *Thiobacillus ferrooxidans*. *In* W. R. Strohl and O. H. Tuovinen (eds.), Microbial chemoautotrophy, pp. 31–46. Ohio State University Press, Columbus.

Cox, J. C., D. G. Nicholls, and W. J. Ingledew. 1979. Transmembrane electrical potential and transmembrane pH gradient in the acidophile *Thiobacillus ferro-oxidans*. Biochem. J. 178:195–200.

Guffanti, A. A., P. Susman, R. Blanco, and T. A. Krulwich. 1978. The protonmotive force and γ-aminoisobutyric acid transport in an obligately alkalophilic bacterium. J. Biol. Chem. 253:708–15.

Harold, F. M. 1972. Conservation and transformation of energy by bacterial membranes. Bacteriol. Rev. 36:172–230.

Hsung, J. C., and A. Haug. 1977. Membrane potential in *Thermoplasma acidophila*. FEBS Lett. 73:47–50.

Krulwich, T. A., L. F. Davidson, S. J. Filip, R. S. Zuckerman, and A. A. Guffanti. 1978. The proton motive force and β-galactoside transport in *Bacillus acidocaldarius*. J. Biol. Chem. 253:4599–4603.

Matin, A., B. Wilson, E. Zychlinsky, and M. Matin. 1982. Proton motive force and the physiological basis of delta pH maintenance in *Thiobacillus acidophilus*. J. Bacteriol. 150:582–91.

Measures, J. C. 1975. Role of amino acids in osmoregulation of non-halophilic bacteria. Nature (London) 257:398–400.

Mitchell, P. 1966. Chemiosmotic coupling in oxidative and photosynthetic phosphorylation. Biol. Rev. 41:445–502.

Searcy, D. G. 1976. *Thermoplasma acidophilum:* intracellular pH and potassium concentration. Biochim. Biophys. Acta 451:278–86.

ALAN B. HOOPER

Ammonia Oxidation and Energy Transduction in the Nitrifying Bacteria

9

INTRODUCTION

Nitrosomonas species grow with the oxidation of ammonia to nitrite as the sole source of reducing power and basis for establishments of an electrochemical gradient for ATP synthesis.[1]

This chapter will consider the present state of knowledge regarding the mechanism of oxidation of ammonia to nitrite by *Nitrosomonas* and *Nitrosocystis:* the nature of N-containing intermediates; the origin of nitrite oxygen; the nature of N-transforming and O-activing enzyme systems (hydroxylamine oxidoreductase, HAO, and ammonia oxidase); and the mechanism of proton pumping for ATP synthesis. The chapter supplements recent reviews (Suzuki, 1974; Aleem, 1977; Hooper, 1978; Suzuki et al., 1981b).

N-INTERMEDIATES

Evidence exists for the following intermediates (oxidation states indicated):

$$NH_3(-3) \rightarrow NH_2OH(-1) \rightarrow [HNO](+1) \rightarrow NO \text{ or } NO_2NHOH(+2) \rightarrow HNO_2(+3) \;. \quad (1)$$

NH_2OH

Hydroxylamine is aerobically oxidized to nitrite by cells at rates the same as, or greater than, the rate of ammonia oxidation (Lees, 1952; Hofman and Lees, 1953; Engel and Alexander, 1958). Production of small amounts of hydroxylamine from ammonia is achieved by *Nitrosomonas* cells in the presence of approximately 1 mM hydrazine (Hofman and Lees, 1953; Yoshida and Alexander, 1964). Hydrazine inhibits oxidation of hydroxylamine by cells (Hofman and Lees, 1953). Hydrazine, a substrate for the hydroxylamine dehydrogenase (Nicholas and Jones, 1960) with a K_m similar to NH_2OH (≈ 1 μM, Hooper and Nason, 1965), is thought to be a competitive inhibitor of hydroxylamine oxidation. It is hypothesized that N-containing intermediates of ammonia oxidation are enzyme bound (Hooper, 1978).

HNO

Evidence for an intermediate of the oxidation state of HNO, nitroxyl, comes from studies of hydroxylamine oxidation by extracts. Falcone and colleagues (1962, 1963) reported anaerobic reduction of 2 mol mammalian cytochrome *c* per mol NH_2OH added and production of N_2O as indicated by mass spectroscopy. N_2O would arise chemically from HNO:

$$2\ HNO \rightarrow N_2O + H_2O\ . \qquad (2)$$

Results by Anderson (1965b) indicate that $H_2N_2O_2$ is not an intermediate in N_2O production. Two-electron dehydrogenation of NH_2OH is presumed coupled to O_2 reduction via a terminal oxidase.

NO

Nitric oxide would be the product of dehydrogenation of nitroxyl. With purified HAO, Hooper and Nason (1965) reported values for the ratio of mammalian cytochrome *c* reduced:NH_2OH utilized as high as 3, consistent with a product of oxidation state +2. Anderson (1964, 1965a) reported production of NO from NH_2OH by cells or extracts anaerobically in presence of methylene blue. With pure HAO, Hooper and Terry (1979) confirmed NO production from NH_2OH in the presence of phenazine methosulfate (PMS). Use of ^{15}N-nitrite or ^{15}N-hydroxylamine allowed demonstration that NO was not produced by reduction of nitrite. Mn ion inhibits dehydrogenation of HNO to NO. Anderson (1964) reported anaerobic uptake of NO in presence of extracts and mammalian cytochrome *c*.

Nitrohydroxylamine

NO_2NHOH, or HONONOH, has the same formal oxidation state as NO. Aleem and Lees (1963) reported that cells or extracts oxidized nitrohydroxylamine to nitrite in the absence of added electron acceptor. Assuming the absence of terminal oxidase activity in the partially purified extracts used, this observation is consistent with oxygenation of nitrohydroxylamine (or a decomposition product) at a point subsequent to NH_2OH dehydrogenation. Aleem and Lees (1963) suggested a dismutation:

$$(NOH) + HNO_2 \rightarrow NO_2NHOH \qquad (3)$$

$$NO_2NHOH + \tfrac{1}{2}\, O_2 \rightarrow 2\, HNO_2 \,. \qquad (4)$$

Experimental use of this substrate has not been pursued.

Anderson (1965b) noted that if schemes for N_2O (equation 2) and NO_2NHOH (equation 3) production are correct, the presence of nitrite might well inhibit enzymatic production of N_2O from NH_2OH. This inhibition was not observed.

ORIGIN(S) OF OXYGEN IN NITRITE

Oxygens of nitrite may originate from water or dioxygen (fig. 1A). Each step may be a simple oxygenation or a sequential hydrogen (or electron) removal and reaction with water.

NH_2OH

The oxygen of NH_2OH produced from NH_3 in the presence of NH_2NH_2 comes from O_2. Dua and colleagues (1979) reported incorporation of $^{18}O_2$ (0.3% atom percent excess) but not $H_2{}^{18}O$ into hydroxylamine formed from NH_3 by cells in the presence of NH_2NH_2. Oxygen of hydroxylamine was isolated in the cyclohexanone oxime and analyzed by mass spectrometry. This experiment was repeated definitively with yields of $> 92\%$ atom percent enrichment by Hollocher and colleagues (1981). Exchange of oxygen between H_2O and NH_2OH was shown to be extremely slow.

HNO_2

Cells. At least one oxygen of nitrite probably comes from O_2. Uncertainty is based on the possibilities that (1) NH_2OH might not, in fact, be an intermediate but is produced in small amounts under special circumstances by a reaction mechanism different from the predominant reaction or (2) oxidation of NH_2OH involves loss of oxygen (fig. 1B). Measurement of the ^{18}O content of nitrite by mass spectrometry is difficult because of the necessity to convert NO_2^- oxygens to CO_2 for gas analysis. With this technique, Rees and Nason (1966) reported that cells incorporated 7% of one oxygen of ammonia-derived

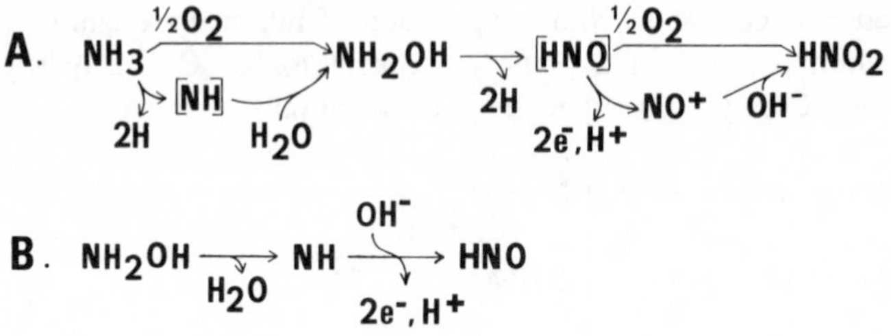

A. $NH_3 \xrightarrow{\frac{1}{2}O_2} NH_2OH \xrightarrow{2H} [HNO] \xrightarrow{\frac{1}{2}O_2} HNO_2$; $NH_3 \xrightarrow{2H} [NH] \xrightarrow{H_2O} NH_2OH$; $[HNO] \xrightarrow{2e^-, H^+} NO^+ \xrightarrow{OH^-} HNO_2$

B. $NH_2OH \xrightarrow{H_2O} NH \xrightarrow[2e^-, H^+]{OH^-} HNO$

Fig. 1. A. Hypothetical pathways in which NO_2^--oxygen may arise from O_2 or H_2O. B. Hypothetical scheme in which the NH_2OH-oxygen may not appear in nitrite.

nitrite from O_2. Recently we have taken advantage of the ^{18}O-isotope shift in ^{15}N NMR to analyze the ^{18}O composition of nitrite (Andersson et al., 1982). As shown in figure 2A, peaks representing $^{15}N^{18}O_2^-$, $^{15}N^{18}O^{16}O_2^-$, and $^{15}N^{16}O_2$ are readily quantified. Figure 2B also illustrates that the technique is applicable to measurement of nitrate. The technique has the advantage that analysis is made directly on the reaction solution after removal of cells. A potential disadvantage is that ≥ 1 mM nitrite concentrations are required for NMR analysis. Utilizing this technique, we have shown that nitrite produced by cells from NH_3 or NH_2OH has the isotope oxygen composition of H_2O. To our surprise, cells were found to catalyze the rapid exchange with water of both oxygen atoms of nitrite:

$$2\ H_2{}^{18}O + H^{15}N^{16}O_2 \rightarrow H^{15}N^{18}O_2 + 2\ H_2{}^{16}O\ . \quad (5)$$

Significantly, the exchange reaction required simultaneous oxidation of ammonia (fig. 3A), although the amount of nitrite exchanged with water could exceed the amount of ammonia oxidized by a factor of three. The mechanism of the exchange is unknown. The reaction suggests (1) that nitrite is in rapid chemical equilibrium with an intermediate in the oxidation of hydroxylamine (the intermediate would contain H_2O-exchangeable oxygens and may have incorporated nitrite-N) or (2) that the cell has regions of high acidity (possibly a result of the energy-transducing proton gradient) in which the rapid H_2O-HNO_2 chemical oxygen exchange occurs. The exchange reaction obviously obscures the origin of oxygen in the nitrite as initially synthesized. Inclusion of a great excess of $^{14}NO_2^-$ during cellular oxidation of $^{15}NH_3$ in $^{18}O_2$ allowed incorporation of 20% of one oxygen from dioxygen. To summarize, studies with intact cells have not clearly identified the origin of the oxygen incorporated during oxidation of hydroxylamine.

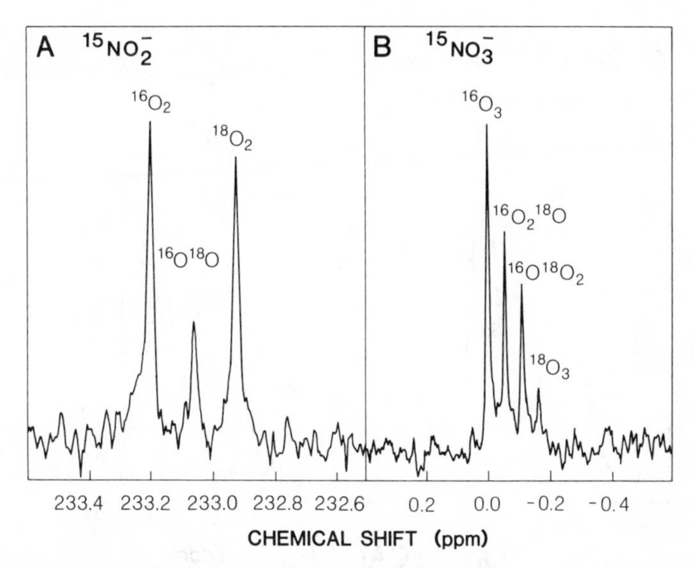

Fig. 2. $^{18}O/^{16}O$ derivatives of nitrite and nitrate resolved by high resolution ^{15}N-NMR (Andersson et al., 1982). A. 100 mM ^{15}N nitrite in 0.25 M phosphate, 0.5 mM carbonate, and 6% $H_2{}^{18}O$, pH 7.5. 180 scans, 12 mm tube (4 ml). B. Solution of ^{15}N-nitrate in 0.25 M phosphate, 5 mM carbonate, and 10% $H_2{}^{18}O$, pH 7.0. 724 scans, 12 mm tube.

Extracts. In reactions catalyzed by extracts in the presence of substrate quantities of electron acceptor, nitrite production, but not hydroxylamine oxidation, is inhibited by anaerobiosis (see Falcone et al., 1963). Likewise, conversion of the putative intermediate nitrohydroxylamine to nitrite required oxygen (Aleem and Lees, 1963). In keeping with those observations, purified HAO catalyzes incorporation of $^{18}O_2$ into NO_2^- produced by oxidation of NH_2OH in presence of PMS as electron acceptor (Andersson et al., 1982) (fig. 3B). Pure HAO does not catalyze the $NO_2^- - H_2O$ oxygen exchange reaction. These observations suggest that O_2 is required for nitrite production and may be incorporated into nitrite.

Support for water as the source of the second nitrite oxygen comes from the report by Yamanaka and Sakano (1980) that nitrite is produced anaerobically by HAO in the presence of substrate amounts of mammalian cytochrome *c*. We have observed what was apparently the same phenomenon but are uncertain of the interpretation. It is possible that a ferrous cytochrome *c* N-oxide complex was formed that subsequently formed nitrite in a chemical reaction taking place in the oxygen-containing sulfanilic acid solution used for assay of nitrite (Hooper and Terry, 1979).

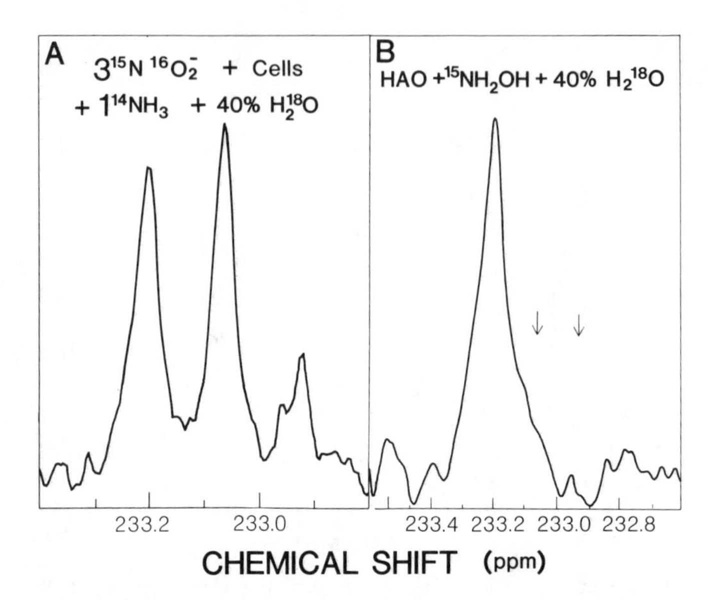

Fig. 3. A. Exchange of oxygen between NO_2 and H_2O during oxidation of ammonia (Andersson et al., 1982). 12 mM $^{15}NO_2{}^-$ has taken on the $^{16}O/^{18}O$ isotopic composition of water during the oxidation of 4 mM $^{14}NH_4$. B. Nitrite formed from the oxidation by HAO of mM $^{15}NH_2OH$ in 40% $H_2{}^{18}O$ has the $^{16}O/^{18}O$ composition of $^{16}O_2$. The aerobic reaction mixture contained 10 μM phenazine methosulfate. Arrows indicate the position of $^{15}N^{16/18}O_2$ and $^{15}N^{18}O_2$.

It is important to keep in mind that some or all of the activities catalyzed by HAO might not occur *in vivo*. Production of NO_3^- by HAO is a clear example (Hooper et al., 1977). In the absence of an enzyme assay involving the appropriate intermediate as substrate, it is not possible to state with certainty that the terminal part of observed nitrite synthesis, ''nitrite synthetase'' as differentiated from ''hydroxylamine dehydrogenase,'' (Hooper, 1978) is enzyme-catalyzed.

To summarize, current evidence suggests that ammonia oxidation begins with an oxygenase (utilizing dioxygen). Evidence for the origin of the second oxygen is much less clear but may favor dioxygen. For simplicity, this article will discuss ammonia oxidation assuming that both nitrite oxygens are from dioxygen (fig. 4A).

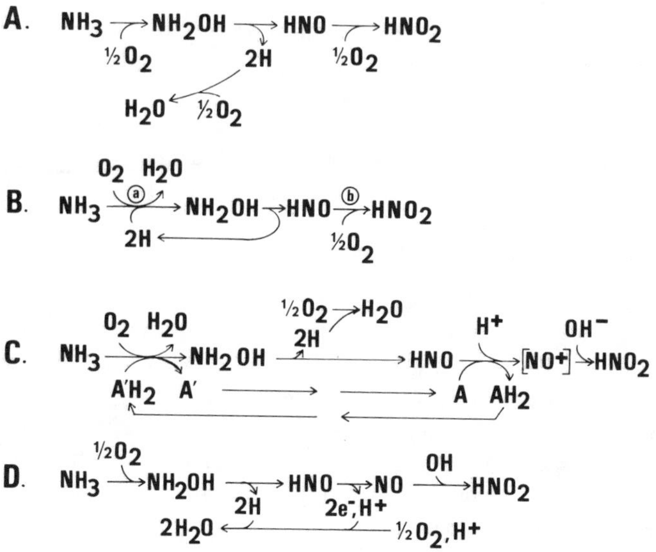

A. $NH_3 \rightarrow NH_2OH \rightarrow HNO \rightarrow HNO_2$
$\frac{1}{2}O_2$ 2H $\frac{1}{2}O_2$
$H_2O \leftarrow \frac{1}{2}O_2$

B. O_2 H_2O
$NH_3 \xrightarrow{(a)} NH_2OH \rightarrow HNO \xrightarrow{(b)} HNO_2$
2H ← $\frac{1}{2}O_2$

C. $\frac{1}{2}O_2 \rightarrow H_2O$
O_2 H_2O 2H H^+ OH^-
$NH_3 \rightarrow NH_2OH \rightarrow HNO \rightarrow [NO^+] \rightarrow HNO_2$
$A'H_2$ A' ⟶ ⟶ A AH_2

D. $\frac{1}{2}O_2$ OH
$NH_3 \rightarrow NH_2OH \rightarrow HNO \rightarrow NO \rightarrow HNO_2$
2H $2e^-, H^+$
$2H_2O \leftarrow \frac{1}{2}O_2, H^+$

Fig. 4. Possible mechanisms of oxidation of ammonia to nitrite by *Nitrosomonas*.

IMPLICATIONS OF THE ORIGIN OF OXYGEN

If Both Oxygens Are from O_2

In this case ammonia oxygenase cannot be water-producing. If, as depicted in figure 4A, (1) both nitrite oxygens are from dioxygen, (2) 2 H's from the dehydrogenation of NH_2OH are consumed by a water-producing terminal oxidase (with associated establishment of a proton gradient), then water cannot also be produced in a "mixed function" oxygenation of ammonia;

$$AH_2 + O_2 + NH_3 \rightarrow A + H_2O + NH_2OH \ . \quad (6)$$

There are not enough valence electrons and protons in ammonia (the sole source of reductant) to incorporate two oxygens, lose two hydrogens in NH_2OH-dehydrogenase, and donate an additional pair of electrons ($+H^+$) to oxygen in a mixed-function ammonia oxygenase.

A water-producing ammonia oxygenase is, however, possible if (1) the ammonia oxygenase is a unique membrane-associated, proton gradient–

forming enzyme (fig. 4B[a]), and/or (2) formation of a proton gradient is associated with a "nitroxyl oxygenase" (fig. 4B[b]). The fact that the ratio of protons pumped:N-substrate oxidized is the same in the oxidation of either NH_3 or NH_2OH (Hollocher et al., 1982) suggests that proton pumping is not (at least obligatorily) associated with ammonia oxidation. This argues against a coupling site associated with ammonia oxygenation.

If the Second Oxygen Is from H_2O

In this case ammonia oxidation could occur with a water-producing monooxygenase (fig. 4C; see Suzuki, 1974). If so, the source of reductant for the ammonia oxygenase must be hydroxylamine or an oxidation product of hydroxylamine. Alternatively, if, as depicted in figure 4D, the ammonia oxygenase is not water-producing, then NH_2OH may pass 4 electrons to the terminal oxidase. This has relevance to the number of reaction cycles of terminal oxidase(s) coupled to hydroxylamine oxidation and the attendant number of protons translocated.

REDUCTION OF NITRITE

Extracts

Soluble enzyme(s) catalyze the reduction of nitrite to a mixture of N_2O and NO (Hooper, 1968). NH_2OH or leucopyocyanine but not NADH or NADPH serves as electron donor:

$$NH_2OH + HNO_2 \rightarrow N_2O + H_2O \tag{7}$$

$$\text{reduced dye} + HNO_2 \rightarrow N_2O + NO + \text{oxidized dye} . \tag{8}$$

The activity has a pH optimum of 6 and is inhibited by diethyldithiocarbamate (DTC). Ritchie and Nicholas (1972) noted that the ratio of NO:N_2O produced is greater with PMS or ferricyanide as electron donor than with 2,6-dichlorophenolindophenol (DClP) or methylene blue. They also found (Ritchie and Nicholas, 1974) that some nitrite reductase (NADH-PMS as electron donor) was closely associated with NAO upon purification. Erickson (1971) had found nitrite reductase activity (leucopyocyanine as electron donor) associated with soluble fractions containing p-phenylenediamine oxidase activity but not HAO.

The nature and physiological significance of nitrite-reducing activities is unclear. They may be reactions of HAO, terminal oxidases, or specialized nitrite-reducing systems. Their activities must be taken into account in studies of N-oxidation pathways in extracts.

Cells

Yoshida and Alexander (1970) reported production of N_2O from ammonia by cell suspensions of *Nitrosomonas*. The yields were as high as 26% of N-produced. Ritchie and Nicholas (1972) showed that $^{15}NO_2^-$ was reduced to $^{15}N_2O$ by cells of *Nitrosomonas*. Subsequently the ammonia-oxidizing chemoautotrophic bacteria have been shown to account for production of significant amounts of N_2O (Bremner and Blackmer, 1978; Blackmer et al., 1980) or N_2O and NO (Goreau et al., 1980). N_2O and NO production by cells is favored at low O_2 concentrations (Goreau et al., 1980). It appears that, under limiting O_2 concentrations, cells oxidize some ammonia only as far as the oxidation state of HNO and/or use nitrite as a terminal electron acceptor in place of oxygen.

Nitrite Assimilation

Wallace and Nicholas (1968) reported assimilatory nitrite- and hydroxylamine-reductases in *Nitrosomonas*. The product was ammonia; NADH or NADPH served as electron donors in the presence of FMN or FAD. ^{15}N of nitrite, hydroxylamine, or ammonia was incorporated into protein. Significantly, nitrate reductase activity was absent in *Nitrosomonas*. Hooper et al. (1967) characterized an NADP-specific glutamate dehydrogenase from *Nitrosomonas*. The *in vitro* rate of glutamate synthesis was similar to that of cellular protein synthesis.

HYDROXYLAMINE OXIDOREDUCTASE (HAO)

HAO constitutes 5% of the soluble protein and 40% of the *c*-type heme of *Nitrosomonas* (Hooper et al., 1978).

Preliminary Characterization

The oxidation of hydroxylamine to nitrite in the presence of substrate amounts of electron acceptor was reported in particles (Nicholas and Jones, 1960; Falcone et al., 1962, 1963) isolated from *Nitrosomonas*. Soluble HAO was purified and shown to contain substrate-reducible hemes and to have a high affinity ($K_m \approx 1$ μM) for hydroxylamine and hydrazine (Hooper and Nason, 1965).

P460. Rees and Nason (1965) reported an unusual absorption maximum at 465 nm resulting from dithionite-reduction of extracts of *Nitrosomonas*. The presence of CO resulted in a shift in absorbancy to approximately 450 nm, suggesting a cytochrome P450-like pigment. Absorbancy at 450 nm was observed in soluble fractions reduced with dithionite (Hooper et al., 1972) and in fractions containing HAO (Ritchie and Nicholas, 1974). Neither NH_2OH nor NH_2NH_2 was able to act as reducing agent in the system. A small portion of the total cellular amount of the pigment was purified by Erickson and Hooper (1972) and was shown to have absorption maxima at 435, 460, and 450 nm in the oxidized, dithionite-reduced or dithionite-reduced-plus-CO forms, respectively. Based on its ligand-binding properties, it appeared to be a heme and was given the name heme P460. Essentially all of the cellular heme P460 was subsequently shown to be associated with HAO (Hooper et al., 1978). Selective destruction of heme P460 of HAO with stoichiometric amounts of H_2O_2 results in loss of hydroxylamine dehydrogenase activity and hydroxylamine reducibility of *c* hemes, indicating that heme P460 is a part of, or near, the substrate binding site (Hooper and Terry, 1977). Electrons may pass from substrate to *c* hemes via the P460 center:

$$NH_2OH \rightarrow P460 \rightarrow c \text{ hemes} . \qquad (9)$$

Prosthetic Groups

Although inhibition by atebrin suggested flavin involvement (Falcone et al., 1963; Hooper and Nason, 1965), chemical analysis did not support the presence of flavin (Hooper et al., 1978). HAO contains > 20 mol Fe per mol enzyme ($M_r \approx 200{,}000$). Al and Ca (2.3 and 1 mol per mol enzyme, respectively) are not thought to be biologically significant. Cu, Zn, Mg, Mo, Co, Mn, Ni, Cd, or Cr are less than 1 per 200,000 M_r (Hooper et al., 1978). All hemes except P460 are *c*-type and account for $\approx$ 18 Fe per enzyme; the remaining 2–3 Fe are assumed to belong to heme P460. The ratio of protein per heme reported as 10,000 (Hooper et al., 1978) or 17,500 (Yamanaka et al., 1979) is not unusual as compared with other hemoproteins. Estimates of the ratio of P460:*c* heme range from 1:6 to 1:9 (Hooper et al., 1978), 1:8 (Lipscomb and Hooper, 1982), and 1:7 (Lipscomb et al., 1982); a value of 1:7 is used as a working number.

Subunit Composition

The relative molecular mass of HAO in aqueous solution has been estimated as 200,000 by sedimentation velocity (Rees, 1968; Maxwell, 1976) or

175,000–180,000 by chromatography on Sephadex or by SDS-polyacrylamide gel electrophoresis (SDS-PAGE) (Yamanaka, et al., 1979). As determined by SDS-PAGE analysis, the subunit structure and denaturation properties of HAO are unusual and poorly understood (Terry and Hooper, 1981). There appear to be 3 peptides detected by SDS-PAGE, two of which have enzymatic significance. Without denaturation HAO has an apparent mass of 200,000. In the presence of SDS, an 11,000 M_r mono-heme cytochrome (called band V) is released. With more stringent denaturation (boiling), a band (IV) that does not contain heme is released. Band IV is not always present in HAO and may be a tightly bound contaminant. Boiling for longer periods in presence of 2-mercaptoethanol progressively converts a third heme-containing protein from apparent mass 125,000 to 195,000 and then to 225,000 (band I). The latter 3 bands have small amounts of hydroxylamine dehydrogenase activity and thus may contain P460. Removal of *c* hemes with 2-nitrophenylsulfenylchloride converts band I into a 63,000 M_+ protein that is present in native HAO in approximately the same ratio as the mono heme cytochrome (band V). As a working model, we suggest that HAO in solution has an $\alpha_3\beta_3$ structure (fig. 5). Subunit α has 6 *c* hemes and heme P460. The α subunits undergo an unusual polymerization during denaturation in SDS or are, in fact, polymerized in the native form of HAO and undergo a sequential denaturation to a form with slower mobility in PAGE (225,000 M_r form). Subunit β is not essential for catalytic activity but imparts stability to HAO. The latter observation suggests that P460 is part of the α subunit.

Native HAO or the 225,000 M_r form are resistant to proteases. Removal of hemes causes the resulting subunits to be easily digestible.

Nature of c *Hemes of HAO*

HAO *c* hemes can be divided into 4 categories based on their oxidation-reduction potentials (fig. 6; C. Larroque and A.B. Hooper, unpublished). One *c*-552 has a midpoint potential greater than +100 mV; one *c*-552 and one *c*-559 have potentials of approximately 0. One *c*-552 and one *c*-559 have potentials in the range −100 to −250 mV. Values for the latter hemes were difficult to determine, suggesting a possible change in redox potential with the oxidation state of the enzyme. Two heme *c*-552 moities and heme P460 have redox potentials of −320 mV. These unusual low-potential *c* hemes could participate in reduction of pyridine nucleotide, activation of dioxygen for oxygenation of substrate, or the initiation of an energy-coupled electron transport chain. In the steady state of oxidation of hydroxylamine by HAO (A. B. Hooper, unpublished) or cells (Drozd, 1976), roughly 1/3 of the heme *c* is reduced.

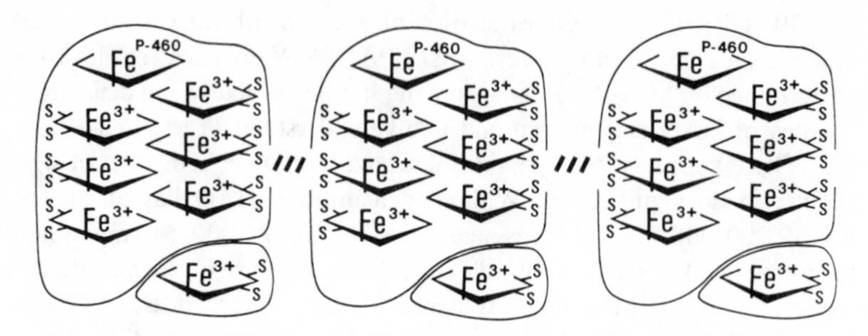

Fig. 5. Schematic view of subunit and heme composition of HAO of *Nitrosomonas*.

EPR. Electron paramagnetic resonance (EPR) measurements on resting HAO by Vickery and Hooper (1981) indicated the presence of at least 4 categories of low-spin (S = 1/2) ferric hemes and the absence of high-spin hemes, iron-sulfur centers, or copper. Subsequent quantification by Lipscomb and Hooper (1982) of the signals during dithionite titration clearly indicated species of g = 3.06, 2.14, 1.35 and g = 2.98, 2.24, 1.44 in a ratio of 3:2, accounting for 5 of the 7 (or 8) *c* hemes (fig. 7). Heme P460 is apparently EPR silent on resting HAO. Two (or three) additional *c* hemes with resonances at 3.38, 2.7, 1.86, and 1.67 have unusual EPR signals, reduce coordinately, and are possibly spin-coupled. EPR spectra made at X, P, and S band (9.2, 15, and 2 GHz, respectively) exhibit shifts in the g = 3.38, 2.7, 1.86, and 1.67 resonances, suggesting a unique electronic coupling between the 2 or 3 *c* hemes (J. D. Lipscomb, K. K. Andersson, A. B. Hooper, and R. Dunham, unpublished). There is no evidence to indicate the role of these hemes in the action of the enzyme.

A preliminary attempt at correlation of reduced absorption maxima, redox potentials (C. Larroque and A. B. Hooper, unpublished), and g values (Liscomb and Hooper, 1982) of *c* hemes of HAO is shown in table 1.

Changes in Structure of HAO. Four examples of inducible changes in spin state of hemes of HAO reflect changes in enzyme structure (Lipscomb and Hooper, 1982): (1) reduction of ≈ 45% of HAO *c* heme by NH_2OH results in shift of a g = 3.06 resonance to g = 2.98 and the appearance of new signals at 2.85 and 2.3 (fig. 7); (2) reduction by dithionite of ≈ 30% of HAO hemes results in a shift of 1 g = 3.06 resonance to g = 2.98; (3) in the pH range 6 to 7 changes in the g = 3.06 and 2.98 resonances occur; and (4) upon reduction of ≈ 70% of HAO heme a g = 6.4 resonance appears.

To summarize, the repeating unit of HAO is a protein containing 6 *c* hemes

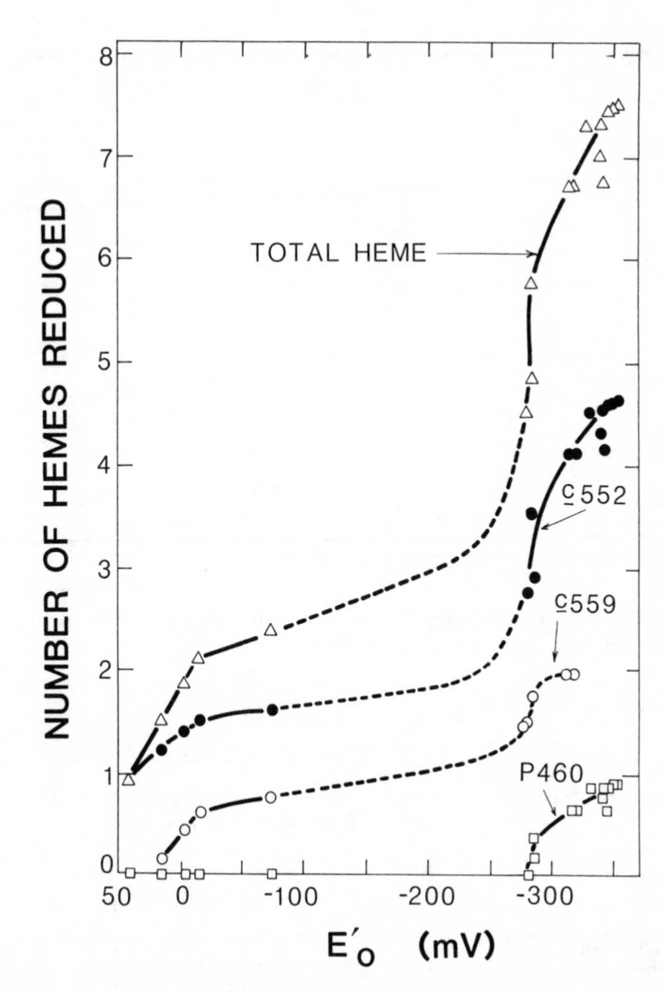

Fig. 6. Oxidation reduction potentials of hemes *c*-552, *c*-559, and P460 of HAO of *Nitrosomonas* (C. Larroque and A. B. Hooper, unpublished). Reduction was achieved by illumination of HAO in presence of acridine orange and EDTA. Redox potential of the solution was estimated by reference to benzyl viologen or methylene blue.

and 1 heme P460 and a second mono heme cytochrome *c*. The *c* hemes of this capacitor-like enzyme range in redox potential from +100 to −320 mV and undergo changes in EPR properties dependent on the oxidation state of the enzyme.

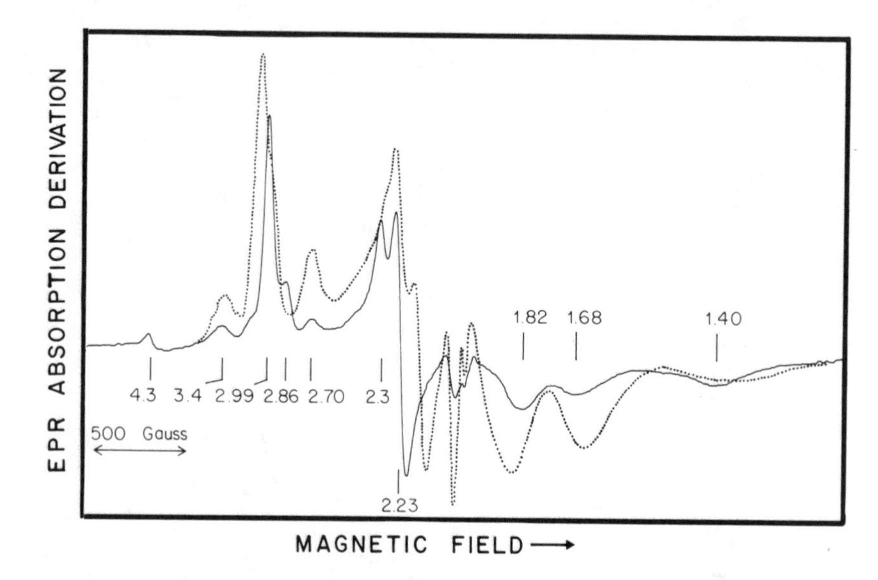

Fig. 7. EPR spectra of HAO (560 μM heme; Lipscomb and Hooper, 1982). Resting enzyme (——); reduced in presence of 1 mM NH_2OH (-----).

TABLE 1

TENTATIVE CORRELATION OF ABSORPTION MAXIMUM, REDOX POTENTIAL, AND EPR PARAMETER OF HEMES OF NITROSOMONAS

	Hemes in Order of Reduction							
	1	2	3	4	5	6	7	8
Wavelength maximum of ferrous form (nm)	552	552,	559	552,	559	552,	552,	460
Approximate redox potential (mV)	+100	0		−120 to −250		−320		
EPR parameter of predominant heme species (g_Z)	3.06	3.06	2.7	3.4	2.98	2.98,	2.98,	6.4

Nature of Heme P460

Almost all heme P460 is found on HAO (Hooper et al., 1978), although a small amount is in the form of the smaller polypeptide "P460 fragment" characterized by Erickson and Hooper (1972). Antibody made against HAO reacts with the P460 Fragment (W. F. Melms, K. K. Andersson, and A. B. Hooper, unpublished). The fragment may be a proteolysis product of HAO. Ferrous P460 of the "fragment" binds CO and CN^-. Reaction of HAO heme P460 with CO is slow ($K_1 = 10^{-3} M^{-1} sec^{-1}$) and the affinity is low ($K_1 =$ 12 μM) as compared with other CO-binding hemoproteins, suggesting the presence of an atypical, tightly bound ligand in the sixth position of iron (Hooper et al., 1983). P460 forms an unusual pyridine derivative with absorption peaks at 449 or 433 nm in the oxidized or reduced forms, respectively. The reduced pyridine derivative binds CO (Erickson and Hooper, 1972). "P460 fragment," M_r 17,000, has one iron (A. B. Hooper and K. R. Terry, unpublished). Mössbauer spectroscopy provides the best evidence that P460 of HAO is a heme (Lipscomb et al., 1982). In reduced HAO approximately 1/8 of the iron (the fraction expected for P460) was accounted for by an ill-defined absorption pattern not previously observed in hemoproteins. Reaction with CO produced a spectrum characteristic of a low-spin ferrous heme CO complex.

Because of the lability of heme P460 the ring structure and method of attachment to HAO is not known. Heme soret resonance Raman spectra (G. Babcock, K. K. Andersson, and A. B. Hooper, unpublished) attributed to heme P460 of the "fragment" or HAO are very unusual for heme proteins. The heme appears not to contain a carbonyl group. Determination of the structure of P460 is an important unresolved problem.

Heme P460 of the fragment and HAO have significantly different properties (table 2). On the resting fragment, P460 is high-spin, g = 6.45 (Hooper

TABLE 2

COMPARISON OF PROPERTIES OF P460 ON HAO WITH P460 OF "P460 FRAGMENT"

Source	EPR Parameter	ABSORPTION MAXIMUM (nm)			
		Oxidized	Reduced	Reduced + CO	Change with NH_2OH
HAO	EPR silent; g ≈ 6 when *c* hemes 70% reduced[a]	415, 566[c]	460[e]	450[e]	none[e]
P460 fragment	g ≈ 6[b]	530[d]	460[d]	448[d]	≈ 442 μm[d], putative reduced and liganded form

[a]Lipscomb and Hooper, 1982.
[b]Hooper and Vickery, 1981.
[c]Andersson and Hooper, 1982.
[d]Erickson and Hooper, 1972.
[e]Hooper et al., 1978.

and Vickery, 1981). On HAO, P460 is EPR-silent but apparently becomes detectable as a high-spin (g = 6.4) signal when the *c* hemes of HAO are 70% reduced (Lipscomb and Hooper, 1982). Although the absorption maxima of the reduced- and reduced-plus-CO forms are similar on the "fragment" and HAO, the values for the oxidized form differ (415 and 435 nm, respectively; Andersson and Hooper, 1982). Reaction of P460 "fragment" with NH_2OH results in an absorbancy shift from 430 to 446 nm and disappearance of the g = 6.45 signal. This suggests that P460 is reduced and liganded by an N-oxide. In contrast, when ≈ 40% of the hemes of HAO are reduced in the presence of NH_2OH, heme P460 apparently remains in the oxidized form (a 446 or 460 nm absorption maximum does not appear). These observations fit a model whereby heme P460 of HAO is reduced by NH_2OH but passes electrons to *c* hemes (equation 9) and remains oxidized at equilibrium. This would be predicted by the relative redox potentials of the hemes *c* and P460.

NATURE OF REACTION CATALYZED BY HAO

Current information suggests that the reaction catalyzed by HAO consists of the dehydrogenation of NH_2OH followed by the oxygenation of an N-intermediate to form nitrite.

NH_2OH Dehydrogenase

Electrons extracted from NH_2OH are thought to initially reduce P460 (fig. 8a). At equilibrium or in the steady state *in vitro* or *in vivo* 30–45% of the *c* hemes are reduced. Electrons may exit from the enzyme from *c* hemes to artificial electron acceptors (e.g., PMS) or to an electron transport chain containing the terminal oxidase (fig. 8d). Alternatively, electrons may exit from HAO directly via P460 (fig. 8e) with the reduced *c* hemes serving only to establish a redox state of the enzyme necessary for catalysis. Under special conditions where the cellular concentration of NADPH is limited, changes in the enzyme may cause exit of electrons through the low potential *c* hemes. Stopped-flow kinetic measurements at 2° (A. B. Hooper and C. Balny, unpublished) indicate that ≈ 90% of the substrate-reducible *c* hemes are reduced in a monophasic manner with first order rate constants of ≈ 30 sec^{-1} (≈ 120 sec^{-1} at 20°). P460 was not reduced nor were transient spectral intermediates detected, indicating that the velocity b is faster than velocity a (fig. 8). Since ≈ 3 *c* hemes are substrate-reduced per P460 center, a reasonable estimate for the rate of turnover of P460 is 360 sec^{-1}. This compares favorably with the turnover of P460 on HAO *in vivo* or *in vitro* (130 molecules NH_2OH oxidized/molecule P460 sec^{-1} or 260 electrons/molecule P460 sec^{-1}) and indicates that electron pathway a → b → d (fig. 8) is feasible.

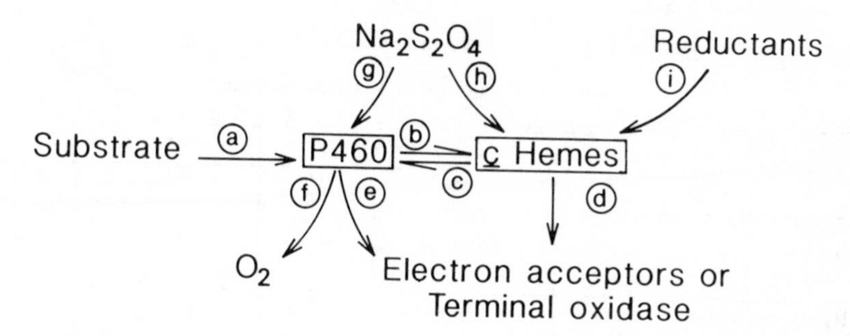

Fig. 8. Pathways of electrons in hydroxylamine oxidoreductase of *Nitrosomonas*.

HAO has at least 2 sites of reduction of *c* hemes (A. B. Hooper and V. M. Tran, unpublished). One site, which contains heme P460, reacts with substrates NH_2OH or NH_2NH_2 and is destroyed by H_2O_2 (table 3). Destruction of the substrate binding site does not affect the level of equilibrium reduction by nonsubstrate reductants at a second type of site(s) (fig. 8i). Reductants are grouped arbitrarily into 5 categories based on the equilibrium level of reduction of *c* hemes (table 3).

Reduction of *c* hemes by dithionite is a multiphasic process (A. B. Hooper and C. Balny, unpublished). In the first phase, one (high potential?) heme *c*-552 is reduced with an apparent first-order rate constant of 7 sec^{-1} at 20°. This value is similar to the rate of reduction of mammalian cytochrome *c* in presence of dithionite. The remaining *c* hemes of HAO are reduced with a rate constant of 0.07 sec^{-1}. The result suggests that one *c* heme has a relatively more exposed site and that others may be more buried. It is puzzling that rapid equilibration of hemes does not occur in presence of dithionite (otherwise all hemes would be reduced with a rate constant of 7 sec^{-1}) although rapid equilibration ($K > 30$ sec^{-1}) occurs during reduction by substrate or oxidation by O_2 or H_2O_2 (see below). Bearing in mind the structural changes observed by EPR, the results suggest changes in the configuration of HAO depending on the nature of the reductant and, in particular, whether the substrate-binding site is occupied.

Oxygen Activation and Nitrite Synthesis

Figure 9 depicts a working model for oxygen activation based on corresponding schemes for cytochrome P450 (White and Coon, 1980). Direct evidence does not exist for any intermediates hypothesized. The scheme is presented to facilitate discussion of unique aspects of activation of dioxygen by *Nitrosomonas*. Cleavage of the dioxygen bond may involve the initial

TABLE 3

EQUILIBRIUM REDUCTION OF C HEMES OF HAO[a]

Reductant	Active HAO		H_2O_2-Inactivated HAO[b]	
	C-552	*C*-559	*C*-552	*C*-559
Ascorbate H_2O_2 NaN_3	0.2[c]	0	0.2	0
Hydroquinone Pyrogallol N-methyl hydroxylamine EDTA NADH-pyocyanine Ascorbate-methylene blue	0.7–1.1	0	0.7–1.1	0
NH_2OH	2	1	0	0
NH_2NH_2	2.8	1.7	0	0
$Na_2S_2O_4$	5	2	5	2

[a]V. M. Tran and A. B. Hooper, unpublished. Solutions of HAO were titrated anaerobically with the indicated reductant.
[b]Inactivated according to Hooper and Terry (1977).
[c]Approximated number of hemes reduced.

2-electron reduction of O_2: $2e^- + O_2 \rightarrow OO^=$. One oxygen would subsequently react with an N-compound. In contrast to reactions of cytochrome P450, the second atom of oxygen does not form water but must be retained by the enzyme and subsequently react with a second N-compound (fig. 9, reactions d and e). Alternatively, both oxygens may react simultaneously with 2 N-compounds. In figure 9, electrons for the activation of oxygen are depicted as arising from HNO and NH_3 (reactions g and h) passing to enzyme metal centers (through P460 to *c* hemes of HAO,) and then (possibly via P460) to O_2 (reactions a and c). Dehydrogenation of HNO by HAO is consistent with the observation of NO production by the enzyme. In fact, electrons may come from NH_3, NH_2OH, or HNO and equilibrate in the pool of heme *c* centers for use in O_2-activation as well as electron-transport or reduction of pyridine nucleotide.

Because heme P460 binds CO, it is likely to bind O_2 as well and may be involved in O_2 activation. We have recently observed that electrons of ferrous *c* hemes of HAO rapidly exit via P460 to oxygen. The product is H_2O_2 and/or O_2^- (Hooper and Balny, 1982). The observations that P460 is oxidized more rapidly than *c* hemes and that oxidation is inhibited by CO suggest the role for

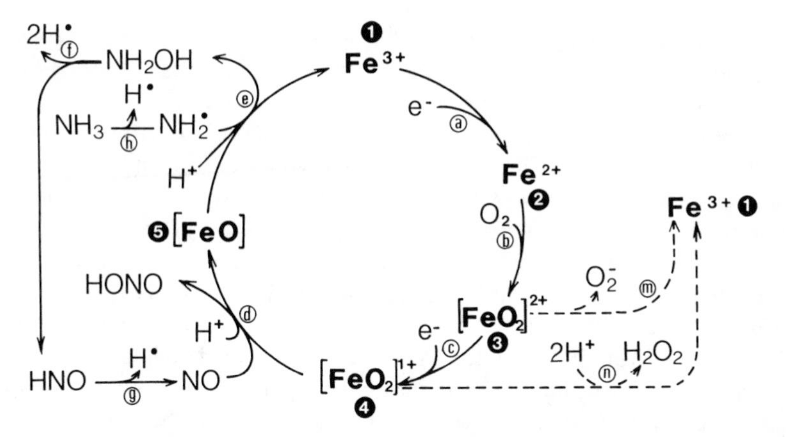

Fig. 9. Hypothetical mechanism illustrating a possible mode of activation of dioxygen by HAO. The model is analogous to corresponding models for P450. The O_2 binding heme is pictured as P460. Electrons for reductive steps (a and c) may come from the pool of *c* hemes. Electrons in the *c* heme pool may have originated from dehydrogenation of NH_3, NH_2OH, or HNO (steps h, f, g, respectively). The scheme differs from P450 in that after the first substrate oxygenation (step d) the second O is not lost as H_2O but oxygenates the substrate. Reaction sequences a, b, m or a, b, c, n illustrate a possible mode of reoxidation of ferrous HAO by O_2 in absence of N-containing substrate (Hooper and Balny, 1982).

TABLE 4

KINETIC DATA FOR OXIDATION BY DIOXYGEN OF FERROUS HEMES P460, *c*-552, AND *c*-558 OF HAO[a]

Component	"PHASE I"		"PHASE II"	"PHASE III"
	0.17–0.2[b]	0.4[b]	70–100[b]	400–500[b]
P460 K_{obs} (sec^{-1})	20	2.5	. . .	. . .
(%)[c]	(60)	(40)	. . .	. . .
c-552 K_{obs} (sec^{-1})	14	. . .	0.03	0.005
(%)[c]	(58)	. . .	(26)	(16)
c-560 K_{obs} (sec^{-1})	16	. . .	0.033	0.008
(%)[c]	(26)	. . .	(58)	(16)

[a]Hooper and Balny, 1982.
[b]Elapsed time (sec).
[c]Percentage of heme oxidized during phase.

P460. The reaction is multiphasic (table 4). In the rapid phase, 1 P460, ≈ 3 *c*-552, and ≈ 0.5 *c*-559 hemes are oxidized with apparent first-order rate constants of 20, 14, and 16 sec^{-1}, respectively. The biological significance of the reaction with O_2 is not certain. It may be a "partial reaction" of oxygen activation: reactions b and c of figure 9 are possible examples; a ferrous P460-O_2 complex may be formed (reaction b), which may undergo a 1 electron reduction from the reservoir of ferrous *c* hemes of HAO to form a hydroxylating intermediate. *In vitro,* in the absence of a nitrogenous substrate, either of the two oxygenated intermediates could decay with the formation of O_2^- (reaction m) or H_2O_2 (reaction n). The rate of oxygen reduction by fully reduced HAO is great enough to account for a possible role in activation of O_2 for N oxygenation. The turnover number for HAO is 130 mol NH_2OH oxidized/mol P460 sec^{-1} (or 130 mol 1 electron-reduced-O/mol P460 sec^{-1}). The first-order rate constant for heme oxidation is $\approx$ 17 sec^{-1}. Because $\approx$ 5 hemes are oxidized during the rapid phase ($\approx$ 4 *c* and 1 P460) and all electrons must exist through P460 (heme P460 is therefore in a transient steady state), a reasonable value for turnover of P460 is $\approx$ 85 mol electrons/mol P460 sec^{-1}.

Electrons also exit rapidly from *c* hemes of HAO via P460 to H_2O_2. The reaction is monophasic and pseudo first order with a rate constant of $\approx 10^2$ M^{-1} sec^{-1} and does not occur in presence of CO or with H_2O_2 inactivated enzyme (Hooper et al., 1983). The reaction of HAO with H_2O_2 can also be interpreted in terms of a model for the biologically significant reactions of oxygen activation. Ferrous P460 may react with H_2O_2 to form a monooxygenated P460 derivative that may, in turn, undergo reduction by *c* hemes and decay to H_2O and ferric P460.

Production of Nitrate

Under certain circumstances, nitrate is a product of the oxidation of hydroxylamine by HAO. The reaction, which does not occur *in vivo*, was first observed by Anderson (1964) and characterized by Hooper et al. (1977). In the presence of O_2 and PMS, HAO catalyzes the oxidation of hydroxylamine to nitrite, nitrate, and gases:

$$1\ NH_2OH + PMS + O_2 \rightarrow \approx 0.4\ NO_2^- + \approx 0.4\ NO_3^- + \text{gases}\ . \qquad (10)$$

In the presence of the sulfhydryl compound diethyldithiocarbamate (DTC), nitrite and the disulfide bis(diethyldithiocarbamoyl)disulfide (Disulfiram) are produced:

$$1\ NH_2OH + 2\ DTC + PMS, O_2 \rightarrow \approx 0.8\ NO_2^- + \text{Disulfiram} + \text{gases}\ . \qquad (11)$$

Nitrate production occurs during hydroxylamine oxidation coupled to a variety of electron acceptors (Hooper et al., 1977).

Nitrate is not always a product. The experimental conditions promoting increased production of nitrite are not understood. They may include limiting O_2 (Andersson et al., 1982) and/or a high concentration of electron acceptor (Yamanaka and Sakano, 1980). Significantly, in the latter case, mammalian cytochrome *c* was reduced and then reoxidized during the course of nitrite production.

Nitrate production by HAO is an example of an artifactual reaction catalyzed by the isolated enzyme. Two possible mechanisms for the reaction are suggested. Nitrate may result from the chemical reaction of enzymatically produced NO with O_2 and H_2O (fig. 10A). According to this model, the presence of a reductant such as DTC (Hooper et al., 1977) or ferrous cytochrome *c* (Yamanaka and Sakano, 1980) may allow a chemical dismutation to nitrite (fig. 10B). An alternative scheme involves the generation, during oxidation of NH_2OH, of a hypothetical active dioxygen that reacts with an N-containing intermediate to produce nitrate. According to this scheme, utilization of a reductant such as DTC allows reduction of one oxygen atom to water rather than incorporation into nitrate (fig. 10C). The second hypothesis is more interesting biologically and is consistent with the

A.

$$2NO + O_2 \longrightarrow 2NO_2 \longrightarrow N_2O_4$$

$$N_2O_4 + H_2O \longrightarrow HNO_2 + HNO_3$$

B.

$$2NO + O_2 + XH_2 \longrightarrow 2HNO_2 + X$$

C.

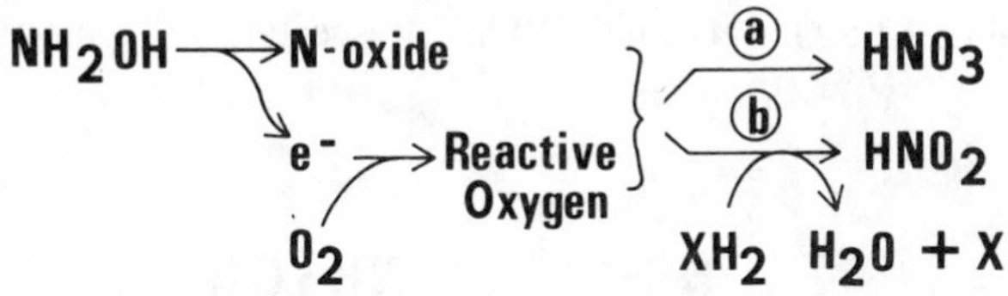

$$NH_2OH \longrightarrow \text{N-oxide}, \quad e^-$$

$$e^- + O_2 \longrightarrow \text{Reactive Oxygen}$$

$$\text{N-oxide} + \text{Reactive Oxygen} \xrightarrow{(a)} HNO_3$$

$$\text{N-oxide} + \text{Reactive Oxygen} + XH_2 \xrightarrow{(b)} HNO_2 + H_2O + X$$

Fig. 10. Chemical reactions of NO producing nitrite and/or nitrate. A. Sequence leading to a mixture of nitrite and nitrate. B. Dismutation in presence of reductant resulting in production of nitrite. C. Model for nitrate production by HAO. Activation of O_2 to reactive oxygen species if achieved with electrons from dehydrogenation of hydroxylamine or subsequent N-intermediates. The reactive oxygen either (1) reacts with a N-intermediate to form nitrate or (2) reacts with an N-intermediate and a reductant (such as diethyldithiocarbonate) to form nitrite.

concept of a chemical coupling between the oxidation of ammonia and HNO (Hooper, 1978) illustrated in figure 11. By this model, HAO lacks ammonia-activating factors required for utilization of the second atom of O_2 to produce NH_2OH (fig. 9, reaction e); DTC substitutes for (NH_2). The latter hypothesis is consistent with the inhibition by DTC of the oxidation of NH_3 but not the oxidation of NH_2OH in cells of *Nitrosomonas*.

Suzuki and colleagues (1981), citing the models of Hughes and Nicklin (1970), noted that the production of nitrate by HAO is consistent with peroxynitrite as an intermediate in the oxidation of hydroxylamine to nitrite:

$$NH_2OH \rightarrow NO^- + 2\ H + H^+ \tag{12}$$

$$H^+ + NO^- + O_2 \rightarrow ONOOH \tag{13}$$

Peroxynitrite may then, in effect, fill the role of an active form of oxygen:

$$NH_3 + ONOOH \rightarrow NH_2OH + HNO_2 \tag{14}$$

Coupling of Hydroxylamine Dehydrogenase with Terminal Oxidase

Cytochrome c-554. A large diheme *c*-type cytochrome (absorbance maximum 553 nm) has been reported by Tronson and colleagues (1973); M_r, pI, and redox potential were 52,000, 4.68, and 500 mV, respectively. The cytochrome binds CO and was half reduced as isolated. Cytochrome *c*-554, purified by Yamanaka and Shinra (1974), is an autooxidizable di-heme *c*-type cytochrome with an unusual soret peak and dithionite-reduced absorbancy maximum at 554 nm. The M_r and pI are 25,000 and 10.4, respectively. These two cytochromes are possibly the same.

Cytochrome c-552. *Nitrosomonas* contains a monoheme *c* cytochrome with a dithionite-reduced absorbance maximum at 551.5, and M_r, pI, and redox potential of 12,500, 4.42, and 240 mV, respectively (Tronson et al., 1973).

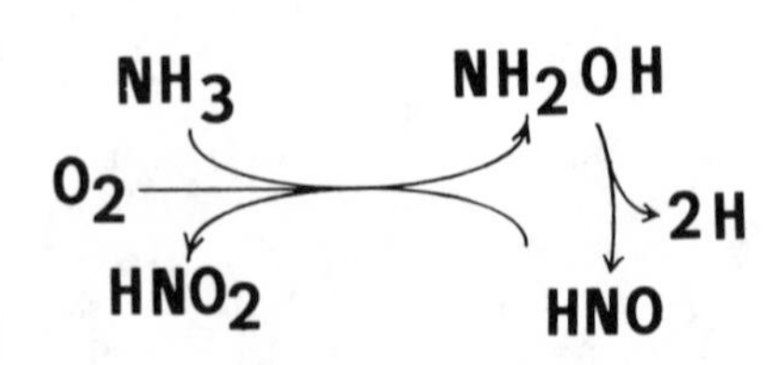

Fig. 11. Coupling of oxygenation of NH_3 and HNO.

This cytochrome may be the same as cytochrome *c*-552, a monoheme *c*-type cytochrome with a dithionite-reduced absorbance maximum at 552 nm and with M_r, pI, and midpoint potential values of 10,000, 3.7, and +250 mV, respectively (Yamanaka and Shinra, 1974).

Oxidases. Soluble oxygen-reducing enzymes (Erickson, 1971) and CO-binding cytochromes (possibly cytochrome *o*; Rees and Nason, 1965) of unknown significance have been reported in *Nitrosomonas*. To be involved in energy coupling, a terminal oxidase is assumed to be membrane-bound. A cytochrome a_1 with low levels of cytochrome *c* oxidase activity has been solubilized and purified from membranes of *Nitrosomonas* (Erickson et al., 1972). Conditions for studying the oxygen-reducing activity and the significance of the reaction to coupled N-oxidation have not been determined.

Coupling. Cytochrome *c*-554 is 50% reduced anaerobically in the presence of HAO and NH_2OH; cytochrome *c*-552 is reduced in the presence of HAO, cytochrome *c*-554, and NH_2OH (Yamanaka and Shinra, 1974). These equilibrium data are consistent with a path of electrons:

$$NH_2OH \rightarrow \text{HAO} \rightarrow \text{cytochrome } c\text{-554} \rightarrow \text{cytochrome } c\text{-552} \rightarrow \text{terminal oxidase} \; . \tag{15}$$

Reconstitution of HAO together with possible electron transport cytochromes and a membrane-associated terminal oxidase has not been systematically explored. Stimulation by cytochromes *c*-554 and *c*-552 of NH_2OH-dependent O_2 utilization catalyzed by a crude membrane fraction has been reported (Suzuki and Kwok, 1981). Although HAO and cytochromes *c*-552 and *c*-554 are very soluble in aqueous solution, the question of whether they are soluble, extrinsically membrane-associated, or intrinsically membrane bound *in vivo* is important (especially for HAO) and unanswered.

AMMONIA OXIDATION

Studies with Cells

The K_m value for ammonia oxidation decreases with increasing pH (Suzuki et al., 1974), suggesting that NH_3 (rather than NH_4^+) is the favored substrate for an ammonia-uptake system (for which there is no evidence to date) and/or the ammonia-oxidizing enzyme system.

Oxidation of ammonia is much more sensitive to inhibition by a great variety of compounds or reaction conditions compared with oxidation of hydroxylamine (Lees, 1952; Hoffman and Lees, 1953; Campbell and Aleem, 1965; Hooper and Terry, 1973; Hynes and Knowles, 1978). Great sensitivity to chelating agents, especially the copper ion binding agents diethyldithio-carbamate, thiourea, and allylthiourea (Lees, 1952), suggests involvement of a metal such as copper. Inhibition by CO (Hooper and Terry, 1973) indicated a possible role for heme P460. Inhibition by methanol (Hooper and Terry, 1973) or acetylene (Hynes and Knowles, 1978) has been taken to indicate the involvement of a free radical. However, inhibition by methanol, methane, or CO is competitive, suggesting that the compounds simply exclude NH_3 from the active site (Suzuki et al., 1976). Ammonia-oxidation activity in cells is specifically destroyed by light in the 400–430 nm range (Hooper and Terry, 1974). Recovery is inhibited by chloramphenicol and thus probably requires protein synthesis, suggesting that an (unidentified) enzyme had been destroyed. Cells are less photosensitive in the absence of ammonia oxidation or during very rapid ammonia oxidation. This was rationalized by suggesting that a photoreactive intermediate or enzyme state present during ammonia oxidation is required for photoinactivation but that protective compounds (e.g., reductants) may be present during rapid oxidation of ammonia.

Inhibition of ammonia oxidation by uncouplers (Hooper and Terry, 1973) suggests involvement of ATP or of an appropriate alignment in the membrane of component proteins. PMS or methylene blue may inhibit by bleeding off reductant (i.e., electrons required for oxygen activation for ammonia oxidation) or by quenching free radical intermediates.

A lag in the time course of ammonia oxidation that is not seen with hydroxylamine oxidation is eliminated in the presence of catalytic amounts of hydroxylamine (Hooper, 1969; Suzuki and Kwok, 1969, 1970). The observation suggests that, for maximum rates of oxidation, the cell must be poised in an appropriate NH_2OH-dependent state (e.g., redox state, enzyme aggregation, or membrane configuration). Coupling of hydroxylamine oxidation with ammonia oxidation is also indirectly suggested by the fact that many inhibitors or treatments (e.g., aging of cells at 4°C) that decrease rates of ammonia oxidation simultaneously increase rates of oxidation of hydroxylamine (Hooper, 1978).

Studies with Extracts

Achievement of high rates of ammonia oxidation in extracts has proved to be extremely difficult. Aerobic oxidation of ammonia to nitrite has been studied in the absence of added electron acceptor (which would inhibit ammonia oxidation). The difficulty in obtaining high activity is understandable since as many as 5 possible components must be together in an appropriate aggregate: ammonia oxidase protein(s), HAO, cytochromes *c*-554 and *c*-552, and a terminal oxidase. In fact, methods of preparing active extracts involved the use of factors that could have the effect of promoting aggregation: Tris, ATP, phosphate, and Mg^{2+} (Watson et al., 1970) or polyamines, Mg^{2+}, and bovine serum albumin (Suzuki and Kwok, 1970) and phosphate, bovine serum albumin, spermine, and Mg^{2+} (Suzuki et al., 1981a).

A particulate fraction of extracts was reported to contain ammonia- , methane- , methanol-, or CO-stimulated NADH activity and ammonia-dependent O_2 utilization that was stimulated by bovine serum albumin, NADPH, or NADH (Suzuki et al., 1976). NADPH or NADH were hypothesized to be electron donors (directly or indirectly) for the mixed-function oxygenation of NH_3 (Suzuki, 1974). In extracts, *c*-type cytochromes were shown to reach a steady-state of reduction prior to the achievement of maximum rate of NH_3-stimulated O_2 utilization (Suzuki et al., 1981). This observation is consistent with the requirement for a low redox potential and a possible role of *c* hemes as electron donors for oxygen activation. Current speculation states that ammonia (fig. 9) or "endogenous" compounds are the electron donors to *c*-type cytochromes for the phase of the reaction preceding rapid oxygen utilization (when hydroxylamine oxidation occurs).

Suzuki and Kwok (1981) have initiated studies of separation and reconstitution of the ammonia-oxidizing system. Crude fractions were separated by chromatography on Sepharose 6B. A membraneous fraction 1, which contained HAO, cytochrome *a*, and cytochrome oxidase activity, catalyzed oxidation of NH_2OH. Fractions 4 and 7 contained HAO and cytochrome *c*-552, respectively. Maximum rates of ammonia oxidation were observed in the presence of the 3 fractions or Fraction 1 plus purified cytochrome *c*-554. Cytochrome *c*-554 (K_m 3.3 μM) was hypothesized to be an electron donor for ammonia hydroxylation (Suzuki and Kwok, 1981).

Tsang and Suzuki (1982) have recently reported an interesting result utilizing the reconstituted membrane fraction with reduced cytochrome *c*-554. Reoxidation of cytochrome *c*-554 and O_2 utilization takes place following the addition of CO or NH_3, suggesting the presence of CO and NH_3 oxygenase activity.

Summary. Ammonia oxidation in *Nitrosomonas* involves a labile, metal-containing multicomponent system. The ammonia oxygenase appears tightly coupled to, and dependent on, hydroxylamine oxidation. Hydroxylamine oxidation may be a source of electrons for activation of dioxygen for ammonia hydroxylation (fig. 4C, fig. 9, fig. 11). It is worth noting that oxygen activation is not likely to involve NADH as electron donor. (1) Reduction of NAD^+ does not occur during oxidation of hydroylamine by HAO. In fact, NAD^+ reduction in *Nitrosomonas* may require electrons from NH_2OH and expenditure of ATP independently generated in the oxidation of an additional

NH_2OH (Aleem, 1966). If this is true, NADH-dependent oxidation of ammonia would require more NADH than is generated in subsequent oxidation of NH_2OH. (2) Addition of NAD^+ or NADH is not required for ammonia oxidation in extracts (Tsang and Suzuki, 1982).

Parallels with Methane-oxidizing Organisms

Several factors suggest similarities in the mechanism of oxidation of ammonia by *Nitrosomonas* and of methane by methanotrophic bacteria. (1) The substrates are similar in structure. Methanotrophs can oxidize ammonia (Ferenci et al., 1975) in addition to CH_3, CH_2OH, and CO. The compounds CO, CH_3, and CH_2OH are competitive inhibitors of ammonia oxidation in *Nitrosomonas* (Suzuki et al., 1976), and extracts can, apparently, oxidize CO (Tsang and Suzuki, 1982). (2) Oxidation of methane by methanotrophic bacteria is sensitive to the same compounds that inhibit ammonia oxidation in *Nitrosomonas* (Ferenci et al., 1975; Topp and Knowles, 1982). (3) The methanotrophs and *Nitrosomonas* both have extensive internal membranes (Murray and Watson, 1965; Davies and Whittenbury, 1970). (4) *Nitrosocystis* (Blumer et al., 1969) and the type I methanotroph (Smith et al., 1970; Maukla, 1978) have predominantly 16:0 and 16:1 fatty acids.

Two differences between the methane oxidizers and ammonia oxidizers appear likely. (1) Whereas methane oxidation appears to involve a water-producing mixed-function oxygenase (Higgins et al., 1981), ammonia oxidation may involve the incorporation of both oxygen atoms of dioxygen into N-compounds. (2) In contrast to methylotrophs (Higgins et al., 1981), in *Nitrosomonas* the electron donor for activation of dioxygen is probably not NADH.

Methane oxidation appears to involve either a 20,000 M_r enzyme containing 2 non-heme irons and 2 acid labile sulfides (Dalton, 1980) or a combination of a soluble, CO-binding, autooxidizable *c*-type cytochrome and a copper protein (Tonge et al., 1977). The latter possibility is interesting considering (1) the presence, in *Nitrosomonas,* of a CO-binding diheme cytochrome *c*-554 (Tronson et al., 1973; Yamanaka and Shinra, 1975) that stimulates ammonia oxidation in extracts (Suzuki and Kwok, 1981); and (2) the sensitivity of ammonia oxidation to copper ion binding agents (Lees, 1952). Significantly we have recently observed that heme of cytochrome *c*-554 can readily exist in a high-spin form in keeping with a possible substrate-binding role (K. K. Andersson, J. D. Lipscomb, and A. B. Hooper, unpublished).

ENERGY COUPLING

Proton gradient-dependent synthesis of ATP has been demonstrated in spheroplasts of *Nitrosomonas* (Kumar and Nicholas, 1982). In *Nitrosomonas* it may be assumed that an energy-linked ATPase with an ADP-binding site on the cytoplasmic side of the membrane is "driven in reverse" by increased proton concentration on the extracellular side of the membrane. Based on morphological studies of *Nitrosomonas,* the noncytoplasmic compartment may consist of extracellular space, periplasmic space, cisternae continuous with periplasmic space, and cisternae non-continuous with periplasmic space (fig. 12). ATP synthesis may be dependent upon increased proton concentration in any or all of the latter 3 compartments. For heterotrophic microorganisms (Haddock and Jones, 1977), proton pumping is thought to involve extraction of hydrogen(s) at a dehydrogenase site on the cytoplasmic membrane face, transfer of hydrogen(s) across the membrane via flavins or quinones, and return of electrons via redox metal centers to a terminal oxidase on the cytoplasmic face. For the present discussion of energy transduction of *Nitrosomonas,* it is assumed that the proton-utilizing and O_2-reactive site of the terminal oxidase is on the cytoplasmic membrane face. This "traditional" scheme is depicted in figure 13A with the N-oxidation reaction scheme (figs. 4A, 9, 11) in which 2 electrons per mole NH_3 oxidized enter an electron transport chain at the level of NH_2OH. Difficulties are encountered with this hypothesis. Oxidation of NH_2OH may involve only electron-transporting

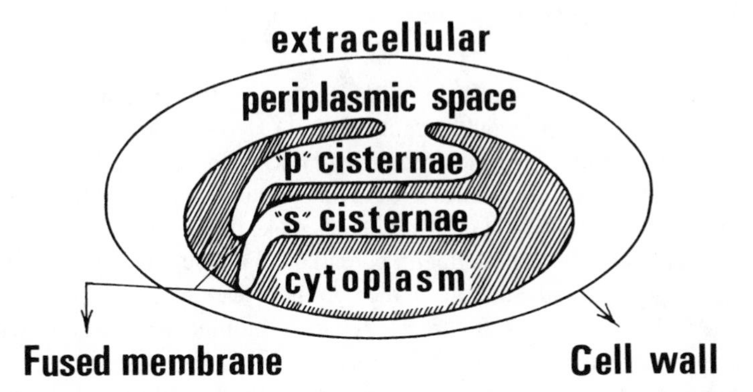

Fig. 12. Hypothesized compartments of *Nitrosomonas* include periplasmic space, cytoplasm, and cisternae of intracellular membranes that may be either continuous with, or separate from, periplasmic space.

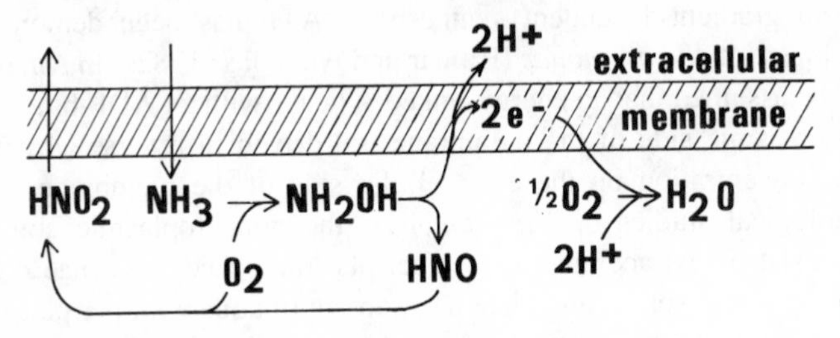

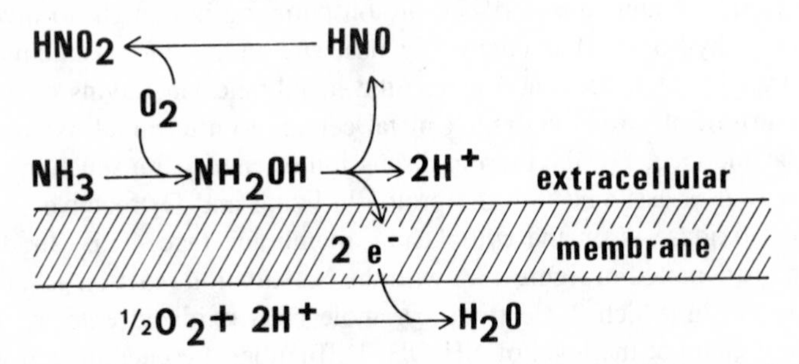

Fig. 13. The dehydrogenase that gives rise to electrons for energy transduction may be either cytoplasmic (A) or extracytoplasmic (B). The terminal oxidase and ATPase (not shown) are assumed to be on the cytoplasmic surface of the membrane. The scheme is illustrated with energy coupling only at the NH_2OH dehydrogenase step (as in fig. 4A).

metal centers (hemes) rather than hydrogen-transporting quinones or flavins. This scheme (fig. 13A) would require that (1) HAO in the membrane (or a membrane factor receiving electrons from HAO) can catalyze hydride transfer, or (2) HAO transfers electrons to flavins or quinones in the membrane. In fact, the low redox potential of HAO hemes makes direct transfer to quinones possible, although the high midpoint potentials of the putative electron shuttle cytochromes *c*-554 and *c*-552 make them unlikely candidates for participation in an HAO-to-quinone electron transfer chain. Ubiquinone (Q_8) is abundant in membranes of *Nitrosomonas* (Hooper et al., 1972).

Serious consideration should be given to the possibility that the dehydrogenase, and indeed, all of N-oxidation may occur in the noncytoplasmic compartment as depicted in figure 13B. This scheme would make HAO (and possibly cytochromes *c*-554 and *c*-552) analogous in location to cytochrome *c* of mitochondria that is extrinsically associated with a terminal oxidase-binding site on the nonmatrix ("noncytoplasmic") face of the membrane. This is also similar to a scheme proposed for iron oxidation in *Thiobacillus* in which the copper protein rusticyanin and soluble cytochrome *c* are hypothesized to be in the periplasmic space (Ingledew et al., 1977). The scheme is attractive in that localization of N oxidation in noncytoplasmic compartments isolates potentially toxic N- or O-containing intermediates. In addition, location of N-oxidation in periplasm or "p" cisternae (fig. 13) eliminates the need for ammonia-uptake and nitrite-secreting systems. Substrate and product transport would be more difficult to and from noncontinuous cisternae unless membranes are fused or "stacked" as in photosynthetic membranes (fig. 12) so as to allow substrates and product to bypass the cytoplasm.

The location of HAO in the compartment of proton accumulation would also provide a novel mechanism of regulation of energy coupling: absence of ADP and the ensuing decrease in the pH of the extracellular compartment will slow the rate of N-oxidation since the pH optimum for HAO is well above 8 (Hooper and Nason, 1965). The possibility of a microenvironment of low pH is also consistent with the observed H_2O-NO_2^- exchange reaction.

It should be noted that the extracellular dehydrogenase scheme might readily apply generally to oxidizers of inorganic compounds and possibly CO, methane, and methanol oxidizers, where, in contrast to organotrophic organisms, the intermediates in oxidation are not necessary for other energy-yielding or biosynthetic reactions.

The existence of separate hypothesis for cytoplasmic and noncytoplasmic energy-coupled dehydrogenases calls for experimental localization of HAO. The unusual reduction of mammalian cytochrome *c* by NH_2OH as catalyzed by intact cells (Campbell and Aleen, 1965) is consistent with an extracellular enzyme. Spheroplasts of *Nitrosomonas* catalyze oxidation of ammonia to nitrite (Bhandari and Nicholas, 1979), indicating that HAO had not been released. This suggests that HAO is intracellular but leaves the possibility that it is inside the noncontinuous cisternae or that spheroplasts, although osmotically fragile, still retained the continuous cisternae in a configuration capable of trapping HAO. We have recently found a procedure for making unusual osmotically shocked "spheroplast-like" preparations of *Nitrosomonas* that have released most HAO activity but retained glutamate dehydrogenase activity (table 5, Olson and Hooper, 1983). The treatment may loosen the wall layer just enough to release HAO but retain enough wall structure to prevent osmotic lysis of the cell membrane and release of glutamate dehydrogenase. This result suggests that HAO is present in the periplasmic space or in cister-

TABLE 5

SELECTIVE RELEASE OF HAO FROM OSMOTICALLY SHOCKED "SPHEROPLAST-LIKE" PREPARATIONS OF NITROSOMONAS[a]

	ENZYME ACTIVITY (%)[b]	
Preparation[a]	HAO	Glutamate dehydrogenase
Particulate	8	83
Supernatant	80	19

[a]Olson and Hooper, 1983. Cells treated by a procedure essentially the same as that used by Bhandari and Nicholas (1979) were suspended in 1 mM phosphate solution containing DNase and certrifuged for 20 min at 20,000 g to yield the particulate and supernatant factions.

[b]Percentage of total activity in cells. Enzyme assays: HAO (Hooper et al., 1978); glutamate dehydrogenase (Hooper et al., 1967).

nae continuous with the periplasmic space and supports the hypothesis depicted in figure 13B.

A hypothesis involving extracellular oxidation of NH_2OH can be modified to account for N-oxidation schemes involving a nitrite-producing, water-utilizing HNO dehydrogenase (fig. 4C), a proton-pumping ammonia oxygenase (fig. 4B), and the existence of proton pumping by the terminal oxidase itself. These modifications may be necessary to account for values of H^+/O ratios of as high as 4 observed with cells of *Nitrosomonas* (Hollocher et al., 1982). Those authors report that values of H^+/O were the same with NH_3 or NH_2OH as substrate. This suggests that a "coupling site" is not associated with NH_3 oxidase and tends to rule out the scheme presented in figure 4B.

ACKNOWLEDGMENTS

This work was supported by Research Grant PCM 8008710, National Science Foundation. I thank K. K. Andersson and J. D. Lipscomb for excellent discussions and collaborators C. Balny, P. Debey, C. Larroque, S. Philson, G. Babcock, T. Olson, and V. M. Tran for the use of unpublished results.

1. Abbreviations: HAO, hydroxylamine oxidoreductase; PMS, phenazine methosulfate; DTC, diethyldithiocarbamate; DCIP, 2,6-dichlorophenolindophenol; SDS, sodium dodecyl sulfate; PAGE, polyacrylamide gel electrophoresis; EPR, electron paramagnetic resonance.

LITERATURE CITED

Aleem, M. I. H., and H. Lees. 1963. Autotrophic enzyme systems. I. Electron transport systems concerned with hydroxylamine oxidation in *Nitrosomonas*. Can. J. Biochem. Physiol. 41:763–78.

Aleen, M. I. H. 1966. Generation of reducing power in chemosynthesis. II. Energy linked reduction of pyridine nucleotides in the chemoautotroph, *Nitrosomonas europaea*. Biochim. Biophys. Acta 113:216–24.

Aleem, M. I. H. 1977. Coupling of energy with electron transfer reactions in chemolithotrophic bacteria. *In* B. A. Haddock and W. A. Hamilton (eds.), Microbial energetics, pp. 351–81. Cambridge University Press, Cambridge.

Anderson, J. H. 1964. The metabolism of hydroxylamine to nitrite by *Nitrosomonas*. Biochem. J. 91:8–17.

Anderson, J. H. 1965a. Estimation of the nitric oxide formed from hydroxylamine by *Nitrosomonas*. Biochem. J. 94:236–39.

Anderson, J. H. 1965b. Studies on the formation of nitrogenous gas from hydroxylamine by *Nitrosomonas*. Biochim. Biophys. Acta 97:337–39.

Andersson, K. K., and A. B. Hooper. 1982. One H_2O_2-sensitive heme *c* is found with heme P-460 at the reaction center of hydroxylamine oxidoreductase (HAO). Fed. Proc. 41:1207.

Andersson, K. K., S. B. Philson, and A. B. Hooper. 1982. ^{18}O isotope shift in ^{15}N NMR analysis of biological N-oxidations: H_2O-NO_2^- exchange in the ammonia-oxidizing bacterium *Nitrosomonas*. Proc. Natl. Acad. Sci. USA 79:5871–75.

Bhandari, B., and D. J. D. Nicholas. 1979. Ammonia, O_2 uptake, and proton extrusion by spheroplasts of *Nitrosomonas europaea*. FEMS Microbiol. Lett. 6:297–300.

Blackmer, A. M., J. M. Bremner, and E. L. Schmidt. 1980. Production of nitrous oxide by ammonia-oxidizing chemoautotrophic microorganisms in soil. Appl. Environ. Microbiol. 40:1060–66.

Blumer, M., T. Chase, and S. W. Watson. 1969. Fatty acids in the lipids of marine and terrestrial nitrifying bacteria. J. Bacteriol. 99:366–70.

Bremner, J. N., and A. M. Blackmer. 1978. Nitrous oxide: emission from soils during nitrification of fertilizer nitrogen. Science 199:295–96.

Campbell, N. E. R., and M. I. H. Aleem. 1965. The effect of 2-chloro, 6-(trichloromethyl) pyridine on the chemoautotrophic metabolism of nitrifying bacteria. I. Ammonia and hydroxylamine oxidation by *Nitrosomonas*. Antonie van Leeuwenhoek 31:124–36.

Dalton, H. 1980. Transformations by methane mono-oxygenase. *In* D. E. F. Harrison, I. J. Higgins, and R. Watkinson (eds.), Hydrocarbons in biotechnology, pp. 85–97. Heyden & Son, London.

Davies, S. L., and R. Whittenbury. 1970. Fine structure of methane and other hydrocarbon utilising bacteria. J. Gen. Microbiol. 61:227–32.

Drozd, J. W. 1976. Energy coupling and respiration in *Nitrosomonas europaea*. Arch. Microbiol. 110:257–62.

Dua, R. D., B. Bhandari, and D. J. D. Nicholas. 1979. Stable isotope studies on the oxidation of ammonia to hydroxylamine by *Nitrosomonas europaea*. FEBS Lett. 106:401–4.

Engel, M. S. and M. J. Alexander. 1958. Growth and autotrophic metabolism of *Nitrosomonas europaea*. J. Bacteriol. 76:217–22.

Erickson, R. H. 1971. Oxygen utilizing enzymes of *Nitrosomonas europaea*. Ph.D. thesis, University of Minnesota, St. Paul.

Erickson, R. H., and A. B. Hooper. 1972. Preliminary characterization of a variant CO-binding heme protein from *Nitrosomonas*. Biochim. Biophys. Acta 275:231–44.

Erickson, R. H., A. B. Hooper, and K. R. Terry. 1972. Solubilization and purification of cytochrome a_1 from *Nitrosomonas*. Biochim. Biophys. Acta 283:155–66.

Falcone, A. B., A. L. Shug, and D. J. D. Nicholas. 1962. Oxidation of hydroxylamine by particles from *Nitrosomonas*. Biochem. Biophys. Res. Commun. 9:126–31.

Falcone, A. B., A. L. Shug, and D. J. D. Nicholas. 1963. Some properties of a hydroxylamine oxidase from *Nitrosomonas europaea*. Biochim. Biophys. Acta 77:199–208.

Ferenci, T., T. Strom, and J. R. Quayle. 1975. Oxidation of carbon monoxide and methane by *Pseudomonas methanica*. J. Gen. Microbiol. 91:79–91.

Goreau, T. J., W. A. Kaplan, S. C. Wofsy, M. B. McElroy, F. W. Valois, and S. W. Watson. 1980. Production of NO_2^- and N_2O by nitrifying bacteria at reduced concentrations of oxygen. Appl. Environ. Microbiol. 40:526–32.

Haddock, B. A., and C. W. Jones. 1977. Bacterial respiration. Bacteriol. Rev. 41:47–99.

Higgins, I. J., D. J. Best, R. C. Hammond, and D. Scott. 1981. Methane-oxidizing microorganisms. Microbiol. Rev. 45:556–90.

Hofman, T., and H. Lees. 1953. The biochemistry of the nitrifying organisms. 4. The respiration and intermediary metabolism of *Nitrosomonas*. Biochem. J. 54:579–83.

Hollocher, T. C., M. E. Tate, and D. J. D. Nicholas. 1981. Oxidation of ammonia by *Nitrosomonas europaea:* definitive ^{18}O-tracer evidence that hydroxylamine formation involves a monooxygenase. J. Biol. Chem. 256:10834–36.

Hollocher, T. C., S. Kumar, and D. J. D. Nicholas. 1982. Respiration-dependent proton translation in *Nitrosomonas europaea* and its apparent absence in *Nitrobacter agilis* during inorganic oxidations. J. Bacteriol. 149:1013–20.

Hooper, A. B., and A. Nason. 1965. Characterization of hydroxylamine-cytochrome c reductase fromn the chemoautotrophs *Nitrosomonas europaea* and *Nitroscocystis oceanus*. J. Biol. Chem. 240:4044–57.

Hooper, A. B., J. Hansen, and R. Bell. 1967. Characterization of glutamate dehydrogenase from the ammonia-oxidizing chemoautotroph *Nitrosomonas europaea*. J. Biol. Chem. 242:288–96.

Hooper, A. B. 1968. A nitrite-reducing enzyme from *Nitrosomonas europaea*. Preliminary characterization with hydroxylamine as electron donor. Biochim. Biophys. Acta 162:49–65.

Hooper, Alan B. 1969. Lag phase of ammonia oxidation by resting cells of *Nitrosomonas europaea*. J. Bacteriol. 97:968–69.

Hooper, A. B., R. H. Erickson, and K. R. Terry. 1972. Electron transport systems of *Nitrosomonas:* isolation of a membrane-envelope fraction. J. Bacteriol. 110:430–38.

Hooper, A. B., and K. R. Terry. 1973. Specific inhibitors of ammonia oxidation in *Nitrosomonas*. J. Bacteriol. 115:480–85.

Hooper, A. B., and K. R. Terry. 1974. Photoinactivation of ammonia oxidation in *Nitrosomonas*. J. Bacteriol. 119:899–906.

Hooper, A. B., and K. R. Terry. 1977. Hydroxylamine oxidoreductase from *Nitrosomonas:* inactivation by hydrogen peroxide. Biochemistry 16:455–59.

Hooper, A. B., K. R. Terry, and P. C. Maxwell. 1977. Hydroxylamine oxidoreductase of *Nitrosomonas* oxidation of diethyldithiocarbonate concomitant with stimulation of nitrite synthesis. Biochim. Biophys. Acta 462:141–52.

Hooper, A. B. 1978. Nitrogen oxidation and electron transport in ammonia-oxidizing bacteria. *In* D. Schlessinger (ed.), Microbiology, pp. 299–304. American Society for Microbiology, Washington, D.C.

Hooper, A. B., P. C. Maxwell, and K. R. Terry. 1978. Hydroxylamine oxidoreductase from *Nitrosomonas:* absorption spectra and content of heme and metal. Biochemistry 17:2984–89.

Hooper, A. B., and K. R. Terry. 1979. Hydroxylamine oxidoreductase of *Nitrosomonas:* production of nitric oxide from hydroxylamine. Biochim. Biophys. Acta 571:12–20.

Hooper, A. B., and L. E. Vickery. 1981. EPR of the substrate-binding heme P-460 of the NH_2OH-oxidizing enzyme from *Nitrosomonas*. Fed. Proc. 40:1863.

Hooper, A. B., and C. Balny. 1982. Reaction of oxygen with hydroxylamine oxidoreductase of *Nitrosomonas*: fast kinetics. FEBS Lett. 144:299–303.

Hooper, A. B., P. Debey, K. K. Andersson, and C. Balny. 1983. Heme P460 of hydroxylamine oxidoreductase of *Nitrosomonas*: reaction with CO and H_2O_2. Eur. J. Biochem. 134:83–87.

Hughes, M. N., and H. G. Nicklin. 1970. A possible role for the species peroxynitrite in nitrification. Biochim. Biophys. Acta 222:660–61.

Hynes, R. K., and R. Knowles. 1978. Inhibition by acetylene of ammonia oxidation in *Nitrosomonas europaea*. FEMS Microbiol. Lett. 4:319–21.

Ingledew, W. John, J. C. Cox, and P. J. Halling. 1977. A proposed mechanism for energy conservation during Fe^{2+} oxidation by *Thiobacillus ferro-oxidans*: chemiosmotic coupling to net H^+ influx. FEMS Microbiol. Lett. 2:193–97.

Kumar, S., and D. J. D. Nicholas. 1982. A protonmotive force-dependent adenosine-5′ triphosphate synthesis in spheroplasts of *Nitrosomonas europaea*. FEMS Microbiol. Lett. 14:21–25.

Lees, H. 1952. The biochemistry of the nitrifying organisms. I. The ammonia-oxidizing systems of *Nitrosomonas*. Biochem. J. 52:134–39.

Lipscomb, J. D., and A. B. Hooper. 1982. Resolution of multiple heme centers of hydroxylamine oxidoreductase from *Nitrosomonas*. I. Electron paramagnetic resonance spectroscopy. Biochemistry 21:3965–72.

Lipscomb, J. D., K. K. Andersson, E. Munck, T. A. Kent, and A. B. Hooper. 1982. 1. Resolution of multiple heme centers of hydroxylamine oxidoreductase from *Nitrosomonas*. 2. Mössbauer spectroscopy. Biochemistry 21:3973–76.

Makula, R. A. 1978. Phospholipid composition of methane-utilizing bacteria. J. Bacteriol. 134:771–77.

Maxwell, P. C. 1976. Isolation and characterization of hydroxylamine dehydrogenase from the chemoautotroph *Nitrosomonas europaea*. Ph.D. thesis, University of Minnesota, St. Paul.

Murray, R. G. E., and S. W. Watson. 1965. Structure of *Nitrosomonas oceanus* and comparison with *Nitrosomonas* and *Nitrobacter*. J. Bacteriol. 89:1594–1609.

Nicholas, D. J. D., and O. T. G. Jones. 1960. Oxidation of hydroxylamine in cell-free extracts of *Nitrosomonas europaea*. Nature (London) 185:512–14.

Olson, T. C., and A. B. Hooper. 1983. Energy coupling in bacterial oxidation of small molecules: an extracytoplasmic dehydrogenase in *Nitrosomonas*. FEMS Microbiol. Lett. 19:47–50.

Rees, M., and A. Nason. 1965. A P-450-like cytochrome and a soluble terminal oxidase identified as cytochrome *o* from *Nitrosomonas europaea*. Biochem. Biophys. Res. Commun. 21:248–56.

Rees, M., and A. Nason. 1966. Incorporation of atmospheric oxygen into nitrite formed during ammonia oxidation by *Nitrosomonas europaea*. Biochim. Biophys. Acta 113:398–401.

Rees, M. K. 1968. Studies of the hydroxylamine metabolism of *Nitrosomonas europaea*. II. Molecular properties of the electron-transport particle, hydroxylamine oxidase. Biochemistry 7:366–72.

Ritchie, G. A. F., and D. J. D. Nicholas. 1972. Identification of the sources of nitrous oxide produced by oxidative and reductive processes in *Nitrosomonas europaea*. Biochem. J. 126:1181–91.

Ritchie, G. A. F., and D. J. D. Nicholas. 1974. The partial characterization of purified nitrite reductase and hydroxylamine oxidase from *Nitrosomonas europaea*. Biochem. J. 138:471–80.

Smith, U., D. W. Ribbons, and D. S. Smith. 1970. The fine structure of *Methylococcus capsulatus*. Tissue Cell 2:513–20.

Suzuki, I., and S. C. Kwok. 1969. Oxidation of ammonia by spheroplasts of *Nitrosomonas europaea* extracts: effects of polyamines, Mg, and albumin. Biochem. Biophys. Res. Commun. 39:950–55.

Suzuki, I. 1974. Mechanisms of inorganic oxidation and energy coupling. Ann. Rev. Microbiol. 28:85–101.

Suzuki, I., U. Dular, and S. C. Kwok. 1974. Ammonia or ammonium ion as substrate for oxidation by *Nitrosomonas* cells and extracts. J. Bacteriol. 120:556–58.

Suzuki, I., S. C. Kwok, and U. Dular. 1976. Competitive inhibition of ammonia oxidation in *Nitrosomonas europaea* by methane, carbon monoxide, or methanol. FEBS Lett. 72:117–20.

Suzuki, I., and S. C. Kwok. 1981. A partial resolution and reconstitution of the ammonia-oxidizing system of *Nitrosomonas europaea:* role of cytochrome *c*554. Can. J. Biochem. 59:484–88.

Suzuki, I., S. C. Kwok, U. Dular, and D. C. Y. Tsang. 1981a. Cell-free ammonia-oxidizing system of *Nitrosomonas europaea:* general conditions and properties. Can. J. Biochem. 59:477–83.

Suzuki, I., S. C. Kwok, D. C. Y. Ysang, J. K. Oh, and R. S. Bhella. 1981b. Oxidation of ammonia by *Nitrosomonas* and of inorganic sulfur by *Thiobacilli*. *In* H. Bothe and A. Trebst (eds.), Biology of inorganic nitrogen and sulfur, pp. 212–21. Springer-Verlag, Berlin.

Terry, K. R., and A. B. Hooper. 1981. Hydroxylamine oxidoreductase: a 20-heme, 200,000 molecular weight cytochrome c with unusual denaturation properties which forms a 63,000 molecular weight monomer after heme removal. Biochemistry 20:7026–32.

Tonge, G. M., D. E. F. Harrison, and I. J. Higgins. 1977. Purification and properties of the methane mono-oxygenase enzyme system from *Methylosinus trichosporium* OB3b. Biochem. J. 161:333–44.

Topp, E., and R. Knowles. 1982. Nitrapyrin inhibits the obligate methylotrophs *Methylosinus trichosporium* and *Methylococcus capsulatus*. FEMS Microbiol. Lett. 14:47–49.

Tronson, D. A., G. A. F. Ritchie, and D. J. D. Nicholas. 1973. Purification of c-type cytochromes from *Nitrosomonas europaea*. Biochim. Biophys. Acta 310:331–43.

Tsang, D. C. Y., and I. Suzuki. 1982. Cytochrome c_{554} as a possible electron donor in the hydroxylation of ammonia and carbon monoxide in *Nitrosomonas europaea*. Can. J. Biochem. 60:1018–24.

Vickery, L. E., and A. B. Hooper. 1981. EPR of hydroxylamine oxidoreductase from *Nitrosomonas europaea*. Biochim. Biophys. Acta 670:291–93.

Wallace, W., and D. J. D. Nicholas. 1968. Properties of some reductase enzymes in the nitrifying bacteria and their relationship to the oxidase systems. Biochem. J. 109:763–73.

Watson, S. W., M. A. Asbell, and F. W. Valois. 1970. Ammonia oxidation by cell-free extracts of *Nitrosomonas oceanus*. Biochem. Biophys. Res. Commun. 38:1113–19.

White, R. E., and M. J. Coon. 1980. Oxygen activation by cytochrome P-450. Ann. Rev. Biochem. 49:315–56.

Yamanaka, T., and M. Shinra. 1974. Cytochrome c_{552} and cytochrome c_{554} derived from *Nitrosomonas europaea:* purification, properties and their function in hydroxylamine oxidation. J. Biochem. 75:1265–73.

Yamanaka, T., M. Shinra, K. Takahashi, and M. Shibaska. 1979. Highly purified hydroxylamine oxidoreductase derived from *Nitrosomonas europaea:* some physiochemical and enzymatic properties. J. Biochem. 86:1101–8.

Yamanaka, T., and Y. Sakano. 1980. Oxidation of hydroxylamine to nitrite catalyzed by hydroxylamine oxidoreductase purified from *Nitrosomonas europaea*. Curr. Microbiol. 4:239–26.

Yoshida, T., and M. Alexander. 1964. Hydroxylamine formation by *Nitrosomonas europaea*. Can J. Microbiol. 10:923–926.

Yoshida, T., and M. Alexander. 1970. Nitrous oxide formation by *Nitrosomonas europaea* and heterotrophic microorganisms. Soil Sci. Soc. Amer. Proc. 34:880–82.

JOHN G. COBLEY

Oxidation of Nitrite and Formate in *Nitrobacter* Membrane Preparations: Evidence that Both Reactions Are Catalyzed by the Same Enzyme

10

Chemoautotrophs of the genus *Nitrobacter* obtain energy from the oxidation of the weak reductant NO_2^- with O_2 ($\Delta G_o' = -75$ kJ mol^{-1}) and assimilate CO_2 by using the Calvin cycle (Malavolta et al., 1960; Aleem, 1965). The NADH required for CO_2 fixation (Kiesow et al., 1977) is generated by energy-dependent reversed electron transport from NO_2^- (Kiesow, 1967). The discovery that the oxidation of nitrite by O_2 occurs in preparations of broken cells (Aleem and Alexander, 1958) led to the demonstration (Aleem et al., 1963) that the enzymes for NO_2^- oxidation and the enzymes for the reduction of NAD^+ by NO_2^- ($\Delta G_o' = +150$ kJ mol^{-1}) are all associated with the extensively convoluted membrane system found at one pole of the *Nitrobacter* cell (Murray and Watson, 1965). It is also known that the isolated membrane fragments (electron-transporting [ET] particles) will couple the oxidation of NO_2^- to the phosphorylation of ADP (Malavolta et al., 1960; Aleem and Nason, 1960). The oxidations of NADH by O_2 or NO_3^- can also energize ADP phosphorylation (Kiesow, 1967; Cobley, 1976a).

The inner mitochondrial membrane catalyzes thermodynamically spontaneous electron transport reactions that are coupled to the phosphorylation of ADP. Uncoupling agents accelerate these electron transport reactions by removing the constraint imposed when the phosphorylation of ADP cannot take place. It might be expected therefore that uncoupling agents would accelerate the spontaneous electron transport reactions of *Nitrobacter* ET particles. It has been found, however, that while uncoupling agents stimulate the oxidation of NADH by either O_2 or NO_3^- (Kiesow, 1967; Cobley, 1976a), they inhibit the oxidation of NO_2^- by O_2 (Butt and Lees, 1960; Aleem and Nason, 1960). The significance of the inhibition of nitrite oxidation by uncoupling agents was questioned by O'Kelley and colleagues (1970a), who proposed that the metal-chelating properties of the uncoupling agent 2,4-dinitrophenol might well be responsible for the inhibition. This proposal now seems highly unlikely since it has been reported (Cobley, 1976a) that in ET particles the oxidation of nitrite by O_2 is inhibited by ADP in the presence of P_i and is stimulated by oligomycin.

In the chemiosmotic hypothesis (Mitchell, 1966), it is considered that the driving force for ADP phosphorylation is a transmembrane gradient of H^+ activity, $\Delta\bar{\mu}_{H^+}$, and that $\Delta\bar{\mu}_{H^+}$ has two components:

$$\Delta\bar{\mu}_{H^+} = \Delta\psi - 60\Delta pH \qquad (1)$$

where $\Delta\psi$ is the difference in electrical potential across the membrane and where ΔpH represents the difference in pH between the two aqueous phases that the topologically closed membrane separates. ET particles from *Nitrobacter* are at least in part composed of topologically closed membrane vesicles since during the oxidation of NO_2^- by O_2, protons are taken up from the suspension medium and are released on the depletion of O_2 (Cobley, 1976a). By the use of ionophorous antibiotics and other lipid soluble molecules, it can be shown that it is specifically the collapse of $\Delta\psi$ rather than the collapse of ΔpH that causes the inhibition of NO_2^- oxidation (Cobley, 1976a). Further, a mechanism has been proposed (see Discussion below) by which NO_2^- oxidation generates $\Delta\bar{\mu}_{H^+}$ by proton translocation; the same mechanism also explains how the rate of NO_2^- oxidation is stimulated by $\Delta\psi$ (Cobley, 1976b).

The oxidation of formate by *Nitrobacter* was first reported by Silver (1960). This activity is found in ET particles (Malavolta et al., 1962) and has the following stoichiometry (O'Kelley et al., 1970b):

$$HCOOH + \tfrac{1}{2} O_2 \rightarrow H_2O + CO_2 \qquad (2)$$

$$\Delta G_o' = -237 \text{ kJ mol}^{-1}$$

The latter authors proposed that formate oxidation takes place by a pathway distinct from that for nitrite oxidation, and such a proposal appears reasonable since the E_o' values for the two couples are markedly different (for NO_2^-/NO_3^-, $E_o' = +0.42$ V; for $HCOO^-/CO_2$, $E_o' = -0.42$ V).

The results of the present investigation indicate that nitrite oxidation and formate oxidation are strikingly similar, whereas the oxidations of NADH and formate by O_2 (reactions with comparable $\Delta G_o'$ values) are markedly different. It will be proposed that the active site that binds and oxidizes nitrite also binds and oxidizes formate.

MATERIALS AND METHODS

The growth of *N. winogradskyi,* the preparation of ET particles, the measurement of oxygen reduction rates, the measurement of cytochrome reductions, and the sources of chemicals are all as described previously (Cobley 1976a,b). All reactions were conducted at 25°C.

RESULTS

In figure 1 the rates of nitrite and formate oxidation are plotted as a function of the concentration of the uncoupling agent, carbonylcyanide–m-

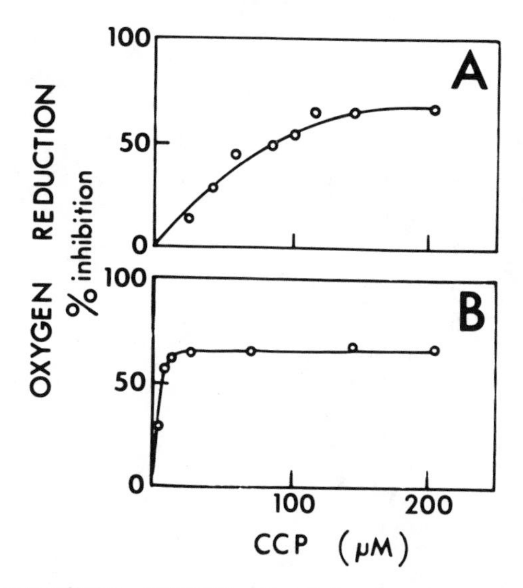

Fig. 1. The relative rates of NO_2^- and $HCOO^-$ oxidation plotted as a function of CCCP concentration. A. *Nitrobacter* ET particles at a final concentration of 0.8 mg ml^{-1} were suspended in 100 mM choline chloride, 5 mM $MgCl_2$, 10 mM tetramethylammonium-P_i, 10 mM tricine at pH 7.8 and at 25°C. B. *Nitrobacter* ET particles (1.6 mg ml^{-1} protein) were suspended in the above medium except that 10 mM TES replaced 10 mM tricine and the pH was 7.0. In A the reactions were initiated by the addition of 3 mM $NaNO_2$ and in B by the addition of 10 mM sodium formate. In the absence of uncoupler, the oxidation rates of NO_2^- and $HCOO^-$ were, respectively, 60, and 10 ng-atoms O min^{-1} mg $protein^{-1}$.

chlorophenylhydrazone (CCCP) (Heytler and Pritchard, 1962). The maximum extent of inhibition for nitrite oxidation is similar to that for formate (approximately 65%; fig. 1). Figure 2 illustrates the pH profile for nitrite oxidation; when the pH is less than 8.2, oxygen reduction is inhibited by ADP in the presence of P_i. Formate oxidation is similarly inhibited by ADP in the presence of P_i, and the pH dependence for this inhibition is similar to that for nitrite oxidation (fig. 3). In contrast the reduction of O_2 with NADH as substrate is stimulated when both ADP and P_i are included in the reaction medium (fig. 4). The uncoupling agent, CCCP, causes a further stimulation

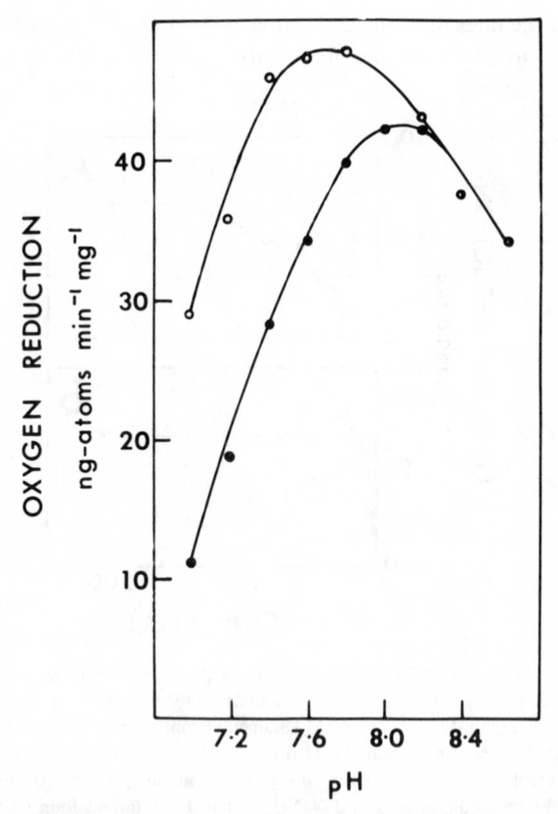

Fig. 2. pH profiles of NO_2^- oxidation. The reaction medium was as described for fig. 1A except that bovine serum albumin (1 mg ml^{-1}) and TES (10 mM) were included. Rates were measured in the presence of 0.8 mM ADP (●) and in the absence of ADP (○). Reactions were initiated by the addition of 1.2 mM $NaNO_2$.

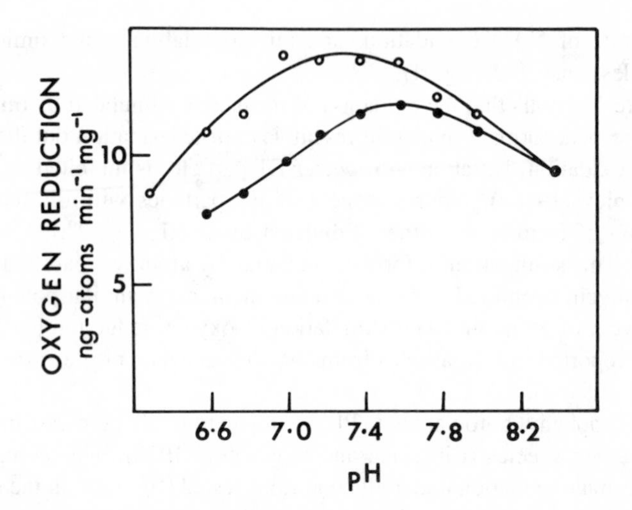

Fig. 3. pH profiles of $HCOO^-$ oxidation. The reaction medium was as described for fig. 2. Rates were measured in the presence (●) and in the absence (○) of ADP (0.4 mM). Reactions were initiated by the addition of 10 mM sodium formate.

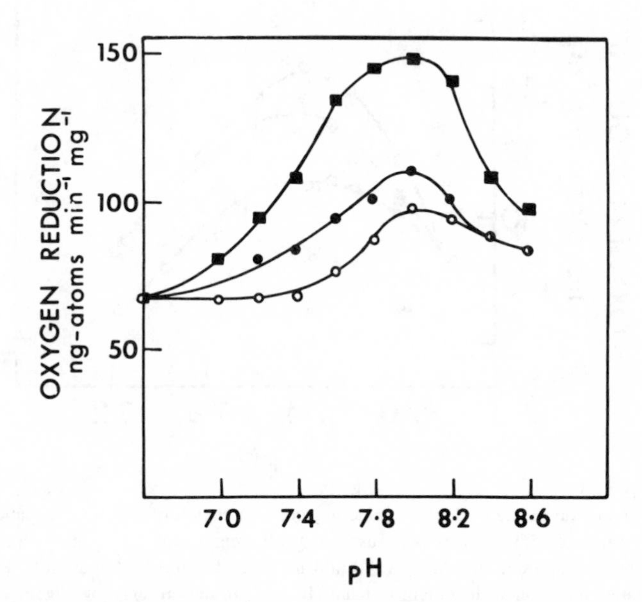

Fig. 4. pH profiles of NADH oxidation. Conditions as described for figs 2 and 3. Rates were measured in the presence of 0.8 mM ADP (●), in the absence of ADP (○), or in the presence of 200 μM CCCP (■). Reactions were initiated by the addition of 1 mM NADH.

of the rate of NADH oxidation, and this stimulation is not limited to pH values less than 8.2 (fig. 4).

Figure 5 reveals that the oxidation of formate is stimulated by oligomycin. In this respect formate oxidation resembles nitrite oxidation but differs from NADH oxidation, which in *Nitrobacter* ET particles is inhibited by oligomycin (Cobley, 1976a). After treatment of ET particles with oligomycin, the oxidation of formate is further stimulated by 1° (fig. 5), 2°, 3°, but not 4° amines. This stimulation of formate oxidation by amines is barely discernible if oligomycin is omitted from the reaction medium. A similar potentiation by oligomycin of amine-induced stimulation of oxygen reduction has been previously reported in ET particles from *Nitrobacter* when nitrite is the substrate (Cobley, 1976a).

The tetraphenyl boron anion (TPB) is known to freely permeate membranes as a charged species (Liberman and Skulachev, 1970). This compound inhibits formate oxidation under all conditions tested (fig. 6) with the exception

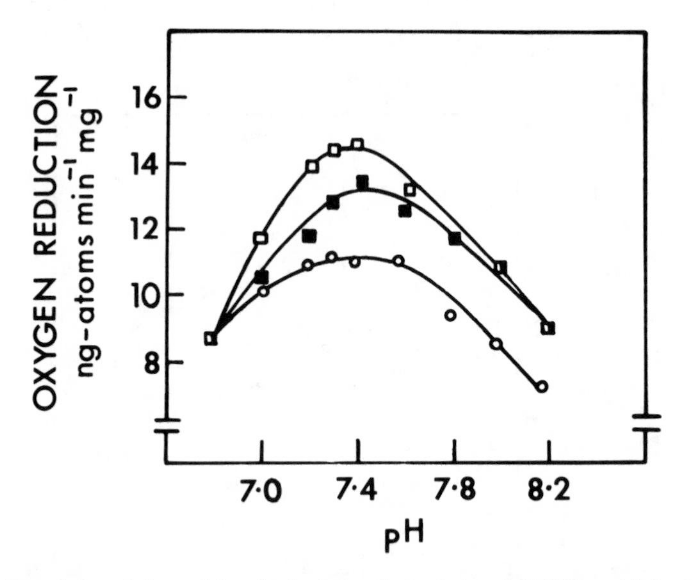

Fig. 5. Potentiation of the cyclohexylamine-induced stimulation of $HCOO^-$ oxidation by oligomycin. *Nitrobacter* ET particles were suspended at 0.9 mg protein ml^{-1} final concentration in 100 mM choline chloride, 10 mM TES, 10 mM tricine and 5 mM $MgCl_2$ at 25°C. Rates were measured after initiation with sodium formate (10 mM) in the presence of oligomycin (5 min preincubation; 5 μg) (■), in the absence of oligomycin (○) and in the presence of both oligomycin and cyclohexylamine (5 mM) (□). In the absence of oligomycin, the stimulation induced by cyclohexylamine was barely detectable.

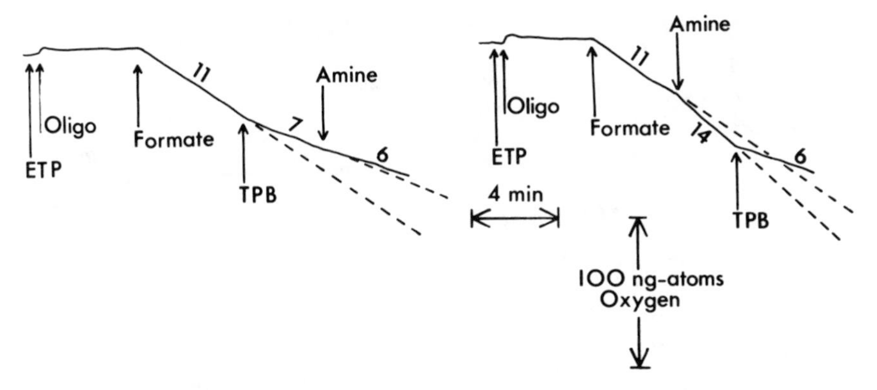

Fig. 6. The influence of cyclohexylamine hydrochloride and sodium tetraphenyl boron on the rate of $HCOO^-$ oxidation. *Nitrobacter* ET particles were suspended to 1.1 mg protein ml^{-1} in the medium described in fig. 5 and at pH 7.2. The additions were as follows: oligomycin (10 μg), sodium formate (10 mM), cyclohexylamine (5 mM) and NaTPB (20 μM). Numbers close to the traces are rates of oxygen reduction expressed as ng-atoms O min^{-1} mg $protein^{-1}$.

that it does not further inhibit the rate of formate oxidation when measured in the presence of an excess of the uncoupling agent, CCCP (200 μM). Nitrite oxidation is similarly inhibited by sodium tetraphenyl boron (Cobley, 1976a). In contrast, the reduction of O_2 by NADH is stimulated by TPB when cyclohexylamine is present in the reaction medium (fig. 7). In the absence of cyclohexylamine, NADH oxidation is little effected by TPB.

In ET particles the reduction of cytochrome *c* (E_m[pH 7.0] + 0.27 V) by NO_2^- is an energy-requiring reaction (Kiesow, 1967). When NO_2^- is being oxidized by O_2, the steady-state level of cytochrome *c* reduction (550–540 nm) is decreased by either CCCP or ADP in the presence of P_i (Cobley, 1976b). Similar responses to the addition of CCCP or ADP are found when O_2 is being reduced by formate (fig. 8). In contrast, when O_2 is being reduced by NADH, the addition of these compounds causes an increase in the steady-state level of cytochrome *c* reduction (fig. 9).

When ET particles from *Nitrobacter* are fully reduced by sodium dithionite, an absorption band appears at 589 nm. Two *a*-type cytochromes contribute equally to this absorption (E_m[pH 7.0] values of +0.14 V and +0.35 V; Sewell et al., 1972). When the steady-state level of the reduced *a*-type cytochromes is monitored (589–574 nm) during formate oxidation, CCCP or ADP in the presence of P_i slightly decreases the level *a*-type cyctochrome reduction (fig. 8). In this respect formate oxidation resembles nitrite oxidation

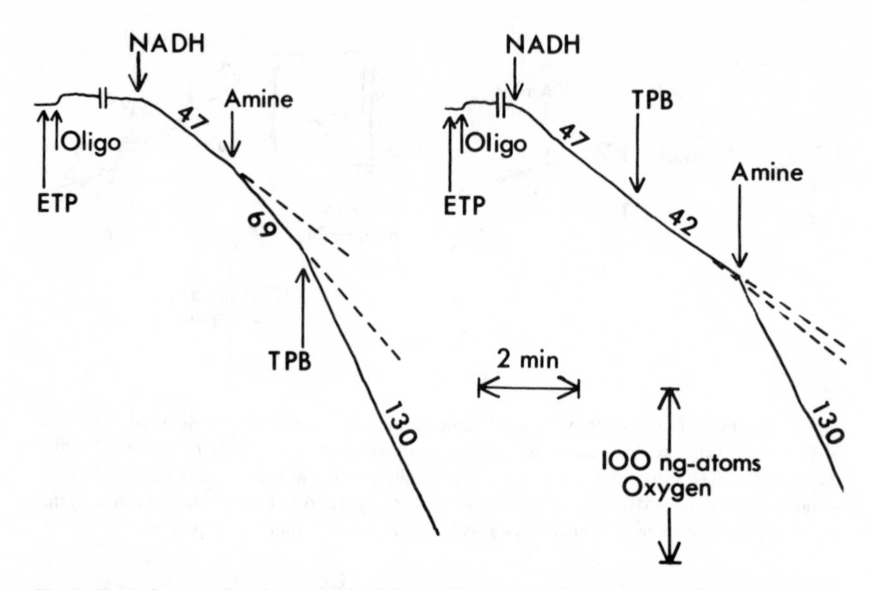

Fig. 7. The influence of cyclohexylamine hydrochloride and sodium tetraphenyl boron on the rate of NADH oxidation. *Nitrobacter* ET particles were suspended to 0.55 mg protein ml^{-1} in the medium described in fig. 5 and at pH 7.8. The additions were NADH (2 mM), oligomycin (10 μg), cyclohexylamine (5 mM) and NaTPB (20 μM).

(Cobley, 1976b). In contrast, the level of reduction of *a*-type cytochromes is increased by ADP or CCCP when NADH is the substrate (fig. 9).

DISCUSSION

The similarity between nitrite and formate oxidation is surprising since formate is a very much stronger reducing agent than nitrite. However, the structure of these two molecules is very similar (fig. 10); the structural information can be found in standard textbooks (Cotton and Wilkinson, 1972; Streitwieser and Heathcock, 1981). Not only are the bond angles and lengths comparable but the molecules are isoelectronic. The dissociation constants for these acids are also similar (HNO_2, pK_a = 3.37; HCOOH, pK_a = 3.75; Weast, 1970). Both nitrite and formate are known to form O-bonded complexes with metals (Basolo and Pearson, 1968; Cotton and Wilkinson, 1972) and to function as two electron-equivalent reducing agents in metal-coordinated systems (Basolo and Pearson, 1968; Green and Sykes, 1970).

Other enzymes are known to interact with both formate and nitrate. Catalase uses both formate and nitrite as peroxidatic substrates (Chance and Her-

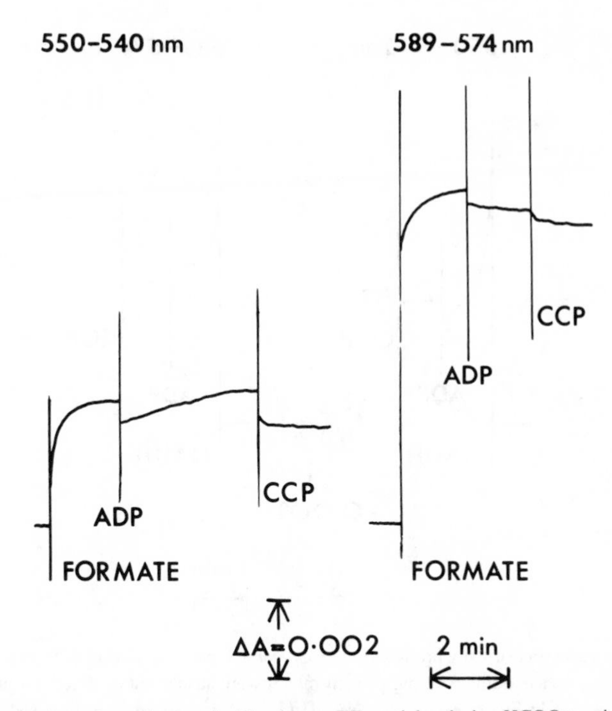

Fig. 8. The reduction of cytochromes in *Nitrobacter* ET particles during $HCOO^-$ oxidation. ET particles (final protein concentration 1.2 mg ml^{-1}) were suspended in 100 mM choline chloride, 20 mM tricine, 10 mM $MgCl_2$, 10 mM tetramethylammonium-P_i and bovine serum albumin (1.6 mg ml^{-1} final concentration). The pH was 7.8, the temperature 25°C, and the final volume 1.25 ml. The additions were sodium formate (10 mM), ADP (32 μM) and CCCP (160 μM).

bert, 1950); in mung bean the formate dehydrogenase is NAD^+-linked and is competitively inhibited by NO_3^- (Peacock and Boulter, 1970). It is clear in *Escherichia coli,* which does not oxidize nitrite with O_2 (Stickland, 1931), that the nitrate reductase and the formate dehydrogenase are not identical enzymes (Ruiz-Herrera and DeMoss, 1969). Lester and DeMoss (1971) have demonstrated for *E. coli* that when selenite is included in the growth medium, the level of formate dehydrogenase activity increases while the level of nitrate reductase activity remains the same. In *Nitrobacter* the inclusion of selenite in the growth medium does not change the ratio of nitrite to formate oxidase activity (W. J. Ingledew and J. G. Cobley, unpublished observation).

Some empirical evidence (O'Kelley and Nason, 1970) seems to suggest that in *Nitrobacter* formate and nitrite are oxidized by different mechanisms. In four strains of *Nitrobacter,* the rates of nitrite and formate oxidation were

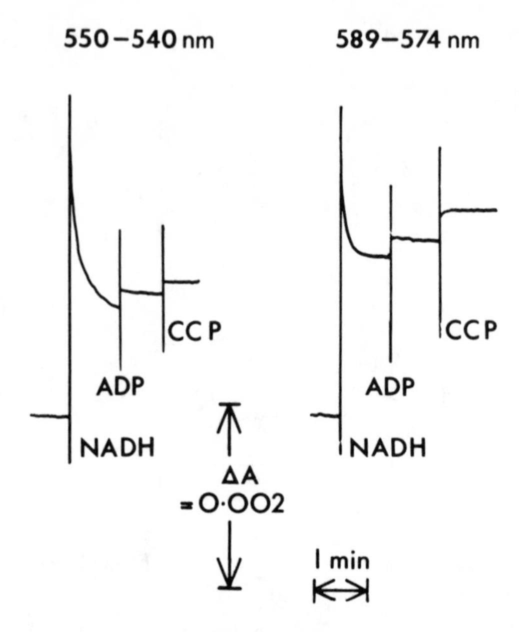

Fig. 9. The reduction of cytochromes in *Nitrobacter* ET particles during NADH oxidation. ET particles (final concentration 0.2 mg protein ml^{-1}) were suspended as described in fig. 8. The additions were as in fig. 8. except that NADH (2 mM) was added instead of sodium formate.

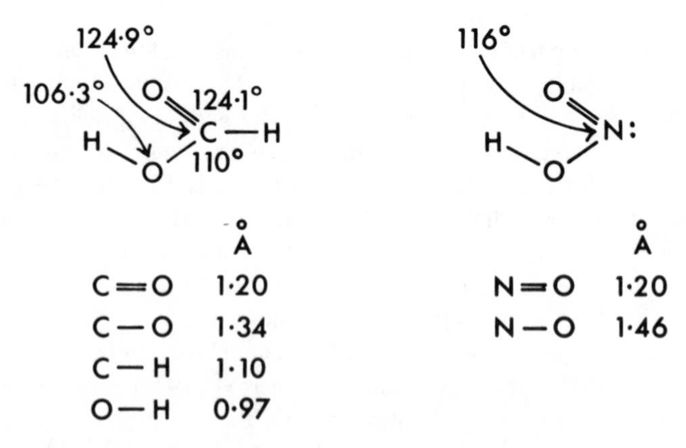

Fig. 10. Comparison of the structures of nitrous and formic acids.

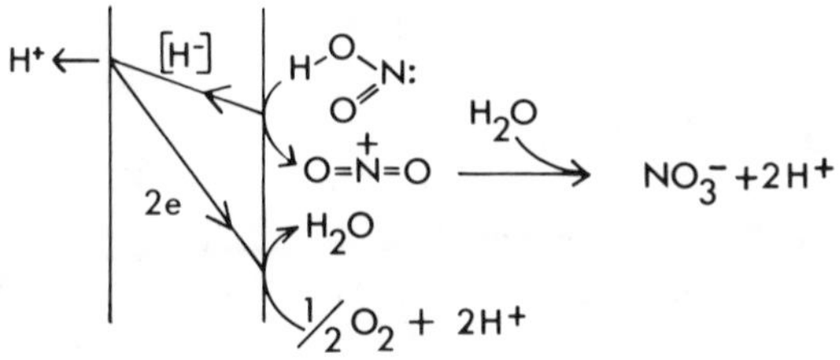

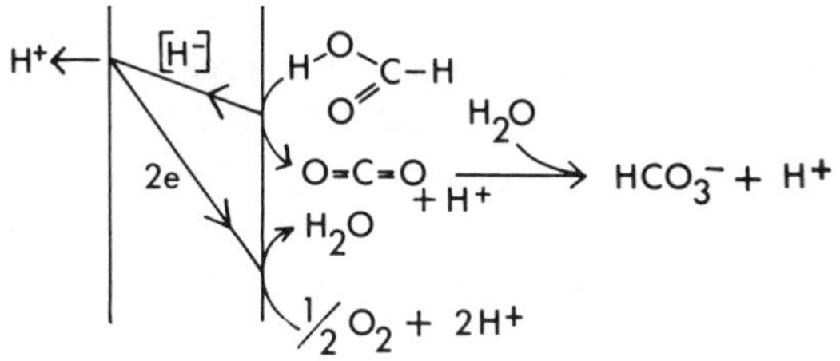

Fig. 11. Oxidations of nitrous and formic acids in ET particles from *Nitrobacter;* representation as analogous reactions.

measured in whole cells and the ratios of the rates were calculated for each strain. The greatest difference between any two strains was not more than a factor of two. This difference may not be significant, especially since the endogenous respiration rate in *Nitrobacter* cells varies from preparation to preparation. The greater sensitivity of formate oxidation to chelating agents as demonstrated by these authors cannot at present be explained. From figure 1 it is clear that formate oxidation is more sensitive to CCCP than nitrite oxidation. The oxidations of formate and nitrite are autocatalytic in the sense that they both generate $\Delta\psi$ by proton translocation and are stimulated in rate by $\Delta\psi$. It would be expected that the concentration of CCCP required to bring about a given percentage of inhibition would be dependent on the absolute value of the uninhibited rate. Hence the difference in sensitivity to CCCP probably reflects the fact that the rate of nitrite oxidation is six times greater than that of formate.

A mechanism has been proposed to explain how the oxidation of nitrite energizes the translocation of H^+ into the ET particle (Cobley, 1976b). This mechanism (fig. 11) also explains why the rate of nitrite oxidation depends on the magnitude of $\Delta\psi$; when $\Delta\psi$ exists (positive within the ET particle), the transfer of a hydride equivalent ($[H^-]$; $2e^-$ *plus* H^+) inward across the membrane would be electrophoretically accelerated. At the inner face of the membrane, the proton is discharged (H^+ translocation), and two electrons move back across the membrane to reduce oxygen at the electrically negative face. Such a flow of reducing equivalents is electrogenic and translocates one proton inward per nitrite oxidized. In this scheme the inward translocation of the hydride equivalent is considered the rate-limiting step.

CCCP inhibits nitrite oxidation, whereas amines stimulate nitrite oxidation. These facts fit the model, since CCCP collapses both $\Delta\psi$ and ΔpH whereas amines decrease ΔpH and indirectly cause an increase in $\Delta\psi$ in compensation for the decrease in ΔpH ($\Delta\bar{\mu}_{H^+} = \Delta\psi - 60\ \Delta pH$). The tetraphenyl boron anion collapses $\Delta\psi$ with an accompanying compensatory increase in ΔpH; under these conditions and in agreement with the model, nitrite oxidation is inhibited. If $\Delta\bar{\mu}_{H^+}$ is the driving force for ADP phosphorylation, then the $\Delta\bar{\mu}_{H^+}$ (and more specifically the $\Delta\psi$) generated during nitrite oxidation should decrease when phosphorylation is taking place and the rate of nitrite oxidation (dependent on the value of $\Delta\psi$) should decrease. The inhibition of nitrite oxidation by ADP together with P_i is thus in full accord with the model.

Although the proposal of a transfer of a hydride equivalent may seem unorthodox as a bioenergetic mechanism, it is not without precedent in enzymology. Edmondson and colleagues (1972) proposed the transfer of a hydride equivalent as a step in the mechanism of the molybdenum-containing xanthine oxidase. In this regard it should be noted that Ingledew and Halling (1976) detected a molybdenum EPR signal in ET particles from *Nitrobacter* and that Zavarin (1960) reported the detection of significant amounts of xanthine and hypoxanthine dehydrogenase activity in *Nitrobacter* whole cells. It is therefore possible that ET particles from *Nitrobacter* contain a molybdenum enzyme of relatively low specificity that is responsible for both nitrite and formate oxidations. Nitrate reductases and formate dehydrogenases are known to contain molybdenum (Coughlan, 1980).

In figure 11 nitrite and formate oxidations are illustrated as analogous reactions. It is tentatively proposed that nitrite oxidation might proceed via the nitronium ion, which is rapidly hydrated in aqueous solutions of neutral pH.

Kiesow (1967) has considered that the inhibition of nitrite oxidation by uncoupling agents occurs because nitrite is not a sufficiently strong reducing agent ($E_o' = +0.42$ V) to reduce cytochrome *c* (E_m[pH 7.0] $= +0.27$ V), an obligatory electron carrier in the flow of reducing equivalents from nitrite to oxygen. If cytochrome *c* in ET particles is located at the electropositive face of the membrane close to the site of proton discharge, then extensive reduction of cytochrome *c* by nitrite can only take place when $\Delta\psi$ lowers the apparent E_m(pH 7.0) for cytochrome *c* by displacing the equilibrium of the hydride equivalent. In this sense the model encompasses Kiesow's proposal. However, this thermodynamic approach cannot be applied to explain formate oxidation since formate causes extensive reduction of cytochrome *c* under anaerobic conditions (not shown).

In conclusion, it seems that the inefficient use by *Nitrobacter* of the energy available from formate oxidation reflects the lack of specificity of the nitrite oxidase system and that formate is not a physiological substrate for the organism. It is interesting to note that attempts to grow *Nitrobacter* anaerobically with formate as electron donor and nitrate as acceptor have been unsuccessful (O'Kelley and Nason, 1970) even though the organism catalyzes this reaction rapidly and the thermodynamics indicate that much free energy would be available. The failure of the organism to grow under these conditions can be explained by the model since the reduction of nitrate by formate occurs at a single active site and would not therefore result in proton translocation.

ACKNOWLEDGEMENTS

I am grateful for financial support from the National Science Foundation and from Research Corporation.

LITERATURE CITED

Aleem, M. I. H. 1965. Path of carbon and assimilatory power in chemosynthetic bacteria. I. *Nitrobacter agilis*. Biochim. Biophys. Acta 107:14–28.

Aleem, M. I. H., H. Lees, and D. J. D. Nicholas. 1963. Adenosinetriphosphate-dependent reduction of nicotinamide adenine dinucleotide by ferro-cytochrome *c* in chemoautotrophic bacteria. Nature (London) 200:759–61.

Aleem, M. I. H., and A. Nason. 1960. Phosphorylation coupled to nitrite oxidation by particles from the chemoautotroph, *Nitrobacter agilis*. Proc. Natl. Acad. Sci. USA 46:763–69

Basolo, F., and R. G. Pearson. 1968. Mechanisms of inorganic reactions. 2d ed. p. 504. Wiley, New York.

Butt, W. D., and H. Lees. 1960. Nitrite oxidation by *Nitrobacter* in the presence of certain phenols. Nature (London) 188:147–48.

Chance, B., and D. Herbert. 1950. The enzyme-substrate compounds of bacterial catalase and peroxides. Biochem. J. 46:402–14.

Chaudhry, G. I., I. Suzuki, H. W. Duckworth, and H. Lees. 1981. Isolation and properties of cytochrome *c-553*, cytochrome *c-550*, and cytochrome *c-549, 554* from *Nitrobacter agilis*. Biochim. Biophys. Acta 637:18–27.

Cobley, J. G. 1976a. Energy-conserving reactions in phosphorylating electron-transport particles from *Nitrobacter winogradskyi*. Biochem. J. 156:481–91.

Cobley, J. G. 1976b. Reduction of cytochromes by nitrite in electron-transport particles from *Nitrobacter winogradskyi*. Biochem. J. 156:493–98.

Cotton, F. A., and G. Wilkinson. 1972. Advanced inorganic chemistry. P. 354 and p. 639. Wiley, New York.

Coughlan, M. P. (ed.). 1980. Molybdenum and molybdenum-containing enzymes. Pergamon Press, New York.

Edmondson, D., V. Massey, G. Palmer, L. Beacham, and G. R. Elion. 1972. The resolution of active and inactive xanthine oxidase by affinity chromatography. J. Biol. Chem. 247:1597–603.

Green, M., and A. G. Sykes. 1970. Kinetic studies on the reaction of nitrous acid with the μ-superoxobis[penta-amminecobalt(III)] complex. J. Chem. Soc. A:3209–14.

Heytler, P. G., and W. W. Pritchard. 1962. A new class of uncoupling agents—carbonyl cyanide phenylhydrazones. Biochem. Biophys. Res. Commun. 7:272–75.

Ingledew, W. J., and P. J. Halling. 1976. Paramagnetic centers of the nitrite oxidizing bacterium, *Nitrobacter*. FEBS Lett. 67:90–93.

Kiesow, L. A. 1964. On the assimilation of energy from autotrophic forms of life. Proc. Natl. Acad. Sci. USA 52:980–88.

Kiesow, L. A. 1967. Energy-linked reactions in chemoautotrophic organisms. Curr. Top. Bioenerg. 2:195–233.

Kiesow, L. A., B. F. Lindsley, and J. W. Bless. 1977. Phosphoribulokinase from *Nitrobacter winogradskyi:* activation by NADH and inhibition by pyridoxal phosphate. J. Bacteriol. 130:20–25.

Lester, R. L., and J. A. DeMoss. 1971. Effects of molybdate and selenite on formate and nitrate metabolism in *Escherichia coli*. J. Bacteriol. 105:1006–14.

Malavolta, E., C. C. Delwiche, and W. D. Burge. 1960. Carbon dioxide fixation and phosphorylation by *Nitrobacter agilis*. Biochem. Biophys. Res. Commun. 2:445–49.

Malavolta, E., C. C. Delwiche, and W. D. Burge. 1962. Formate oxidation by cell-free preparations from *Nitrobacter agilis*. Biochim. Biophys. Acta 57:347–51.

Mitchell, P. 1966. Chemiosmotic coupling in oxidative and photosynthetic phosphorylation. Biol. Rev. 41:445–502.

Murray, R. G. E., and S. W. Watson. 1965. Structure of *Nitrosocystic oceanus* and comparison with *Nitrosomonas* and *Nitrobacter*. J. Bacteriol. 89:1594–609.

O'Kelley, J. C., G. E. Becker, and A. Nason. 1970a. Characterization of the particulate nitrite oxidase and its component activities from the chemoautotroph, *Nitrobacter agilis*. Biochim. Biophys. Acta 205:409–25.

O'Kelley, J. C., and A. Nason. 1970b. Particulate formate oxidase from *Nitrobacter agilis*. Biochim. Biophys. Acta 205:426–36.

Peacock, D., and Boulter, D. 1970. Kinetic studies of formate dehydrogenase. Biochem. J. 120:763–69.

Ruiz-Herrera, H., and J. A. DeMoss. 1969. Nitrate reductase complex from *Escherichia coli* K12: participation of specific formate dehydrogenase and cytochrome b_1, components in nitrate reduction. J. Bacteriol. 99:720–29.

Sewell, D. L., M. I. H. Aleem, and D. E. Wilson. 1972. The oxidation-reduction potentials and rates of oxidation of the cytochromes of *Nitrobacter agilis*. Arch. Biochem. Biophys. 153:312–18.

Silver, W. S. 1960. Endogenous respiration in *Nitrobacter*. Nature (London) 185:555–56.

Stickland, L. M. 1931. The reduction of nitrates by *Bact. coli*. Biochem. J. 25:1543–54.

Streitwieser, A., and C. M. Heathcock. 1981. Introduction to organic chemistry. 2d ed., p. 498. Macmillan Publishing Co., New York.

Weast, R. C. 1970. Handbook of chemistry and physics. 51st ed., pp. D120–21. Chemical Rubber Co., Cleveland, Ohio.

Zavarin, G. A. 1960. The incident of the second phase of nitrification. IV. The dehydrogenase activity of a washed suspension of *Nitrobacter winogradskyi*. Mikrobiologiya 29:476–77.

M. I. H. ALEEM AND DAVID L. SEWELL

Oxidoreductase Systems in *Nitrobacter agilis*

11

INTRODUCTION: THE ELECTRON TRANSPORT COMPLEXES

The bioenergetic activities of chemolithotrophic bacteria are functionally dependent upon the structural aspects and consequent operation of their cytoplasmic membranes. The latter contain electron transfer complexes constituting the oxidoreductase systems as essential components of the respiratory chain that can catalyze ATP synthesis when electrons are transferred from a low-potential to a high-potential complex, or ATP hydrolysis when electrons have to be driven from a high-potential substrate (such as an inorganic nitrogen or sulfur compound) to a low potential complex (usually NAD(P)) under appropriate metabolic conditions requiring active biosynthesis for growth at the expense of CO_2 reduction. The energy generation and utilization in the respective downhill and uphill transfer of electrons is driven, in principle, by the electrochemical proton motive force ($\Delta\bar{\mu}_{H^+}$).[1]

It is the purpose of this paper to discuss the salient features of the electron transfer complexes or oxidoreductase systems in a typical chemoautotrophic bacterium, which is unique in electron transfer and energy transduction mechanisms. This unusual chemolithotroph is *Nitrobacter agilis,* which plays an ecologically indispensable role in the biological nitrogen cycle by oxidizing the toxic nitrite ions to nitrates, the useful nitrogen form for plant growth. Nitrite oxidation is the sole source of energy that is utilized for cell biosynthesis from CO_2; the latter is the sole carbon source for growth. The organism is an obligate aerobe and synthesizes a membrane-bound electron transport system composed of NADH-dehydrogenase, ubiquinone-10, cytochrome *b*, cytochrome *c* (E_m [pH 7.0] + 274 mV), cytochrome *a* (E_m [pH 7.0] + 240 mV), cytochrome a_3 (E_m [pH 7.0] + 400 mV), cytochrome *o*, and two cytochromes (E_m [pH 7.0] + 352 mV and E_m [pH 7.0] + 100 mV) with absorbance bands at 590 nm (α-peak) and 438 nm (γ-peak) attributed to cytochrome a_1. The cytochromes c, a_1, a, and a_3 are rapidly oxidized when O_2 is added to an anaerobic suspension of cell-free extracts. The properties of cytochromes c, a, and a_3 are similar to those of mitochondrial cytochromes c, a, and a_3. Difference spectra of *Nitrobacter* cell-free extracts do not clearly show the presence of cytochrome *b* because of its being masked by the large absorbance of cytochrome *c*. However, the acid-acetone extraction of intact cells separates protoheme and heme *b* from cytochrome *c*. The intact *Nitrobacter* cells contain approximately 0.1 nmol cytochrome *b* per mg dry weight. We have observed a ratio of cytochromes *b*:*c*:*o*:heme *a* in intact cells as 0.1:1.5:1:1. Cytochrome a_3 acts as the terminal oxidase as evidenced by inhibition of respiration by CO, and release of inhibition by light. Although *Nitrobacter agilis* contains cytochrome *o*, it reacts slowly with CO, and low concentrations of nitrite either do not reduce cytochrome *o* or CO cannot bind with reduced cytochrome *o* in the presence of nitrite. Although cytochrome *o* exhibits much lower affinity toward O_2 as compared with cytochromes *a* and a_3, further work is needed to elucidate the conditions under which cytochrome *o* can function as an oxidase or terminal electron acceptor.

The cytoplasmic membrane of *Nitrobacter* contains nitrite: O_2 oxidoreductase (NO_2^- oxidase) as well as dissimilatory NADH:NO_3^- oxidoreductase (respiratory NO_3^- reductase) systems. Although the requirements of these two enzyme systems with respect to the NO_2^-/NO_3^- couple are opposite, the functional aspects of the two systems appear to show resemblance. The EPR profile of *Nitrobacter* subcellular particles resembles that of the nitrate reductase enzyme of *Paracoccus denitrificans* (Lam and Nicholas, 1969), and of *Escherichia coli* (Devertanian and Forget, 1975). All paramagnetic centers in *Nitrobacter* exhibit relatively high midpoint potentials ranging from 32 mV to 340 mV. These signals constitute ferredoxin-type, HiPIP-type Fe-S, and Mo-V-type paramagnetic centers (Ingledew and Halling, 1976).

The unique features of the chemoautotrophic metabolism in *Nitrobacter* are the mechanisms or electron transfer reactions involved in the oxidation of nitrite, and the coupled generation of energy (e.g., ATP) and reducing power (e.g., NAD(P)H); all of these cellular events are essential to drive the reductive and endergonic CO_2-based biosynthetic metabolism. Thus, during growth of *Nitrobacter,* nitrite serves as the sole electron donor for the reduction of O_2 as well as for the reduction of NAD(P) as shown by the following reactions:

$$NO_2^- + H_2O \rightarrow NO_3^- + 2\ H^+ + 2\ e^- \quad (1)$$

$$2\,H^{+} + 2\,e^{-} + \tfrac{1}{2}\,O_2 \rightarrow H_2O \qquad (2)$$

$$NO_2^{-} + \tfrac{1}{2}\,O_2 \rightarrow NO_3^{-}$$

$$\Delta G_o' = -18 \text{ kcal mol}^{-1} \qquad (3)$$

$$NO_2^{-} + H_2O \rightarrow NO_3^{-} + 2\,H^{+} + 2\,e^{-} \qquad (4)$$

$$2H^{+} + 2e^{-} + NAD(P)^{+} \rightarrow NAD(P)H + H^{+} \qquad (5)$$

$$NO_2^{-} + H_2O + NAD(P)^{+} \rightarrow NO_3^{-} + NAD(P)H + H^{+} \qquad (6)$$

$$\Delta G_o' = +35 \text{ kcal mol}^{-1}$$

Reaction 3 represents the energy-yielding process in *Nitrobacter* and is catalyzed by the enzyme system NO_2^-:O_2 oxidoreductase. The latter must of necessity drive the highly endergonic reaction 6 by a tight coupling during growth because of the obligate requirement of reduced pyridine nucleotides for CO_2 reduction involving the Calvin-Benson cycle (Aleem, 1965). The reaction 6, which in essence is the reversal of the dissimilatory nitrate reductase (e.g., NADH:NO_3^- oxidoreductase), is catalyzed by the energy-linked NO_2^-:NAD^+ oxidoreductase enzyme system. Two additional enzyme systems, NADH:O_2 oxidoreductase and NADH:NO_3^- oxidoreductase (Aleem, 1968; Sewell and Aleem, 1979; Kiesow, 1964) appear to play an important role in the energy metabolism of this chemoautotroph. Since the cytoplasmic membrane contains additional electron transfer complexes comprising several other oxidoreductase systems (table 1), this paper will consider their characteristics, mechanism of action, and possible roles in the overall energy metabolism in *Nitrobacter*.

DESCRIPTION OF OXIDOREDUCTASE SYSTEMS

NO_2^-:O_2 Oxidoreductase

This enzyme system catalyzes the oxidation of nitrite and is sedimentable at 144,000 *g* for 1 h, as first reported by Aleem and Alexander (1958). Attempts to solubilize nitrite oxidase have been unsuccessful in part because of the extreme lability of the enzyme system after treatment with detergents. The nitrite oxidation rate as measured polarographically by O_2 uptake by intact cells is linear up to 2 min with NO_2^- concentrations of 2.5 mM to 60 mM and

TABLE 1

CHARACTERISTICS OF THE NITROBACTER OXIDOREDUCTASE SYSTEMS
(% INHIBITION)

Oxidoreductase	0.01 mM Rotenone	1 mM Amytal	5 mM Salicylaldoxime	5 mM *o*-phenanthroline	5 μg mg protein^{-1} Antimycin A	10 μg mg protein^{-1} HOQNO	0.05 mM CN^-	0.1 mM N_3^-	0.01 mM CCCP	0.1 mM 2,4-DNP
$NO_2^-:O_2$	10	5	18	11	0	0	100	70	60	40
$NO_2^-:K_3Fe(CN)_6$	17	13	35	24	0	0	26	92	46	21
$NO_2^-:NAD^+$ (ATP-dependent)	100	40	70	50	80	85	—	—	75	100
Ascorbate:O_2	18	15	31	45	0	0	100	28	4	8
Formate:O_2	12	27	38	38	65	75	100	75	86	33
NADH:O_2	80	74	64	31	62	60	75	23	(21*)	8
NADH:$K_3Fe(CN)_6$	11	13	65	29	0	0	0	0	11	22
NADH:cyt.*c*	91	48	85	59	50	45	0	0	0	0
NADH:NO_3^-	90	44	87	76	52	30	0	0	(78*)	(9*)

*% stimulation.

increases with higher concentrations of nitrite. In the case of cell-free extracts, however, a 50 mM NO_2^- concentration becomes inhibitory after 30 sec. With membrane particles the O_2-uptake activity increases with NO_2^- concentrations of up to 40 mM. The apparent K_m for NO_2^- at pH 8.0 is estimated to be 2.5 mM for 20,000 *g* supernatant fraction and 1.7 mM for 144,000*g* pellet fraction containing virtually all of the nitrite oxidase activity. The optimal activity occurs at pH 7.5–8.0 with 20 mM NO_2^- and at pH 6.5–7.0 with 2 mM NO_2^-. Tris-HCl is inhibitory at concentrations higher than 0.1 M (approx. 50% and 70% inhibition in presence of 0.5 and 1.0 M Tris-HCl, respectively).

Involvement of flavoproteins in nitrite oxidation was suggested by Van Gool and Laudelout (1966) because atabrine inhibited the process. However, the inhibition was not reversed by added FMN or FAD (O'Kelley et al., 1970). Since other classical inhibitors of flavoproteins, such as rotenone, amytal, and TTFA, have little or no effect on NO_2^- oxidation, and since atabrine also inhibits ascorbate oxidation that does not involve the flavoprotein system (Aleem, 1968), the inhibition of nitrite oxidation by atabrine may be due partly to its effect on cytochrome oxidase and partly to its uncoupling action on the energy-linked nitrite coupling with the electron transport chain. Atabrine has been reported by Weinbach and Garbus (1969) as an uncoupling agent. That the flavoproteins and cytochrome *b* do not participate in nitrite oxidation is further supported by the lack of inhibition by antimycin A or HOQNO (Aleem, 1968). The enzyme system is relatively insensitive to the metal-binding agents such as 8-hydroxyquinoline, *o*-phenanthroline, α,α-bipyridyl, and salicylaldoxime. However, low concentrations of cyanide, CO, and azide cause a potent inhibition implicating the involvement of aa_3-type cytochromes in the oxidoreductase system (Aleem and Sewell, 1981).

The NO_2^- oxidase system is extremely sensitive to uncouplers of oxidative phosphorylation due most probably to the energy-linked cytochrome *c* reduction by NO_2^- (Kiesow, 1967; Aleem, 1967; 1968; 1977). The marked inhibition of the enzyme system by 10 μM *m*-chlorocarbonyl cyanide phenylhydrazone (CCCP) appears to be due to its uncoupling action on the energy-linked cytochrome *c* reduction by cytochrome a_1. This conclusion is supported by the fact that nitrite reduces cytochrome a_1 directly in the presence of CCCP but the reduction of cytochrome *c* does not occur. Thus, unless cytochrome *c* is reduced and oxidized by the cytochrome *c*:O_2 oxidoreductase system containing cytochromes *a* and a_3, nitrite oxidation and concomitant ATP formation do not take place. The suggestion by O'Kelley and colleagues (1970) that cytochrome *c* reduction by nitrite is non-energy-dependent and that inhibition of nitrite oxidation by uncouplers is due to metal chelation in the

NO_2^-:cytochrome *c* oxidoreductase is not in harmony with our findings. Our work has revealed that at pH 8.0 the $NO_2^-:O_2$ oxidoreductase system exhibits 5-fold greater affinity for nitrite as compared with NO_2^-:ferricyanide or NO_2^-:cytochrome *c* oxidoreductases. Based on electron equivalents transferred, the activity of the $NO_2^-:O_2$ oxidoreductase is 3- to 4-fold greater than the activity of either of the reductases, indicating that probably the "nitrite dehydrogenase" is not readily accessible to the artificial electron acceptors such as ferricyanide or cytochrome *c*. This is at odds with the report by O'Kelley and colleagues (1970) that the NO_2^-:cytochrome *c* reductase activity of the particles is 2- to 3-fold greater than that of the $NO_2^-:O_2$ oxidoreductase. The reaction mixture used by these investigators contained 50 mM nitrite, 160 μM mammalian cytochrome *c*, and enzyme preparation. Since the very rapid leveling off of the reaction after 90 sec (and even faster at higher enzyme concentrations) results in the reduction of only one-thirtieth (5 μM) of the added cytochrome *c*, the reaction can hardly be free from experimental artifacts, especially when the measurements were made in a single-beam instrument using rather high concentrations of nitrite and exogenously added mammalian cytochrome *c*.

We have observed that the freezing and thawing of intact cells causes 20% inhibition of NO_2^- oxidase and 50% inhibition of the steady-state anaerobic cytochrome *c* reduction by nitrite. The inhibition of nitrite oxidase by freezing and thawing is partially reversed by incubation at 30°C for 20 min probably due to the restoration of the structural organization of the membrane structure; the addition of Tween 80 prevents this restoration. Upon sonication of *Nitrobacter* cells, reduction of cytochrome a_1 by nitrite increases 50%, whereas cytochrome *c* reduction decreases 70%, indicating that membrane integrity is essential for cytochrome *c* reduction. A similar effect was observed by Van Gool and Laudelout (1967), but they postulated that the breakage of intact cells slows the rate of electron flow from nitrite to O_2, and therefore the steady-state reduction of the cytochrome nearest O_2 increases. This postulate does not conform to the experimental observations (Aleem, 1977; Aleem and Sewell, 1981) that cytochrome a_1 is the first electron acceptor from nitrite and that reduction of cytochrome *c* requires ATP (Aleem, 1967; Kiesow, 1967; Sewell and Aleem, 1969) or membrane energization involving most probably the electrical potential component of the protonmotive force. This view is in harmony with the observations by Ingledew and Chappell (1975) that cytochrome *c* is energized in the presence of ATP and the midpoint potential at pH 7 shifts from 274 mV to 360 mV, which is the midpoint potential of cytochrome a_1 as well as of the NO_2^-/NO_3^- couple (Aleem, 1977; Aleem and Sewell, 1981). It is quite apparent that the energization of cytochrome *c* facilitates the overall process of electron transfer from nitrite to molecular oxygen and coupled energy generation; however, nitrite cannot be oxidized in the absence of cytochrome a_1. Interesting studies by Sundermeyer and Bock (1981) have clearly demonstrated that cytochrome a_1 as well as the ability to oxidize nitrite disappeared in *Nitrobacter* X_{14} (a facultative chemolithotroph) when grown heterotrophically, although the bacterial cells contained cytochromes *b, c, a,* and a_3. Earlier studies of Dessers and colleagues (1970) show that *Nitrobacter winogradskyi* conserves almost 50% of the total available free energy from the oxidation of nitrite, that is, at best one mole of ATP can be generated per mole of nitrite oxidized. Slowing the growth due to depletion of nitrite depresses cytochrome a_1 content, and its turnover rate decreases during the first 5–6 generations and then remains constant (Tsien and Laudelout, 1970). However, synthesis of poly-β-hydroxybutyrate continues until nitrite is depleted, at which time the poly-β-hydroxybutyrate is depolymerized (Van Gool et al., 1970). Cytochrome a_1 is reduced at a faster rate than cytochrome *c* when nitrite is added to cell-free preparations at 5°C (Kiesow, 1967). The rate of reduction of cytochrome *c* by nitrite is greatly enhanced in *Nitrobacter* particles in the presence of added ATP; during this process cytochrome a_1 undergoes cyclic oxidation/reduction patterns (Sewell and Aleem, 1969). The addition of an uncoupler inhibits the reduction of cytochrome *c* by nitrite without affecting the reduction of cytochrome a_1. All of these experimental observations support the conclusion that cytochrome a_1 is the primary electron acceptor from nitrite and that the coupling of electrons from nitrite to cytochrome *c* is an energy-linked process requiring perhaps energization of cytochrome *c*, whose participation appears to be essential to act as electron acceptor from nitrite via cytochrome a_1 and to serve as an electron donor for the *Nitrobacter* cytochrome $c{:}O_2$ oxidoreductase system mediated by cytochromes *a* and a_3 for ATP generation (Aleem, 1967; 1968; Sewell et al., 1972). Kiesow (1967) made the interesting observation that the reduction of ferricyanide or chlorate by nitrite is not affected by the uncoupler 2,4-DNP, which inhibits aerobic nitrite oxidation. It appears, therefore, that cytochrome *c* is not involved in the electron transfer from nitrite to either ferricyanide or chlorate. The oxidation of nitrite can proceed in the absence of O_2 if the latter is replaced by an artificial electron acceptor such as ferricyanide (Aleem et al., 1965; these workers have demonstrated unequivocally that the formation of nitrate from nitrite oxidation occurs at the expense of oxygen atom of water and not molecular oxygen).

In view of the foregoing description of the $NO_2^-{:}O_2$ oxidoreductase system, it appears to contain several electron transfer complexes catalyzing the following reactions:

$$NO_2^- + H_2{}^{18}O \rightarrow N^{18}O_3^- + 2\ H^+ + 2\ e^- \quad (7)$$

$$2\ e^- + 2\ Cyt.a_1{\cdot}Fe^{3+} \rightarrow 2\ Cyt.a_1{\cdot}Fe^{2+} \quad (8)$$

$$2\ Cyt.a_1{\cdot}Fe^{2+} + 2\ Cyt.c{\cdot}Fe^{3+} \rightarrow 2\ Cyt.a_1{\cdot}Fe^{3+} + 2\ Cyt.c{\cdot}Fe^{2+} \quad (9)$$

$$2\ Cyt.\ c{\cdot}Fe^{2+} + 2\ Cyt.aa_3{\cdot}Fe^{3+} \rightarrow 2\ Cyt.c{\cdot}Fe^{3+} + 2\ Cyt.aa_3{\cdot}Fe^{2+} \quad (10)$$

$$2\ Cyt.aa_3{\cdot}Fe^{2+} + 2\ H^+ + \tfrac{1}{2}\ O_2 \rightarrow 2\ Cyt.aa_3{\cdot}Fe^{3+} + H_2O \quad (11)$$

$$\text{Sum: } NO_2^- + \tfrac{1}{2}\ O_2 \rightarrow NO_3^-\ . \quad (12)$$

This scheme shows the removal of electrons from nitrite and protons and oxygen from water, incorporation into nitrite of the oxygen atom from water, transfer of electrons to cytochrome a_1 driven by the electrochemical proton gradient, and finally the transfer of electrons and protons to molecular oxygen with the mediation of the cytochrome $c{:}O_2$ oxidoreductase components identified in *Nitrobacter* as cytochromes a and a_3. The overall operation of this enzyme system yields energy by oxidative phosphorylation, and P/O ratios approaching 1 have been obtained during nitrite oxidation (Aleem, 1968).

The kinetics of nitrite oxidation has been studied by Boon and Laudelout (1962); they suggested that the inhibition of nitrite oxidation by high nitrite concentrations is due to undissociated nitrous acid since the inhibition varies with pH. In addition, the pH and nitrite concentration may also affect nitrite oxidation by altering the midpoint potential of the nitrite/nitrate couple (Aleem and Sewell, 1981); however, the oxidation of ascorbate and NADH is not inhibited by high levels of nitrite. O'Kelley and colleagues (1970) proposed that HNO_2 is both the substrate and the inhibitor of nitrite oxidation. If this is the case, the site of inhibition does not appear to occur in that portion of the $NO_2^-{:}O_2$ oxidoreductase complex which is essential for the oxidation of ascorbate and NADH. Nitrite oxidation is also inhibited by high concentrations of nitrate, which inhibits the enzyme system noncompetitively (Boon and Laudelout, 1962; Sewell, 1972); these results are not in harmony with those of O'Kelly et al. (1970) that nitrate competitively inhibits nitrite oxidation.

NO_2^-:Ferricyanide Oxidoreductase

Since the reduction of mammalian cytochrome c (E_m [pH 7.0] + 250 mV) by nitrite (E_m [pH 8.0] + 360 mV) is energy-dependent, ferricyanide (E_m [pH 8.0] + 360 mV) can be used as the primary electron acceptor to study the enzyme system catalyzing the removal of electrons from nitrite. Ferricyanide reduction can be measured spectrophotometrically by following absorbance change at 425 nm. Reaction mixture in a total volume of 2.0 ml contains 1.0 mM ferricyanide, 50 μM cyanide, 90 mM Tris-HCl (pH 8.0), and 200–600 μg enzyme protein. The reaction is started by the addition of 10–50 mM nitrite. Maximal activity occurs at pH 8.0 and in the presence of 1 mM ferricyanide and 60 mM nitrite; however, the reaction is linear with nitrite concentrations up to 20 mM. The apparent K_m for nitrite and ferricyanide is calculated to be 9.1 mM and 190 μM, respectively. The rate of electrons transferred from nitrite to ferricyanide is approximately 25% of the rate of electron transferred to O_2.

Neither cyanide nor anaerobiosis has any effect on ferricyanide reduction. The enzyme activity is markedly sensitive, however, to azide (65% inhibition at 5 μM NaN_3). Approximately 50% inhibition is caused by 0.1 mM CCCP or atabrine. Amytal, rotenone, and TTFA have little or no effect. There is no significant effect of metal-binding agents, as well as antimycin A or HOQNO. Nitrate (5 mM) causes approximately 40% inhibition. Further examination of the data using Lineweaver-Burk plots indicates that the inhibition by nitrate is noncompetitive. Thus, although the enzyme system exhibits some similarities with the NO_2^-:O_2 oxidoreductase and ascorbate:O_2 oxidoreductase systems, the latter system is not affected by CCCP whereas the former system is much more sensitive to CCCP at one-tenth of its concentration.

NO_2^-:NAD^+ Oxidoreductase

The electrons from nitrite originate with a rather high electropositive potential (E_m [pH 7.0] + 420 mV), and thus the NO_2^-:NO_3^- couple possesses therefore a strong oxidizing potential. In *Nitrobacter,* however, the electrons from nitrite must reach a very electronegative potential (E_m [pH 7.0] −340 mV) in order to reduce pyridine nucleotides essential for the chemoautotrophic CO_2 reduction. As reported by Gibbs and Schiff (1961), the reduction of NAD^+ by nitrite, ammonia, or hydroxylamine is highly unfavorable thermodynamically. Nevertheless, *Nitrobacter* grows at the expense of nitrite oxidation and CO_2 reduction, and Kiesow (1963) reported that nitrite oxidation by *Nitrobacter* was coupled to NAD^+ reduction. Simultaneously Aleem and colleagues (1963) demonstrated an ATP-driven reversal of electron transfer from ferrocytochrome *c* to NAD^+ in *Nitrobacter, Nitrosomonas,* and *Ferrobacillus (Thiobacillus).*

The mechanism of NAD^+ reduction by nitrite was elucidated subsequently by Sewell and Aleem (1969). The experimental results show that the aerobic preincubation of the cell-free extracts with 3.3 mM nitrite followed by anaerobiosis results in the reduction of added NAD^+ concomitant with the oxidation of cytochrome *c*. In a typical experiment, 30 nmol of cytochrome *c* are oxidized concomitant with the reduction of 15 nmol of NAD^+. When ADP and phosphate are included in the aerobic preincubation with nitrite followed by anaerobiosis and addition of NAD^+, 24 nmol of NAD^+ are reduced with the concomitant oxidation of 44 nmol of cytochrome *c*. Thus the aerobic oxidation (in the absence or presence of added ADP and phosphate) results in the reversal of electron transfer (e.g., respiratory control) from ferrocytochrome *c* to NAD^+ under anaerobic conditions obviously at the expense of energy generated by the NO_2^-:O_2 oxidoreductase system. In this system the aerobic nitrite oxidation can be replaced by added ATP for pyridine nucleotide reduction. The addition of ATP and nitrite to the membrane preparations containing exogenously added NAD^+ results in the initial reduction of cytochrome *c* but not NAD^+. As cytochrome *c* becomes oxidized, NAD^+ gets reduced; FMN is found to mediate the reversal of electron flow between cytochrome *c* and NAD^+. It is observed that nitrite initially reduces cytochrome a_1 and that ATP is required for the transfer of electrons from cytochrome a_1 to cytochrome *c* (inhibited by uncouplers), from cytochrome *c* to cytochrome *b*-ubiquinone 10 (inhibited by antimycin A and HOQNO), and from ubiquinone 10 to NAD^+ (inhibited by rotenone and amytal). The nitrite-linked NAD^+ reduction increases with increasing ATP concentrations from 0.1 mM to 1.0 mM; higher ATP concentrations are inhibitory.The optimal activity occurs at pH 8.0 and in the presence of 3.4 mM nitrite, 1.0 mM ATP, and 0.7 mM NAD^+. Under the experimental conditions, 4–5 μmol of ATP are hydrolyzed per μmol of NAD^+ reduced, indicating that the energy-coupling efficiency in terms of nitrite-linked ATP-driven reversed electron flow is almost 100%. In addition to the uncouplers, the NO_2^--linked ATP-driven reversal of electron transfer is highly sensitive to oligomycin (70% inhibition by 0.8 μg mg $protein^{-1}$). On the basis of electron transport chain and energy transfer inhibitors, the NO_2^-:NAD^+ oxidoreductase system involves the reversal of the oxidative phosphorylation sequence. Thus, the occurrence of oxidative phosphorylation and its reversal within the same electron transport chain during the growth of *Nitrobacter* offers some challenging research problems dealing with the bioenergetics and control of energy metabolism in the chemoautotroph.

Ascorbate:O_2 Oxidoreductase

Ascorbate oxidation may be considered as a measure of cytochrome oxidase (cytochrome *c*:O_2 oxidoreductase) activity. As opposed to the nitrite oxidase, ascorbate oxidase is relatively insensitive to uncouplers of oxidative phosphorylation. In *Nitrobacter* particles, atabrine inhibits both ascorbate oxidase and nitrite oxidase, indicating that atabrine may interfere with electron transport in the cytochrome oxidase region rather than being a flavoprotein inhibitor. Ascorbate oxidation is more sensitive to cyanide than to azide. Thus, 10 μM CN^- causes 90% inhibition whereas a 10-fold azide concentration causes only 28% inhibition. The enzyme system is also more sensitive to the metal-binding agents such as *o*-phenanthroline and salicylaldoxime as compared with the systems oxidizing formate or nitrite. Ascorbate oxidation is not appreciably inhibited by amytal, TTFA, or rotenone except atabrine, which causes 75% inhibition at 0.17 mM concentration. Antimycin A and NHQNO have no effect on ascorbate oxidation, and unlike NO_2^- oxidation, the process is virtually unaffected by CCCP, PCP, 2,4-DNP, and 2,4-DBP. Although the mechanism of cytochrome *c* reduction by nitrite is quite different from that involving the chemical reduction of cytochrome *c* by ascorbate, the oxidation of cytochrome *c* reduced either by NO_2^- or by ascorbate shares a common electron transport pathway. However, the ascorbate:O_2 oxidoreductase system is relatively much less sensitive to azide as compared with the NO_2^-:O_2 oxidoreductase system. In the latter case, azide would appear to interfere with the establishment of $\Delta\bar{\mu}_{H^+}$ generated as a result of electron transfer from NO_2^- to cytochrome a_1, a process that appears to be linked either with proton translocation (Aleem, 1977) or translocation of a hydride ion (Cobley, 1976). This electrochemical membrane potential is essential for the reduction of cytochrome *c* by NO_2^- (mediated by cytochrome a_1) but not for cytochrome *c* reduction by ascorbate. It is precisely for this reason that NO_2^- oxidation is inhibited by CCCP or 2,4-DNP (e.g., the agents that collapse the ΔpH and Δψ components) but these compounds have virtually no effect on ascorbate oxidation.

Formate:O_2 Oxidoreductase

Formate oxidation by *Nitrobacter* cells was first observed by Silver (1960), who reported that cytochromes *c* and a_1 were reduced in cell suspensions treated with formate and that the O_2 uptake occurred without lag. However, the Q_{O_2} with formate was one-fifth of that with NO_2^-. The enzyme system catalyzing formate oxidation is recovered principally in membrane particles, although 30% of the activity is lost upon centrifugation (Malavolta et al., 1962). The ratio of CO_2 evolved to O_2 consumed is about 2, and that of formate consumed and CO_2 evolved is 1. No H_2 production is detected during formate oxidation (O'Kelley and Nason, 1970). Based primarily on the observations that both formate and nitrite reduce cytochromes *c* and a_1, it has been suggested that a similar electron transport system is utilized between cytochrome *c* and O_2 (Van Gool and Laudelout, 1966; O'Kelley and Nason, 1970). We have conducted further studies on the formate: O_2 oxidoreductase system. The most active fraction containing the enzyme system (100% activity) is 20,000 *g* supernatant. The pellet 144,000 *g* fraction contains only 21% of the total activity, and supernatant 144,000 *g* has a negligible amount (0–5%) of the total activity. The enzyme is particulate, but 70% of the enzyme activity is not recovered after centrifugation in either S_{144} or P_{144} fraction. The optimal pH is 6.5 with 0.1 M phosphate buffer, and pH 7.0 with 0.1 M Tris-HCl buffer. Activity increases with increasing formate concentrations with maximal activity occurring at 20 mM formate. Higher concentrations slightly decrease activity. The potent inhibition by 10 μM CN^- or N_3^- indicates the participation of cytochrome oxidase in the enzyme system. Addition of copper ions partially reversed (about 50%) the inhibition caused by cyanide but not by azide. Amytal, TTFA, and atabrine (1 mM) inhibit approximately 30%, but the system is quite sensitive to antimycin A and NHQNO, and extremely sensitive to very low concentrations of CCCP (86% inhibition at 1 μM and 51% inhibition at 0.1 μM concentration). Pentachlorophenol (10 μM PCP) causes 65% inhibition. All inhibitors exhibit similar effects on whole cells, 20,000 *g* supernatant and 144,000 *g* pellet. Addition of FMN, FAD, and cytochrome *c* has little effect on formate oxidation, and neither FMN nor FAD is able to prevent or relieve the inhibition by amytal, TTFA, atabrine, or rotenone. Cytochromes *c*, a_1, *a*, and a_3 are reduced in the presence of added formate, and addition of nitrate causes the oxidation of the cytochromes under anaerobic conditions. Almost 90% inhibition of the aerobic formate oxidation occurs in the presence of 10 mM added nitrate; the inhibition by nitrate is competitive. More or less similar results are reported by O'Kelley and Nason (1970). Formate oxidation has been observed to couple with CO_2 fixation at a greatly reduced rate (Van Gool and Laudelout, 1966), and the P/O ratios are reported to range between 0.13–0.28 (O'Kelley and Nason, 1970). In highly coupled preparations, we have obtained P/O ratios between 0.5–0.6 with formate as the electron donor (table 2).

The proposed electron transport chain and the vectorial organization of the formate:O_2 and formate:NO_3^- oxidoreductase systems in *Nitrobacter* are shown in figure 1. The coupling of formate with the respiratory chain is shown to take place at the level of low-potential cytochrome a_1 (E_m [pH 7.0] +110 mV); this is based purely on spectroscopic observations. However, the possibility cannot be excluded that the entry of formate takes place at the ubiquinone 10-cytochrome *b* level since formate is able to reduce fumarate

TABLE 2

COMPARATIVE ENERGETICS OF *NITROBACTER* OXIDOREDUCTASE SYSTEMS

Oxidoreductase System	Electron Donor	Electron Acceptor	$\Delta E'_o$ (Volt)	$\Delta G'_o$ kcal mol^{-1}	$P/2e^-$	
					Theoretical	Observed
$NO_2^-:O_2$	NO_2^-	O_2	0.46	−21.2	1.00	0.80
$NO_2^-:K_3Fe(CN)_6$	NO_2^-	$K_3Fe(CN)_6$	0.00	0.00	0.00	0.00
$NO_2^-:NAD(P)^+$	NO_2^-	$NAD(P)^+$	−0.74	+34.1	−4.00*	−4 to −6*
$K_4Fe(CN)_6:O_2$	$K_4Fe(CN)_6$	O_2	0.46	−21.1	1.00	0.80
Ascorbate:O_2	Ascorbate	O_2	0.77	−35.5	1.00	0.55
$HCOO^-:O_2$	$HCOO^-$	O_2	1.27	−58.6	3.00	0.50
$HCOO^-:NO_3^-$	$HCOO^-$	NO_3^-	0.87	−40.1	2.00	N.D.
NADH:O_2	NADH	O_2	1.14	−52.6	3.00	1.20
NADH:NO_3^-	NADH	NO_3^-	0.74	−34.1	2.00	0.82
NADH:Cyt.*c*	NADH	Cyt.*c*	0.57	−28.3	2.00	N.D.
NADH:$K_3Fe(CN)_6$	NADH	$K_3Fe(CN)_6$	0.68	−31.4	2.00	N.D.

*Mol of ATP hydrolyzed/mol $NAD(P)^+$ reduced.

(E_m [pH 7.0] +30 mV) and the oxidation of formate is inhibited by antimycin A or HOQNO. Possibly cytochrome *c* and the high potential cytochrome a_1 (Em 7.0 +357 mV) mediate electron transfer to the respiratory nitrate reductase, and cytochromes *c, a,* and a_3 participate in the aerobic oxidation of formate. There can be two proton-translocating loops, one at the ubiquinone level and the other at the cytochrome oxidase level. Proton transport through the membrane-localized ATPase is shown to couple with ATP synthesis from ADP and P_i (ATPase structure is adopted after Garland, 1977), the F_0 component being sensitive to oligomycin and DCCD and F_1 to aureovertin. The $H^+/2e^-$ ratios using formate as electron donor and O_2 or NO_3^- as electron acceptor have not as yet been determined.

NADH:O_2 Oxidoreductase

The enzyme activity in *Nitrobacter* has been reported by Aleem (1959, 1967, 1968) and Kiesow (1967). The specific activity (nmol NADH oxidized min^{-1} mg $protein^{-1}$) varies from 36 to 110 in crude cell-free extracts and particles sedimented at 144,000 *g* centrifugation. Our data show that maximum activity occurs at pH 6.5–7.0 with 0.1 mM NADH. The apparent K_m for NADH (Lineweaver-Burk plot) is 44 μM. The enzyme activity resides in 144,000 *g* pellet or in the 0–35% saturation with ammonium sulfate. Treatment of the 20,000 *g* supernatant with ammonium sulfate results in 50% loss of activity. The enzyme system exhibits similarities with mitochondrial NADH oxidase. Thus, the enzyme system is sensitive to cyanide and azide

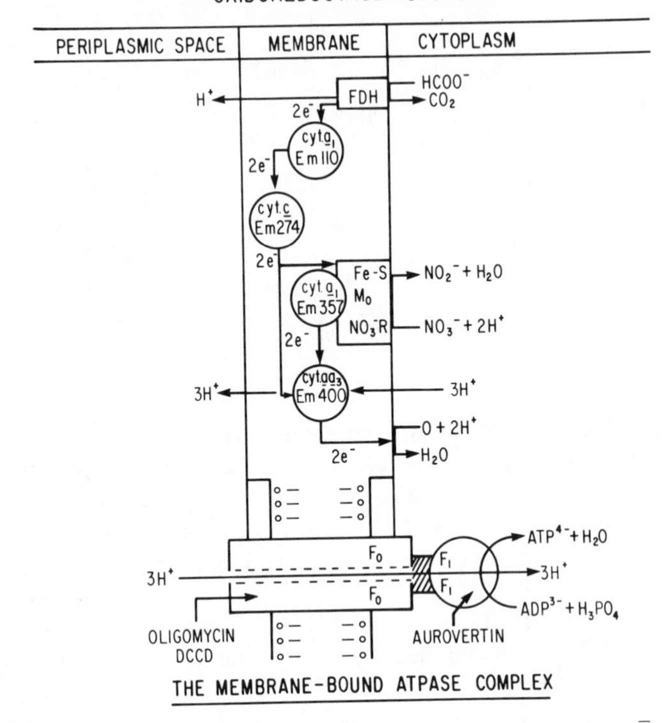

Fig. 1. Proposed vectorial organization of *Nitrobacter* formate:O_2 and formate:NO_3^- oxidoreductase systems.

indicating the participation of cytochrome oxidase. The inhibition by salicylaldoxime implicates the involvement of a metal component, and the complete inhibition by 1.0 mM *p*-hydroxymercuribenzoate strongly indicates that sulfhydryl groups are necessary for activity. Amytal, rotenone, and antimycin A cause a marked inhibition of NADH oxidation, whereas atabrine (0.1 mM) has no effect at all. The uncoupler CCCP causes a 20–30% stimulation of the activity, whereas other uncouplers have little or no effect.

In the cyanide-inhibited system, addition of copper causes a partial restoration of activity and a 20% stimulation in the system inhibited by *o*-phenanthroline. However, the addition of MoO_4^{2-}, Fe^{3+}, and Mg^{2+} to systems inhibited by salicylaldoxime, cyanide, or *o*-phenanthroline has no effect on the prevention or relief of inhibition. The oxidation of NADH is stimulated approximately 30% in the presence of added FMN or FAD; these flavins also cause a relief or bypass of the inhibition caused by cyanide, salicylaldoxime, *o*-phenanthroline, antimycin A, amytal, and rotenone. The addition of NADH to cell-free fractions results in the reduction of cytochromes c, a_1, and aa_3. The cytochrome system is not reduced in the presence of rotenone, salicylaldoxime, and antimycin A; thus, these compounds inhibit on the substrate side of cytochromes and exhibit cross-over points. The O_2-uptake during aerobic NADH oxidation is inhibited with increasing NO_3^- concentrations. Ferricyanide at low concentrations of up to 0.5 mM causes a stimulation in the O_2 uptake, and at higher concentrations ferricyanide inhibits respiration due to competition in electron flux to ferricyanide and O_2.

NADH:Ferricyanide Oxidoreductase

The enzyme measures the NADH-dehydrogenase activity. The maximum rate of ferricyanide reduction occurs at pH 8.0 with 1 mM ferricyanide. The apparent K_m for ferricyanide is approximately 0.3 mM and for NADH 5.1 μM; the substrate becomes non-rate-limiting at about 0.05 mM. The flavoprotein inhibitors antimycin A, azide, cyanide, and uncouplers of oxidation phosphorylation have little or no effect. The sensitivity to metal chelators and *p*-hydroxymercuribenzoate suggests the involvement of a metal component and sulfhydryl groups in the electron transfer from NADH to ferricyanide. The major portion of the enzyme system is not sedimented by centrifugation at 144,000 *g* for 1 h. Ammonium sulfate fractionation (45–60% saturation) of 144,000 *g* supernatant results in recovery of more than 60% of the activity. Optimal pH is 8.0; the apparent K_m for NADH is 73 μM and for ferricyanide 0.23 mM. The activity of the enzyme system is heat labile and 90% of the activity is lost after 3 min at 50°C. The activity of this fraction is also sensitive to the metal chelators salicylaldoxime and *o*-phenanthroline as well as to the sulfhydryl group antagonist *p*-hydroxymercuribenzoate.

NADH:Cytochrome c Oxidoreductase

The activity of the enzyme complex can be correlated with the efficiency of NADH dehydrogenase. Optimal activity occurs at pH 8.0. The apparent K_m for NADH and for mammalian cytochrome *c* is 20 μM and 43 μM respectively (calculated from the Lineweaver-Burk plots). The activity of the enzyme system, unlike the NADH:ferricyanide oxidoreductase, is sedimentable at 144,000 *g* for 1 h, and it is insensitive to 1 mM CN^- or N_3^-. A metal component and sulfhydryl groups appear to be involved in the NADH:cytochrome *c* reductase activity because of the inhibition by the metal chelators and PHMB. The transfer of electrons from NADH to cytochrome *c* is inhibited by flavoprotein inhibitors and by antimycin A. Thus, the enzyme complex involves the mediation of flavoproteins and Q_{10}-cytochrome *b* with cytochrome *c* as the electron acceptor. The electron transfer system is usually intact and tightly coupled since there is little, if any, bypass of electrons.

NADH:NO_3^- Oxidoreductase

The reduction of chlorate and nitrate by *Nitrobacter* cells was first reported by Lees and Simpson (1957). Later investigations confirmed nitrate reduction by cell-free extracts of the bacterium (Kiesow, 1964; Straat and Nason, 1965; Fault et al., 1969; Wallace and Nicholas, 1968). The latter investigators have shown that $^{15}NH_4NO_3$, $^{15}NH_2OH$, $Na^{15}NO_2$, and $K^{15}NO_3$ were incorporated into cellular proteins by intact cells of *N. agilis* and they also detected the presence of hydroxylamine, nitrite, and nitrate reductases in the cell-free extracts. Nitrate reductase has been separated from nitrite oxidase by ammonium sulfate fractionation, and it was suggested that nitrate reductase is an assimilatory enzyme and is thus a separate enzyme from the nitrite oxidase (Straat and Nason, 1965; Wallace and Nicholas, 1968). Based on the stoichiometry 0.5:1.0:0.75 of O_2 uptake, NO_2^- uptake, and NO_3^- extrusion, respectively, by *Nitrobacter,* Nicholas (1981) concluded that 25% of the NO_3^- produced from NO_2^- oxidation was assimilated via the NADH-linked nitrate and nitrite reductase systems. Straat and Nason (1965) suggested, however, that cytochrome a_1 (587 nm) donates electrons either directly to nitrate or to O_2 via cytochrome *a* (605 nm). These investigators identified two components of cytochrome a_1 with absorption maxima at 583 nm and 587 nm in a low-temperature absorption spectrum.

Sewell and Aleem (1979) further characterized the dissimilatory enzyme system and observed that the anaerobic reduction of nitrate in the presence of NADH and cell-free extracts resulted in the accumulation of nitrite. Nitrite absorbs at 355 nm and may interfere with the measurement of the absorbance change at 340 nm due to oxidation of NADH. However, if the ratio of NO_2^-/NADH is less than 15, NO_2^- does not significantly contribute to the absorbance at 340 nm.

The optimal activity occurs at pH 6.5–7.0 with 20,000 *g* supernatant fraction. The apparent K_m for NADH is 6.3 μM and for NO_3^- 0.72 mM. Virtually all of the activity is recovered in the pellet after centrifugation of the S_{20} fraction at 144,000 *g* for 1 h.

Cyanide has little or no effect, whereas N_3^- (1 mM) causes about 70% inhibition. The enzyme system is sensitive to rotenone, amytal, and antimycin A. The uncoupler CCCP (10 μM) causes about 80% stimulation of NADH oxidation by NO_3^-; PCP at the same concentration causes 16% stimulation. The anaerobic electron transport from NADH to NO_3^- is proposed to involve two H^+-translocating loops comprised of an NADH-dehydrogenase-Fe-S center and the protonmotive Q cycle involving Q_{10} and cytochrome *b*. The aerobic NADH oxidation is shown to involve a proton-translocating loop in the cytochrome oxidase region. The stoichiometry of H^+-translocation coupled to NADH oxidation in *Nitrobacter* has not so far been determined. With the exception of the NO_2^-:O_2 oxidoreductase and NADH:NO_3^- oxidoreductase systems, the *Nitrobacter* electron transport chain shares similarities with the respiratory chain in mammalian mitochondria.

CONCLUDING REMARKS

With the exception of the ATP-dependent NO_2^-:NAD^+ oxidoreductase system, all of the oxidoreductases described in *Nitrobacter* catalyze reactions that are exergonic and proceed with a standard free energy change between -20 to -53 kcal mol^{-1} (table 2). The reactions catalyzed by NO_2^-:O_2 oxidoreductase, ferrocyanide:O_2 oxidoreductase, and ascorbate:O_2 oxidoreductase are coupled to ATP generation in the terminal energy-coupling site only and yield P/O ratios of 0.8, 0.8, and 0.55, respectively. NADH oxidation yields P/NO_3^--reduced ratio of 0.7 as opposed to a P/O ratio of 1.1, suggesting that the anaerobic oxidation of NADH by nitrate involves one less phosphorylation site than the aerobic oxidation of NADH. The CN^--insensitivity of the NADH:NO_3^- oxidoreductase-linked NADH oxidation as well as the coupled ATP synthesis suggests that the terminal energy-coupling site involving cytochrome oxidase does not participate in the anaerobic NADH oxidation by nitrate.

The proposed organization of the electron transfer complexes participating in the NADH:O_2 oxidoreductase and NADH:NO_3^- oxidoreductase systems is shown in figure 2. In this scheme the electron transport from NADH to either nitrate or O_2 shares a common (or similar) respiratory chain mediated by flavoproteins, iron-sulfur center(s), ubiquinone-10, cytochrome *b*, and cytochrome *c*; this sequence is supported by the sensitivity of both the systems to rotenone or amytal and to antimycin A or HOQNO. The nitrate reductase complex is proposed to contain the high-potential cytochrome a_1 component along with Fe-S and Mo. In aerobic NADH oxidation, the major portion of the electron flux is mediated by cytochrome *c*, cytochrome *a*, and cytochrome a_3, and this system is highly sensitive to low cyanide concentrations. The

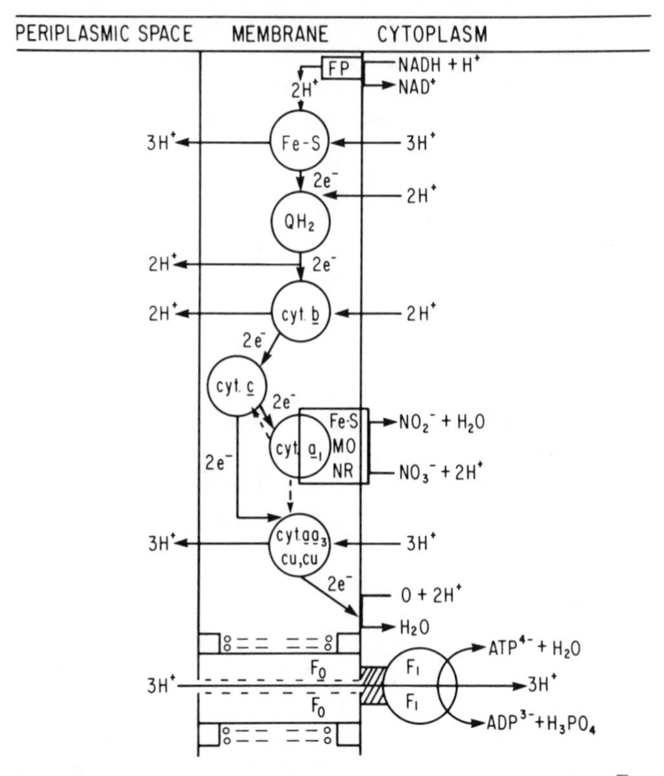

Fig. 2. Proposed vectorial organization of *Nitrobacter* NADH:O_2 and NADH:NO_3^- oxidoreductase systems.

formate:O_2 oxidoreductase system appears to involve energy-coupling sites 2 and 3 because of its marked sensitivity to antimycin A or HOQNO and to cyanide or azide (table 1). Although formate oxidation is coupled to ATP generation, the P/O ratios are in the range of 0.5 even in the case of highly coupled membrane preparations. The NADH:O_2 and NADH:NO_3^- oxidoreductase systems yield P/O and P/NO_3^--reduced ratios of 1.2 and 0.82, respectively. The former system involves all of the 3 energy-coupling sites; the latter system employs coupling sites 1 and 2 for ATP generation. The ATP-dependent NO_2^-:NAD^+ oxidoreductase system involves electron transfer from NO_2^- to cytochrome a_1 and then energy-linked reversal of electron transfer from cytochrome a_1 to NAD^+ with the mediation of cytochrome *c*, cytochrome *b*, Q10, and the flavoprotein system with utilization of 4–6 mol of ATP/mol of pyridine nucleotide reduced. Since the electron transfer reactions in systems catalyzed by NADH:O_2 oxidoreductase, NADH:NO_3^- oxidoreductase, and ATP-dependent NO_2^-:NAD^+ oxidoreductase exhibit similar inhibition patterns with flavoprotein inhibitors (e.g., rotenone and amytal), metal-binding agents (e.g., salicylaldoxime and *o*-phenanthroline), sulfhydryl-group antagonists such as PHMB (data not shown), and antimycin A or HOQNO, it would appear that either the same or similar electron transport chains are involved in these oxidoreductase systems, and that the ATP-driven NAD^+ reduction involves reversal of the oxidative phosphorylation reactions. All of the NADH-linked oxidoreductase systems are much more sensitive to the metal-binding agents and to sulfhydryl groups antagonists. However, the NADH:cytochrome *c* oxidoreductase system is more tightly coupled (by virtue of its sensitivity to flavin inhibitors, metal-binding agents and to antimycin A or HOQNO) than the NADH:ferricyanide oxidoreductase system, which is not affected by rotenone or amytal and by antimycin A or HOQNO. The uncoupler CCCP causes strong inhibition of the NO_2^-:O_2 oxidoreductase and the ATP-dependent NO_2^-:NAD^+ oxidoreductase, indicating the involvement of $\Delta\bar{\mu}_{H^+}$ in both systems; the reduction of cytochrome *c* by NO_2^- via cytochrome a_1 appears to be driven by the electron transport-linked electrochemical proton gradient or membrane potential. CCCP also inhibits the formate:O_2 oxidoreductase system but has no effect on the ascorbate:O_2 oxidoreductase; the uncoupler causes 21 and 78% stimulation in the NADH oxidation by O_2 and NO_3^-, respectively, exhibiting a mitochondrial-type respiratory control in *Nitrobacter*. Since the electrons from formate couple with the respiratory chain at the succinate-fumarate level, the reason for inhibition of formate oxidation by CCCP (table 1) is not clear; perhaps the inhibition may be caused at the formate-dehydrogenase level, but such determinations have not been made.

The oxidoreductase systems display a vital role in the chemoautotrophic metabolism of *Nitrobacter*. The endergonic CO_2-dependent cellular biosynthesis in this organism is tightly coupled to the exergonic oxidation reactions responsible for the generation of energy through the electron transport chain, the efficiency of which is controlled by CO_2 and its reductive assimilation. The latter process is driven at the expense of ATP and NADH, both of which are generated within the same electron transport chain during electron transfer to O_2 catalyzed by NO_2^-:O_2 oxidoreductase, and the electron transfer to NAD^+ catalyzed by the ATP-dependent NO_2^-:NAD^+ oxidoreductase. The latter enzyme system is reversed by the NADH:NO_3^- oxidoreductase system capable of generating energy with nitrate (the product of NO_3^- oxidation) as the terminal electron acceptor. The same enzyme system, when linked to O_2, constitutes the NADH:O_2 oxidoreductase activity exhibiting similarities with that of the mitochondrial systems. As suggested by Kelly (1971), chemolithotrophic bacteria, in the absence of exogenous energy source, may oxidize endogenous materials via NAD^+-linked dehydrogenase. Characteristically, the chemoautotrophic growth of *Nitrobacter* results in the cellular accumulation of the endogenous reserve material as poly-β-hydroxybutyrate. This is an important electron donor for the generation of NADH by the β-hydroxybutyrate dehydrogenase present in the organism. The electron transfer from the resultant NADH to either O_2 catalyzed by the NADH:O_2 oxidoreductase or to NO_3^- (under O_2-limiting environments) catalyzed by the NADH:NO_3^- oxidoreductase yields ATP required for the maintenance energy for survival of *Nitrobacter* in environments defficient in NO_2^- and/or O_2. In this context Richie and Nicholas (1972) also suggest that the observed possession of dissimilatory NO_2^--reductase by *Nitrosomonas* may play an important role in the survival of the organism since microaerophilic or anaerobic conditions often occur in the soil environments.

The Electron Transport Systems in Nitrobacter

The electron transport pathways and energy-coupling sites in *Nitrobacter agilis* are shown in figure 3. The coupling of nitrite at the level of high-potential cytochrome a_1 (E_m [pH 7.0] + 327 mV) is based on the following considerations:

1. Under anaerobic conditions in the presence of ATP, NAD^+, and NO_2^-, cytochrome a_1 is oxidized and cytochrome c is reduced.
2. The uncoupler CCCP inhibits the ATP-dependent nitrite-linked cytochrome c reduction as well as nitrite oxidation.
3. The reduction of cytochrome c by nitrite in intact cells and in cell-free extracts is inhibited by freezing and thawing or by metal chelators more than cytochrome a_1 reduction by nitrite.
4. Sonication of intact cells results in a decrease in cytochrome c reduction and an increase in cytochrome a_1 reduction.
5. Potentiometrically, it is difficult to envision nitrite (E_m [pH 7.0] + 420 mV) coupling at the level of cytochrome c (E_m [pH 7.0] + 274 mV). The reduction of cytochrome c by cytochrome a_1 requires an energy equivalence of approximately 7 kcal normally provided by the hydrolysis of one mole of ATP or $\Delta E_o'$ of 160 mV. This thermodynamic gap can be bridged

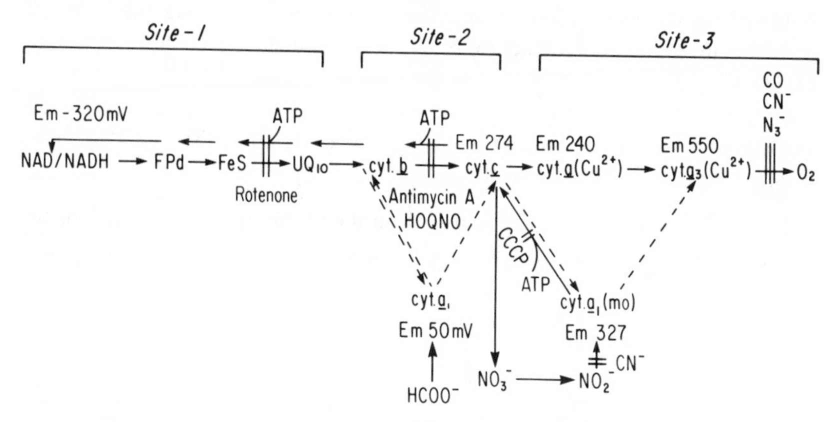

Fig. 3. Electron transport pathways and energy coupling sites in *Nitrobacter agilis*.

by the electron transport-linked proton electrochemical gradient across the energy-transducing membrane as a consequence of the operation of NO_2^-:O_2 oxidoreductase, which requires the reduction of cytochrome *c* and its subsequent oxidation by the cytochrome *c*:O_2 oxidoreductase. Presence of CCCP inhibits cytochrome *c* reduction by cytochrome a_1 because the uncoupler is capable of inhibiting both components (i.e., ΔpH and $\Delta\psi$) of the $\Delta\bar{\mu}_{H^+}$, thereby causing collapse of the membrane potential. Since ascorbate oxidation, unlike nitrite oxidation, does not involve unfavorable thermodynamics, it does not have to depend upon the generation of $\Delta\bar{\mu}_{H^+}$, and therefore the ascorbate:O_2 oxidoreductase system should be unaffected by CCCP, although ascorbate oxidation and nitrite oxidation are mediated by cytochromes *c, a,* and a_3, and ATP generation occurs at coupling site 3 only.

The entry of formate may occur either at the level of low potential cytochrome a_1 (E_m [pH 7.0] +50 mV), and/or cytochrome *b* followed by electron transport to O_2 with the mediation of cytochrome *c* and cytochrome oxidase components; the coupling of ATP synthesis involves the participation of energy conservation sites 2 and 3. Under anaerobic conditions, the formate:NO_3^- oxidoreductase system appears to be mediated by cytochromes *b, c,* and high-potential a_1 and may involve the participation of coupling site 2 only (fig. 1). The NADH:O_2 oxidoreductase catalyzes electron transfer that is mediated by electron transfer components similar to those of mitochondrial systems involving the participation of all 3 energy-coupling sites. The NADH:NO_3^- oxidoreductase complex contains energy-coupling sites 1 and 2 with the high-potential cytochrome a_1 as the immediate electron donor to nitrate (fig. 2).

PROPOSED ACTIVITIES OF THE RESPIRATORY NITRATE REDUCTASE IN *NITROBACTER*: A HYPOTHESIS

Figure 4 represents the proposed structural and functional interrelationship between the *Nitrobacter* NO_2^-:O_2 oxidoreductase, NO_2^-:NAD^+ oxidoreduc-

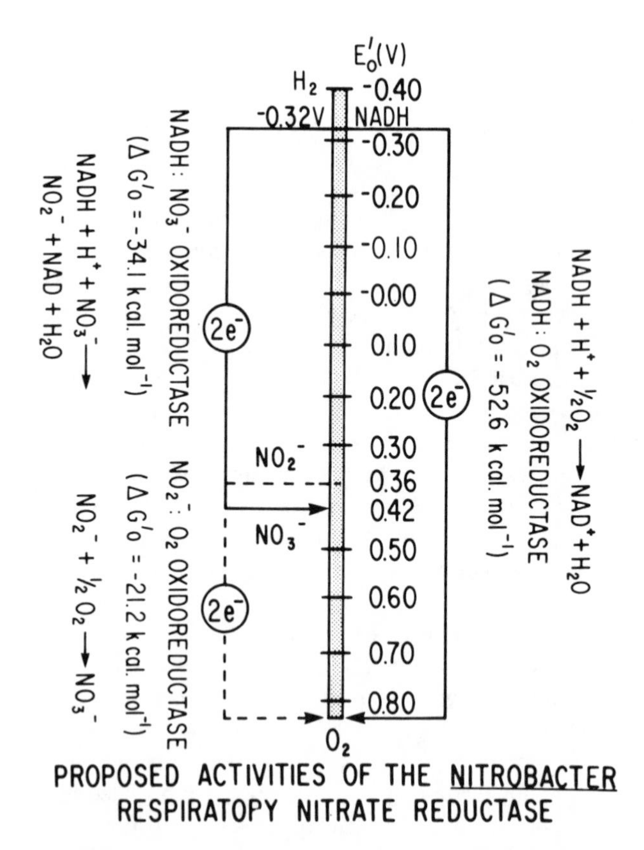

Fig. 4. Proposed activities of the *Nitrobacter* respiratory nitrate reductase.

tase, NADH:NO_3^- oxidoreductase, and NADH:O_2 oxidoreductase systems. It would appear that all of these structural and functional activities have probably originated in, or evolved from, the respiratory enzyme complex constituting NADH:NO_3^- oxidoreductase. This enzyme system, like the NO_2^-:O_2 oxidoreductase, is constitutive; and, therefore, it is always present under aerobic growth conditions that have no inhibitory effect on its biosynthesis and activity, unlike other microbial systems possessing the dissimilatory nitrate reductase. In this respect the system is also unique in *Nitrobacter* because the electron transport from NADH to NO_3^- is mediated by the flavoprotein system and cytochromes *b, c,* and a_1 involving energy-coupling sites 1 and 2. The normal operation of the enzyme complex uses NADH as the electron donor and nitrate as the electron acceptor. This reaction is accompanied with a release of 35 kcal mol^{-1} and is sufficient for the coupled synthesis of 2 mol of ATP. Under physiological growth conditions, the respiratory enzyme system involving NADH:NO_3^- oxidoreductase is reversed in *Nitrobacter* since electrons from nitrite have to reduce NAD^+ for the generation of CO_2-reducing power. It is obvious that no organism can obtain energy by reversing the respiratory nitrate reductase, a process that is thermodynamically unfeasible. In *Nitrobacter* this system appears to have been split, resulting in the evolution of NO_2^-:O_2 oxidoreductase containing new electron transport hemoproteins such as cytochromes a_1, a, and a_3 linking nitrite metabolism with molecular oxygen. The NO_2^-:O_2 oxidoreductase so evolved can greatly diminish the unfavorable thermodynamics by constituting a highly potent energy-generating cytochrome c:O_2 oxidoreductase system, which when linked with the NADH:NO_3^- oxidoreductase gives rise to an actively functional NADH:O_2 oxidoreductose system. Metabolically, the organism is still at a disadvantage to employ the NO_2^-:O_2 oxidoreductase system as the sole energy source because the growth of the bacterium on CO_2 must employ the operation of the highly endergonic NO_2^-:NAD^+ oxidoreductase system. The electron transfer reactions in *Nitrobacter* catalyzed by the inherent component enzyme systems such as NADH:NO_3^- oxidoreductase, NO_2^-:NAD^+ oxidoreductase, and NADH:O_2 oxidoreductase are mediated by the same or similar respiratory carriers, and exhibit similar inhibition patterns with respect to the respiratory chain inhibitors. All of the enzyme systems shown as components of the *Nitrobacter* respiratory nitrate reductase are associated with the energy transduction and transmission processes, and are localized in the same membrane particles. The hypothesis lends further support to our unpublished experimental observations that the enzymic activities are inseparable from the methods employed thus far. Thus, from the evolutionary as well as from the energy transduction and transmission aspects, *Nitrobacter* appears to be a unique bacterium among the chemoautotrophs with respect to their ecological, physiological, and biochemical characteristics.

ACKNOWLEDGMENTS

This investigation was supported in part by grants from the National Science Foundation and the Biomedical Research grant awarded to the University of Kentucky.

1. Abbreviations: $\Delta\bar{\mu}_{H^+}$, electrochemical gradient involving electrical potential difference and proton concentration difference constituting protonmotive force across the biological membrane; TTFA, thenoyltrifluoroacetone; CCCP, carbonyl cyanide m-chlorophenylhydrazone; 2,4-DNP, 2,4-dinitrophenol; PCP, pentachlorophenol; 2,4-DBP, 2,4-dibromophenol; NHQNO, 2-n-nonyl-4-hydroxyquinoline-N-oxide; HOQNO, 2-n-heptyl-4-hydroxyquinoline N-oxide; DCCD, N,N′-dicyclohexylcarbodiimide; PHMB, *p*-hydroxymercuribenzoate.

LITERATURE CITED

Aleem, M. I. H. 1959. The physiology and chemoautotrophic metabolism of *Nitrobacter agilis*. Ph.D. thesis, Cornell University.

Aleem, M. I. H. 1965. Path of carbon and assimilatory power in chemosynthetic bacteria. 1. *Nitrobacter agilis*. Biochim. Biophys. Acta 107:14–28.

Aleem, M. I. H. 1967. Energy conversions in the chemoautotroph *Nitrobacter agilis*. Bacteriol. Proc. 1967:112.

Aleem, M. I. H. 1968. Mechanism of oxidative phosphorylation in the chemoautotroph *Nitrobacter agilis*. Biochim. Biophys. Acta 162:338–47.

Aleem, M. I. H. 1977. Coupling of energy with electron transfer reactions in chemolithotrophic bacteria. *In* B. A. Haddock and W. A. Hamilton (eds.), Microbial energetics, pp. 351–81. Cambridge University Press, Cambridge.

Aleem, M. I. H., and M. Alexander. 1958. Cell-free nitrification by *Nitrobacter*. J. Bacteriol. 76:510–14.

Aleem, M. I. H., H. Lees, and D. J. D. Nicholas. 1963. Adenosine triphosphate-dependent reduction of nicotinamide adenine dinucleotide by ferro-cytochrome *c* in chemoautotrophic bacteria. Nature (London) 200:759–61.

Aleem, M. I. H., G. E. Hoch, and J. E. Varner. 1965. Water as the source of oxidizing and reducing power in bacterial chemosynthesis. Proc. Natl. Acad. Sci. USA 54:869–73.

Aleem, M. I. H., and D. L. Sewell. 1981. Mechanism of nitrite oxidation and oxidoreductase systems in *Nitrobacter agilis*. Curr. Microbiol. 5:267–72.

Boon, B., and H. Laudelout. 1962. Kinetics of nitrite oxidation by *Nitrobacter winogradskyi*. Biochem. J. 85:440–47.

Cobley, J. G. 1976. Energy-conserving reactions in phosphorylating electron-transport particles from *Nitrobacter winogradskyi*. Biochem. J. 156:481–92.

Dervertanian, D. V., and P. Forget. 1975. The bacterial nitrate reductase, EPR studies on the enzyme A of *Escherichia coli* K_{12}. Biochim. Biophys. Acta 379:74–80.

Dessers, A., C. Chiang, and H. Laudelout. 1971. Calorimetric determination of free energy efficiency of *Nitrobacter winogradskyi*. J. Gen. Microbiol. 64:71–76.

Fault, K. F., W. Wallace, and D. J. D. Nicholas. 1969. Nitrite oxidase and nitrate reductase in *Nitrobacter agilis*. Biochem. J. 113:449–55.

Garland, P. B. 1977. Energy transduction and transmission in microbial systems. *In* B. A. Haddock and W. A. Hamilton (eds.), Microbial energetics, pp. 1–21. Cambridge University Press, Cambridge.

Gibbs, M., and J. A. Schiff. 1960. Chemosynthesis: the energy relations of chemoautotrophic organisms. *In* F. C. Steward (ed.), Plant physiology, 1B:279–319. Academic Press, New York.

Ingledew, W. J., and P. Halling. 1976. Paramagnetic centers of the nitrite oxidizing bacterium, *Nitrobacter*. FEBS Lett. 67:90–93.

Kelly, D. P. 1971. Autotrophy: concepts of lithotrophic bacteria and their organic metabolism. Ann. Rev. Microbiol. 25:117–210.

Kiesow, L. 1963. Über die Reduktion von Diphosphopyridinonucleotid bei der Chemosynthese. Biochem. Zeits. 338:400–406.

Kiesow, L. 1964. On the assimilation of energy from inorganic sources in autotrophic forms of life. Proc. Natl. Acad. Sci. USA 52:980–88.

Kiesow, L. 1967. Energy-linked reactions in chemoautotrophic organisms. Curr. Top. Bioenerg. 2:195–233.

Lam, Y., and D. J. D. Nicholas. 1969. A nitrate reductase from *Micrococcus denitrificans*. Biochim. Biophys. Acta 178:225–34.

Lees, H., and J. R. Simpson. 1957. The biochemistry of the nitrifying organisms. 5. Nitrite oxidation by *Nitrobacter*. Biochem. J. 65:297–305.

Malavolta, E., C. C. Delwiche, and W. D. Burge. 1962. Formate oxidation by cell-free preparations from *Nitrobacter agilis*. Biochim. Biophys. Acta 57:347–51.

Nicholas, D. J. D. 1981. Studies with nitrifying bacteria *Nitrosomonas* and *Nitrobacter* using electrode and fluorescence techniques. *In* K. Mageyama, K. Nakamura, T. Oshima, and T. Ochida (eds.), Science and scientists, pp. 267–74. Japanese Scientific Society Press, Tokyo.

O'Kelley, J. C., and A. Nason. 1970. Particulate formate oxidase from *Nitrobacter agilis*. Biochim. Biophys. Acta 205:426–36.

O'Kelley, J. C., G. E. Becker, and A. Nason. 1970. Characterization of the particulate nitrite oxidase and its component activities from the chemoautotroph *Nitrobacter agilis*. Biochim. Biophys. Acta 205:409–25.

Ritchie, G. A. F., and D. J. D. Nicholas. 1972. Identification of the sources of nitrous oxide produced by oxidative and reductive processes in *Nitrosomonas europaea*. Biochem. J. 125:1181–91.

Sewell, D. L., and M. I. H. Aleem. 1969. Generation of reducing power in chemosynthesis. 5. The mechanism of pyridine nucleotide reduction by nitrite in the chemoautotroph *Nitrobacter agilis*. Biochim. Biophys. Acta 172:467–75.

Sewell, D. L., and M. I. H. Aleem. 1979. NADH-linked oxidative phosphorylation in *Nitrobacter agilis*. Curr. Microbiol. 2:235–37.

Sewell, D. L., M. I. H. Aleem, and D. F. Wilson. 1972. The oxidation-reduction potentials and rates of oxidation of the cytochromes of *Nitrobacter agilis*. Arch. Biochem. Biophys. 153:312–19.

Sewell, D. L. 1972. Aspects of the electron transport system of *Nitrobacter agilis*. Ph.D. thesis, University of Kentucky, Lexington.

Silver, W. S. 1960. Exogenous respiration in *Nitrobacter*. Nature (London) 185:555–56.

Straat, P. A., and A. Nason. 1965. Characterization of a nitrate reductase from the chemoautotroph *Nitrobacter agilis*. J. Biol. Chem. 240:1412–26.

Sundermeyer, H., and E. Bock. 1981. Characterization of nitrite oxidizing system in Nitrobacter. *In* H. Bothe and A. Trebst (eds.), Biology of inorganic nitrogen and sulfur, pp. 317–24. Springer-Verlag, Berlin.

Tsien, H. C., and H. Laudelout. 1971. Changes in cytochrome content and turnover number during growth of *Nitrobacter*. Arch. Mikrobiol. 75:266–68.

Van Gool, A., and H. Laudelot. 1966. The mechanism of nitrite oxidation by *Nitrobacter winogradskyi*. Biochim. Biophys. Acta 113:41–50.

Van Gool, A., and H. Laudelot. 1967. Spectrophotometric and kinetic study of nitrite and formate oxidation in *Nitrobacter winogradskyi*. J. Bacteriol.93:215–20.

Van Gool, A. P., P. P. Tobback, and I. Fischer. 1971. Autotrophic growth and synthesis of reserve polymers in *Nitrobacter winogradskyi*. Arch. Mikrobiol. 76:252–64.

Wallace, W., and D. J. D. Nicholas. 1969. The biochemistry of nitrifying microorganisms. Biol. Rev. 44:359–91.

Weinbach, E. C., and J. Garbus. 1969. Mechanism of action of reagents that uncouple oxidative phosphorylation. Nature (London) 221:1016–18.

YOUNG PARK AND GEORGE HEGEMAN

The Oxidation of Carbon Monoxide by Bacteria

12

Carbon monoxide is an atmospheric trace gas found in concentrations (0.1–0.3 ppm) of the order of those of N_2O, H_2, and CH_4 (Seiler and Schmidt, 1975). The ambient concentration of CO in the northern hemisphere is three times that found in the less-industrialized southern hemisphere, which testifies to the major role of incomplete combustion of fossil fuels, principally by automobile motors, as a source of CO. Natural biological processes contribute to the atmospheric CO, for example, during oxidation of methane and in the decomposition of porphyrins (Seiler and Schmidt, 1975). Local soil and air concentrations can reach appreciable but transient levels, but the mean residence time for CO in the atmosphere is relatively short (0.3–0.4 year), and the concentration of CO is not rising with time. These latter facts indicate that worldwide consumption of CO exceeds production (Bortner et al., 1972; Seiler and Schmidt, 1975; Warneck, 1975; Bartholomew and Alexander, 1981).

Oxidation of CO by ·OH radicals in the troposphere and stratosphere is a major CO sink (table 1), but oxidation in soils and surface waters constitutes a sink nearly as great as this photophysical process. Oxidation in soils is principally a microbiological process (Seiler and Schmidt, 1975; Conrad and Seiler, 1982). Bacteria are the principal agents of CO oxidation, and many types that may participate in the process have been isolated (Kim and Hegeman, 1983). Some of these bacteria evidently use CO as an energy and, possibly, as a carbon source following oxidation of CO to CO_2 (utilitarian CO oxidizers), whereas others (e.g., the methane-oxidizing bacteria) perform a gratuitous oxidation of CO by means of enzymes presumably evolved under selection to catalyze other processes and which yield the cell neither carbon nor energy. Both aerobes and anaerobes of several different major bacterial groups carry out both gratuitous and utilitarian CO oxidation.

TABLE 1

THE CO CYCLE: A BUDGET IN OUTLINE

Process	Nature (CO Source or Sink)	Estimate of Amount of CO involved ($\times 10^{14}$g yr^{-1})[a]
Human activities	Source	6.4
CH_4 oxidation in the upper atmosphere	Source	4.0
All other sources (plants, forest fires, oceans, etc.)	Source	3.5
Metabolism in soils (microbes)	Sink	4.0
Oxidation in the upper atmosphere	Sink	7.1

SOURCE: Adapted from Seiler and Schmidt (1975).

[a]Although estimates given here indicate that CO produced exceeds CO consumed, this cannot be true at the surface of the earth since the CO concentration of the atmosphere is not rising with time (Bortner et al., 1972; Seiler and Schmidt, 1975).

The oxidation of CO by anaerobic bacteria has recently been reviewed (Zavarzin and Nozhevnikova, 1977; Kim and Hegeman, 1983). The present paper describes the presence of plasmids in various carboxydobacteria.

Our laboratory has been interested in the CO oxidizing process of the carboxydobacteria during the past several years. We have focused on the electron transport processes (Kim and Hegeman, 1981a), CO oxidizing enzyme (Kim and Hegeman, 1981b), and the genetic basis for CO oxidation in *Pseudomonas carboxydohydrogena* (formerly *Seliberia carboxydohydrogena*; Meyer and Schlegel, 1978) and strains of *Azomonas* and *Azotobacter* isolated in our laboratory (Kim et al., 1982; S. Kirkconnell and G. Hegeman, unpublished results).

P. carboxydohydrogena has been shown to have a branched cytochrome system that contains cytochrome *o*, a terminal oxidase usually relatively resistant to CO inhibition (Kim and Hegeman, 1981a). UV destruction and restoration experiments indicated that the electrons from CO oxidation feed into the electron transport chain at the level of quinone. It should be noted that reverse electron flow would be required to provide reduced pyridine nucleotides for biosynthesis and for CO_2 fixation, a process which occurs in all aerobic carboxydobacteria so far examined by means of the reductive pentose cycle (Zavarzin and Nozhevnikova, 1977).

Our interest has been arrested by the apparent similarity in the cross-reactivity observed among the CO dehydrogenases (COD) from independently isolated carboxydobacteria. A gene for hydrogenase in some hydrogen bacteria is evidently encoded by a plasmid (e.g., Pootjes, 1977). The basis for similarity among different CODHs may be that the structural gene for the enzyme responsible for CO oxidation is encoded by a broadly transferrable genetic element. What follows is a description of our search for such an element.

MATERIALS AND METHODS

P. carboxydohydrogena DSM 1083 (Meyer et al., 1980; obtained as *Seliberia carboxydohydrogena* Z-1062 from Dr. J. Schmidt, University of Arizona, Tempe) and *P. carboxydovorans* DSM 1227 (gift of Dr. O. Meyer, Institut für Mikrobiologie der Universität, Göttingen, FRG) were grown on CO in liquid mineral medium under conditions described previously (Kim and Hegeman, 1981b). *Azotobacter* sp. 1 and *Azomonas* spp. 1 and 2 (S. Kirkconnell and G. Hegeman, unpublished) and unidentified strain 3B (Park, 1982) were grown on the same medium solidified with 1.5% agar (Difco) in an atmosphere of 30% (v/v) CO in air.

Soluble cell extracts were prepared from cells by the method described previously (Kim and Hegeman, 1981b) up to, but not including, the stage of protamine sulfate treatment.

DNA isolation for detection of small plasmids was performed by the alkaline extraction procedure (Birnboim and Doly, 1979) with some modifications (Park, 1982). Large plasmid DNA was isolated by the Eckhardt (1978) procedure with some modification (Pootjes, 1977; Park, 1982).

Agarose gel electrophoresis was performed in 0.7 to 1.2% agarose (HGT, FMC Corp.) dissolved in Tris-phosphate buffer (36 mM Tris [hydroxymethyl] aminomethane base, 30 mM NaH_2PO_4 and 1 mM EDTA) in either vertical or horizontal apparatus. Following electrophoresis the gels were stained with ethidium bromide and photographed under UV light to reveal the presence of DNA-ethidium complexes by fluorescence (Park, 1982).

RESULTS AND DISCUSSION

Antiserum prepared against purified *P. carboxydohydrogena* CODH reacted to some degree with extracts of each of the five CO-grown carboxydobacteria examined, but not with extracts from heterotrophically grown cells. Pre-immune serum from the animal used to make the antibodies did not react. *Azotobacter* sp. 1 and *Azomonas* sp. 1 reacted more weakly than other strains, unidentified strain 3B less weakly, and strong cross-reaction with spurring was observed with *P. carboxydovorans* extract (fig. 1). This level of cross-reaction implies as much as 70% homology of amino acid sequence among the CODHs of independently isolated (and evidently different) carboxydobacteria (Wilson et al., 1977). These findings extend and confirm earlier work (Kim et al., 1982).

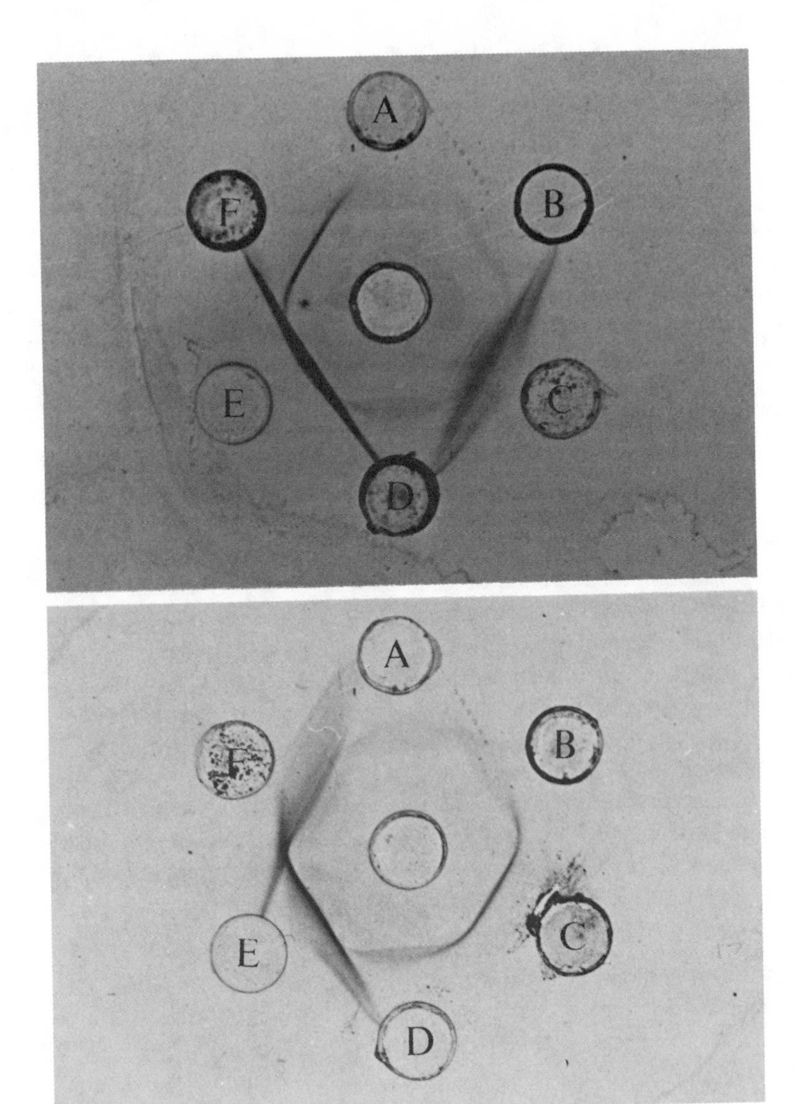

Fig. 1a (top) and 1b (bottom). Stained double immunodiffusion patterns using crude extracts of CO-grown aerobic carboxydobacteria as antigens. Assays were performed in 1.2% agarose gel as described by Kim et al. (1982). Centerwells contained antiserum prepared against purified CODH from *P. carboxydohydrogena.* 1a. Wells designated by letter contain extracts of *Azotobacter* sp. 1 (A), *Azomonas* sp. 1 (D), *Azomonas* sp. 2 (F), *P. carboxydohydrogena* (C), *P. carboxydovorans* (E), and strain 3B (B). 1b. Wells contained *Azotobacter* sp. 1 (A), *Azomonas* sp. 1 (B), *Azomonas* sp. 2 (C), strain 3B (D), *P. carboxydovorans* (E), and *P. carboxydohydrogena* (F) extracts. Pre-immune serum and extracts of cells grown on succinate or ribose did not yield visible precipitates.

Initial attempts to demonstrate plasmids from the five strains examined were successful in all strains except *P. carboxydohydrogena*. The first indication of the presence of plasmid DNA was attained at the ethidium bromide-cesium chloride centrifugation step. Subsequent agarose gel electrophoresis (fig. 2) confirmed these results and permitted estimation of the molecular masses and number of plasmid types in each strain (Meyers et al., 1976). Five plasmids of known size were used as molecular mass references (fig. 2). A plot of migration distance vs. log molecular mass for the covalently closed circular form yielded a linear relationship among the reference plasmids, and it was used to estimate the size of the single size class of plasmid found in the carboxydobacteria (Park, 1982). Within error of measurement, all four were the same size, ca. 3.5 kilobase pairs (kb). This striking similarity prompted a

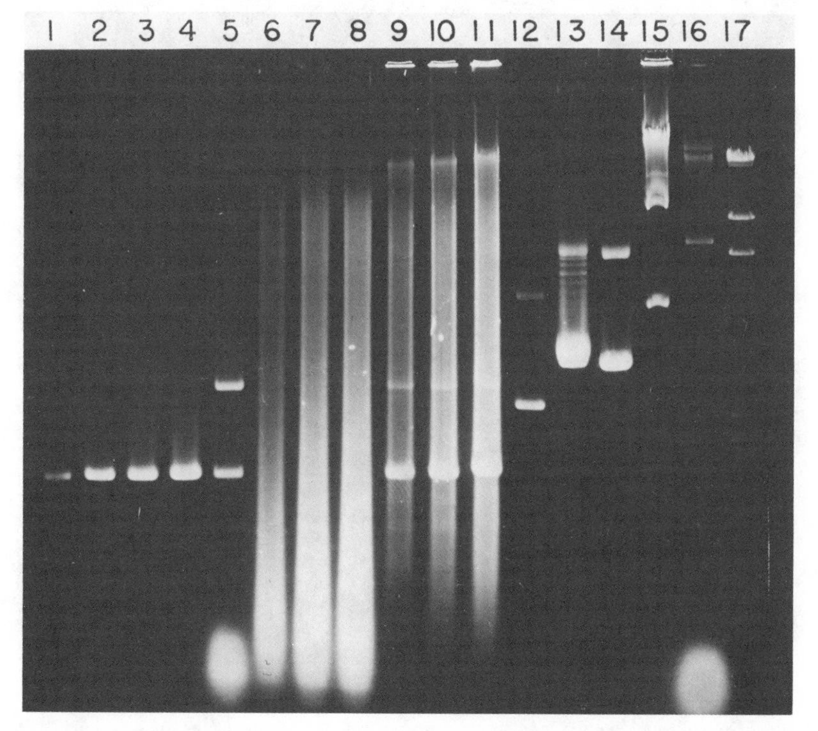

Fig. 2. Agarose gel electropherogram stained with ethidium bromide and photographed by fluorescence of this dye as a stain for DNA. Sample wells contained small plasmid preparations from *Azotobacter* sp. 1 (well 1), *Azomonas* sp. 2 (2–4), *Azomonas* sp. 1 (5), *P. carboxydohydrogena* (6–8), *P. carboxydovorans* (9–11), pBR322 (4.36 kb) (12), pKPG 16 (5.65 kb) (13), pSM 1 (5.4 kb) (14), pHuB4 (7.6 kb) (15), pRK248 (9.6 kb) (16), and λ treated with *Hind*III (17).

further test of the identity of this small plasmid from the four strains. Plasmids were digested with various restriction enzymes and the resulting fragments were separated by agarose gel electrophoresis (fig. 3).

A 3.5 kb length of DNA could theoretically encode a polypeptide of almost 144,000 molecular mass. However, some coding capacity is unavailable owing to the need to encode the origin of replication and possibly regulatory elements associated with the putative plasmid-borne CODH gene. If the origin of the 3.5 kb plasmid is about the same size as the small plasmid cloning vector pBR322 (4.36 kb), i.e., about 580 base pairs, then the amount of DNA left for encoding CODH protein is reduced to an amount sufficient to encode a

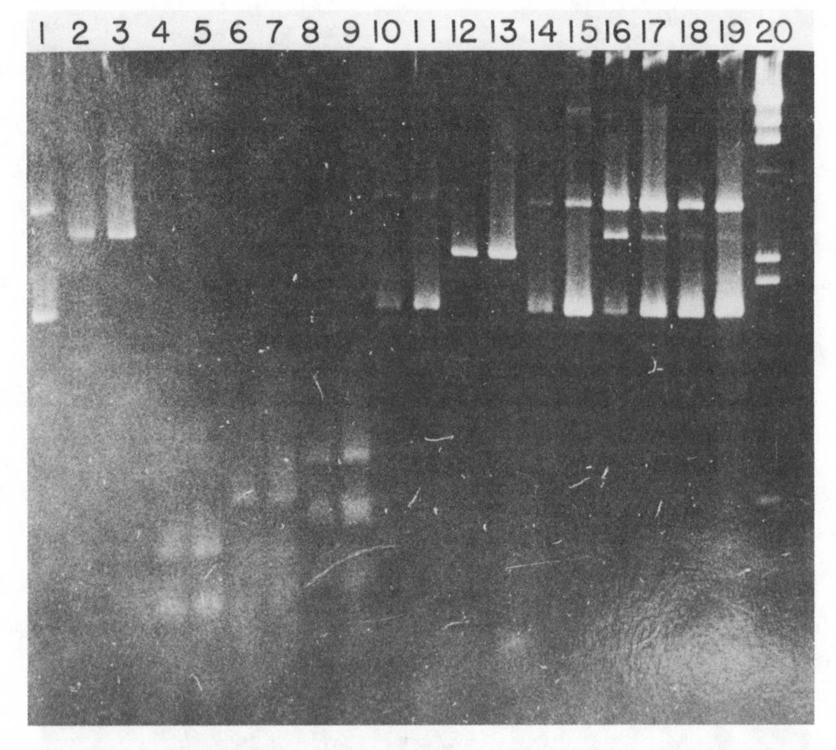

Fig. 3. Agarose electropherogram of endonuclease digestion patterns of fragments of the small plasmid DNAs from *Azotobacter* sp. 1 and *Azomonas* sp. 2. Lane 1 contains small plasmid DNA from *Azomonas* sp. 2 without digestion. Odd-numbered lanes 3–19 contain digested DNA samples from *Azotobacter* sp. 1 and even-numbered lanes 2–18 from *Azomonas* sp. 2. Restriction enzymes used were: *Alu* (2,3), *Hae*III (4,5), *Hha*I (6,7), *Taq*1 (8,9), *Eco*RI (10,11), *Hind*II (12,13), *Hind*III (14,15), *Bgl*II (16,17), and *Pst*I (18,19). Lane 20 contained lambda DNA treated with *Hind*III.

protein of about 200,000 molecular mass. This estimate ignores regulatory elements. Cypionka and colleagues (1980) reported that the native CODHs of all the aerobic carboxydobacteria tested by sucrose density gradient centrifugation shared a common molecular mass of 230,000 except for a 170,000 M_r CODH from *Achromobacter carboxydus*. Meyer and Schlegel (1978) previously showed the CODH from *P. carboxydovorans* to be composed of two equivalent subunits of 107,000 molecular mass. The molecular mass of the *P. carboxydohydrogena* CODH apoenzyme was found to be 400,000 by Kim and Hegeman (1981b). This CODH was a trimer whose monomeric peptide complement calculates to a molecular mass of 133,000. Given these measurements of the molecular masses of CODHs from different aerobic carboxydobacteria, there seems barely enough DNA to encode the CODHs on a 3.5 kb plasmid.

The small size of the common 3.5 kb plasmid provoked a search for larger plasmids by use of the Eckhardt (1978) procedure. Large plasmids of a similar size were revealed in *Azomonas* sp. 2, *Azotobacter* sp. 1, and unidentified strain 3B, but their molecular masses were not determined. These large plasmids could be detected only in cells grown with CO. Experiments presently under way using curing and cloning techniques should reveal more directly the relationship of plasmid(s) to CODH synthesis and function in the aerobic carboxydobacteria.

ACKNOWLEDGEMENTS

Work in the authors' laboratory is supported by Research Grant No. PCM 78-12482 from the National Science Foundation. We thank Dr. O. Meyer and Dr. H. G. Wood for sharing unpublished findings with us and B. Polisky's laboratory for hospitality and suggestions.

LITERATURE CITED

Anderson, K., R. C. Tait, and W. R. King. 1981. Plasmids required for the utilization of molecular hydrogen by *Alcaligenes eutrophus*. Arch. Microbiol. 129:384–90.

Bartholomew, G. W., and M. Alexander. 1981. Soil as a sink for atmospheric carbon monoxide. Science 212:1389–91.

Birnboim, H. C., and J. Doly. 1979. A rapid alkaline extraction procedure for screening recombinant plasmid DNA. Nucleic Acid Res. 7:1513–23.

Bortner, M. H., R. M. Kummler, and L. S. Jaffe. 1972. A review of carbon monoxide sources, sinks, and concentrations in the Earth's atmosphere. NASA Report CR-2081, Washington, D.C.

Conrad, R., and W. Seiler. 1982. Utilization of traces of carbon monoxide by aerobic, oligotrophic microorganisms in oceans, lakes, and soil. Arch. Microbiol. 132:41–46.

Cypionka, H., O. Meyer, and H. G. Schlegel. 1980. Physiological characteristics of various types of carboxydobacteria. Arch. Microbiol. 127:301–7.

Eckhardt, T. 1978. A rapid method for the identification of plasmid desoxyribonucleic acid in bacteria. Plasmid 1:584–88.

Kim, Y. M., and G. D. Hegeman. 1981a. Electron transport system of an aerobic carbon monoxide-oxidizing bacterium. J. Bacteriol. 148:991–94.

Kim, Y. M., and G. D. Hegeman. 1981b. Purification and some properties of carbon monoxide dehydrogenase from *Pseudomonas carboxydohydrogena*. J. Bacteriol.148:904–11.

Kim, Y. M., S. Kirkconnell, and G. D. Hegeman. 1982. Immunological relationships among carbon monoxide dehydrogenases of carboxydobacteria. FEMS Microbiol. Lett. 13:219–23.

Kim, Y. M., and G. D. Hegeman. 1983. Oxidation of carbon monoxide by bacteria. Int. Rev. Cytol. 81:1–32.

Liebl, K. H., and W. Seiler. 1975. CO and H_2 destruction at the soil surface. *In* H. G. Schlegel, G. Gottschalk, and N. Pfennig (eds.), Microbial production and utilization of gases (H_2, CH_4, CO), pp. 215–19. E. Goltze KG, Göttingen.

Meyer, O., and H. G. Schlegel. 1978. Reisolation of the carbon monoxide utilizing hydrogen bacterium *Pseudomonas carboxydovorans* (Kistner) comb. nov. Arch. Microbiol. 118:35–43.

Meyer, O., J. Lalucat, and H. G. Schlegel. 1980. *Pseudomonas carboxydohydrogena* (Sanjieva and Zavarzin) comb. nov., a monotrichous non-budding, strictly aerobic, carbon monoxide-utilizing hydrogen bacterium previously assigned to *Seliberia*. Int. J. System. Bacteriol. 30:189–95.

Meyer, O. 1982. Chemical and spectral properties of carbon monoxide:methylene blue oxidoreductase. J. Biol. Chem. 257:1333–41.

Meyers, J. A., D. Sanchez, L. P. Elwell, and S. Falkow. 1976. Simple agarose gel electrophoresis method for the identification and characterization of plasmid desoxyribonucleic acid. J. Bacteriol. 127:1529–37.

Park, Y. 1982. Evolution of carbon monoxide dehydrogenases among different carboxydobacteria. M.A. thesis, Indiana University, Bloomington.

Pootjes, C. F. 1977. Evidence for a plasmid coding for the ability to use hydrogen gas by *Pseudomonas facilis*. Biochem. Biophys. Res. Commun. 76:1002–6.

Reh, M., and H. G. Schlegel. 1975. Chemolithotrophie als eine übertragbare, autonome Eigenschaft von *Nocardia opaca* lb. Nachr. Akad. Wiss. Göttingen, Math. Phys. K1.II. 12:207–16.

Seiler, W., and U. Schmidt. 1975. The role of microbes in the cycle of atmospheric trace gases, especially of hydrogen and carbon monoxide. *In* H. G. Schlegel, G. Gottschalk, and N. Pfennig (eds.), Microbial production and utilization of gases (H_2, CH_4, CO), pp. 35–45. E. Goltze KG, Göttingen.

Warneck, P. 1975. The role of chemical reactions in the cycles of atmospheric trace gases, especially CH_4. *In* H. G. Schlegel, G. Gottschalk, and N. Pfennig (eds.), Microbial production and utilization of gases (H_2, CH_4, CO), pp. 53–62. E. Goltze KG, Göttingen.

Wilson, A. C., S. S. Carlson, and T. T. White. 1977. Biochemical evolution. Ann. Rev. Biochem. 46:573–639.

Zavarzin, G. A., and A. N. Nozhevnikova. 1977. Aerobic carboxydobacteria. Microb. Ecol. 3:305–26.

MARY E. LIDSTROM-O'CONNOR

Genetics of Methylotrophy and Methanotrophy

13

The ability to grow on reduced one-carbon compounds is generally known as methylotrophy, and the ability to utilize methane as a sole carbon and energy source has been termed methanotrophy. Methylotrophs and methanotrophs possess unique metabolic capabilities related to growth on C-1 compounds, including the dissimilatory pathways that convert reduced C-1 compounds to CO_2 and the assimilatory pathways for incorporating C-1 units into cell material (fig. 1). The physiology of C-1 utilizers has been extensively reviewed in several recent papers (Colby et al., 1979; Hanson, 1980; Higgins et al., 1981). The unique attributes of these microorganisms have significance in several areas. The ability of methanotrophs to convert methane into cell

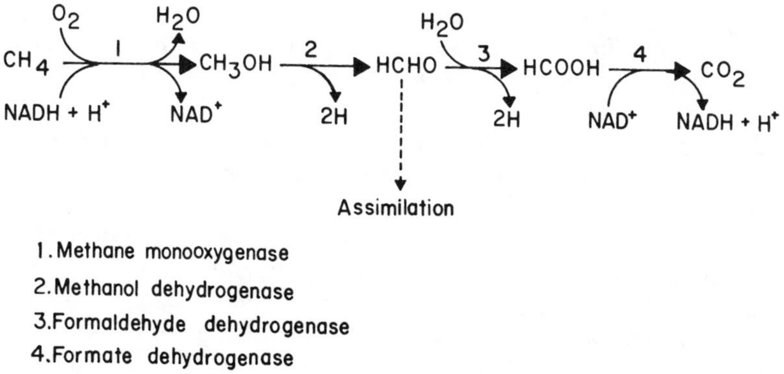

O_2 H_2O H_2O
CH_4 —1→ CH_3OH —2→ HCHO —3→ HCOOH —4→ CO_2
NADH + H^+ NAD^+ 2H 2H NAD^+ NADH + H^+
Assimilation

1. Methane monooxygenase
2. Methanol dehydrogenase
3. Formaldehyde dehydrogenase
4. Formate dehydrogenase

Fig. 1. A summary of the physiology of C-1 utilizers. Methane or methanol is oxidized to CO_2, while carbon is assimilated either at the level of formaldehyde, by the serine or ribulose monophosphate pathways, or at the level of CO_2 by the Calvin-Benson cycle.

material and CO_2 plays an important role in the carbon cycle of aquatic ecosystems (Hanson, 1980; Rudd and Taylor, 1980). It has been estimated that in the global carbon cycle 20–50% of all the methane produced is oxidized microbiologically, and the remainder is oxidized photochemically in the upper atmosphere (Ehhalt and Schmidt, 1978; Reeburgh, 1982).

In addition to their role in nature, C-1 utilizers also have considerable biotechnological potential. The unique oxidative enzymes these organisms contain may be useful in biocatalysis applications (Higgins et al., 1980; Hou et al., 1980), and the ability of these microbes to convert C-1 compounds into multicarbon compounds has attracted attention from the chemical industry.

It will not be possible to fully understand or to utilize the various activities of methylotrophs and methanotrophs without the ability to generally manipulate these microorganisms. Genetic studies will enhance the basic understanding of biochemistry and regulation of C-1 metabolism. In addition, genetic manipulation is essential for the production of strains with altered or novel properties. Until a few years ago, genetic studies in these organisms were hampered by the lack of suitable genetic systems. However, with the advent of recombinant DNA techniques and conjugative plasmid systems, it is now possible to carry out genetics in both methylotrophs and methanotrophs. Several laboratories, including my own, are currently using this approach to study various aspects of genetics in these microorganisms. This paper will review the progress made in this area and highlight some recent work from my own laboratory.

GENETIC TECHNIQUES

In order to carry out genetic studies in microorganisms, it is necessary to have capabilities for mutant isolation and for gene transfer. In addition, cloning vectors are necessary for genetic engineering. All three capabilities are now possible in C-1 utilizers. Mutants have been difficult to isolate in methanotrophs for reasons that are not entirely clear. Some mutagens, including ethylmethanesulfonate and ultraviolet light, have been ineffective in increasing the spontaneous mutation frequency in methanotrophs (Williams et al., 1977). Both auxotrophic and methanol mutants have been isolated in the facultative methanotroph *Methylobacterium organophilum* XX using nitrosoguanidine and ultraviolet light (O'Connor et al., 1977; O'Connor and Hanson, 1977), but only during growth on non-methane substrates. Mutants of some methylotrophs have been readily isolated using nitrosoguanidine and ultraviolet light (Quayle, 1972; Bamforth and O'Connor, 1979; Warner, et al., 1980; Holloway, 1981; Marison and Attwood, 1982). Transposons have now been used to isolate auxotrophic and methanol mutants in methylotrophs

(D. Anderson and M. Lidstrom-O'Connor, unpublished), and attempts to isolate such mutants in methanotrophs are under way in several laboratories. It seems clear that the isolation of mutants remains an important priority in approaching the genetics of both methylotrophy and methanotrophy in any strain.

The lack of gene transfer systems has been a major hurdle in attempts to study genetics in these bacteria. No transducing phages have been identified in methylotrophs or methanotrophs, so transduction has not been possible. Genetic transformation with linear DNA has been described for two methane oxidizers, *Methylococcus capsulatus* (Williams and Bainbridge, 1971) and *M. organophilum* XX (O'Connor et al., 1977), but large amounts of DNA were required in both cases. Moreover, the frequency of transformation of *M. organophilum* XX was low. Low-frequency transformation of plasmid DNA has also been reported for this strain (Haber et al., 1983). Attempts to transform other methylotrophs have so far been unsuccessful (M. Lidstrom-O'Connor, unpublished; R. S. Hanson, personal communication).

More recently, wide-host range conjugative plasmids have been used to transfer both chromosomal and plasmid DNA in methylotrophs and methanotrophs (Jeyaseelan and Guest, 1979; Gautier and Bonewald, 1980; Warner et al., 1980; Windass et al., 1980; Holloway, 1981; O'Connor, 1981b; Hennam et al., 1982; Haber et al., 1983). The plasmid R68.45 has been used to mobilize chromosomal genes (Holloway, 1981; Allen and Hanson, 1982) and nonconjugative plasmids have been mobilized by the plasmids RP4 (Gautier and Bonewald, 1980), R64drdii (Windass et al., 1980), R751 (Hennam et al., 1982), and pRK2013 (M. Lidstrom-O'Connor, unpublished). Plasmid-mediated conjugation systems seem to hold the most promise for gene transfer in these organisms at this time.

In order to carry out genetic engineering in bacteria, it is necessary to identify a cloning vector that will replicate in the prospective host. Two general classes of cloning vectors have been successfully used in methylotrophs and methanotrophs: Inc Q plasmid derivatives, such as R1162 (Gautier and Bonewald, 1980), pTB70 (Windass et al., 1980), and pGSS15 (Hennam et al., 1982) and Inc P1 plasmid derivatives, such as pRK290 (Ditta et al., 1980; M. Lidstrom-O'Connor, unpublished). Each of these cloning vestors is mobilizable by conjugative plasmids.

GENETICS OF METHYLOTROPHY

Very little is currently known concerning the genetics of methylotrophic functions. Table 1 lists possible C-1 specific functions. In some cases the

TABLE 1

POSSIBLE METHYLOTROPHIC FUNCTIONS[a]

Function	Methylotrophic Type[b]
Oxidation pathway	
Methanol dehydrogenase	SP, RuMP, C-B
PQQ (cofactor)[c]	SP, RuMP, C-B
Cytochrome *c*-co	SP, RuMP, C-B
Formaldehyde dehydrogenase	Variable
NAD-linked	
GSH-dependent	
GSH-independent	
Dye-linked	
DCPIP	
PMS	
Formate dehydrogenase	SP, RuMP, C-B
6-phosphogluconate dehydrogenase	Some RuMP
Assimilation pathway	
Hydroxypyruvate reductase	SP
Serine-glyoxylate aminotransferase	SP
Glycerate kinase	SP
Malate thiokinase	SP
Malyl CoA lyase	SP
Serine transhydroxymethylase (C-1 specific)	SP
PEP carboxylase (C-1 specific)	SP
Glyoxylate regenerating enzymes (unknown)	SP
Hexulose phosphate synthase	RuMP
Phosphohexuloisomerase	RuMP
Regulatory genes (unknown)	?

[a]References: Colby et al., 1979; Higgins et al., 1981.

[b]Organisms containing as their major carbon assimilatory pathway either the serine pathway (SP), the ribulose monophosphate pathway (RuMP), or the Calvin-Benson cycle (C-B).

[c]PQQ, polyquinoline quinone; GSH, glutathione; DCPIP, dichlorophenol indolephenol; PMS, phenazine methosulfate.

proposed genes have not been identified as C-1 specific, and their position in this category is merely speculative. In several cases, as for example in the synthesis of the PQQ cofactor, the number of genes for each function is unknown. A list of this sort is useful, however, since it provides a general overview of the number and types of genes that may be involved in genetic studies of C-1 functions. In contrast, table 2 summarizes the known C-1 mutants. Clearly, the task of isolating new C-1 mutants is an important one. However, using the few available mutants and current genetic systems, several laboratories are now beginning to obtain data concerning the genetic organization and regulation of a few of these genes.

One study has been carried out involving methylotrophic genes in *M. organophilum* XX, using the linear DNA transformation system (O'Connor et

TABLE 2

MUTANTS DEFICIENT IN METHYLOTROPHIC FUNCTIONS

Organism	Strain	Activities Missing[a]	Reference[b]
Pseudomonas AMl	M15A	MeDH	1
	20B-L	HPred	1
	PCT57	MaCoA Ly	2
	PCT76	cyt *c*-co	3
	82GT	STHM, SGAT	4
Hyphomicrobium X	1	MeDH	5
Pseudomonas aminovorans	M5	HPred, SGAT, PEPcarb, ForDH, FormaldDH, MGDH, GGMS	6
Methylobacterium organophilum XX	4M	HPred, SCAT, STHM, MeDH, GlycK, MaColy, PEPcarb	7
	8A	HPred, SGAT, MeDH, PEPcarb	7
	17A	HPred, SGAT, STHM, MeDH, GlycK, MaCoLy, PEPcarb, cyt *c*-co	7
	17M	MeDH, PEPcarb	7
	8Z	GlycK	7
	22C	cyt *c*-co	7
	24C	cyt *c*-co	7
	7A	HPred, SGAT, STHM, MeDH, PEPcarb	7

[a]HPred, hydroxypyruvate reductase; SGAT, serine-glyoxylate aminotransferase; STHM, serine transhydroxymethylase; MeDH, methanol dehydrogenase; GlycK, glycerate kinase; MaColy, malyl CoA lyase; PEPcarb, PEP carboxylase; cyt *c*-co, cytochrome *c*-co; ForDH, formate dehydrogenase; FormalDH, formaldehyde dehydrogenase; MGDH, N-methylglutamate dehydrogenase; GGMS, γ-glutamylmethylamide synthetase.

[b]References: 1, Heptinstall and Quayle, 1970; 2, Salem et al, 1974; 3, Anthony, 1975; 4, Harder and Quayle, 1971; 5, Marison and Attwood, 1982; 6, Bamforth and O'Connor, 1979; 7, O'Connor, 1981a.

al., 1977). Studies of regulation demonstrated that several of the C-1 enzymes were coordinately induced by methanol, suggesting the possibility of a C-1 operon (O'Connor and Hanson, 1977). Eight mutants were isolated that were defective in the ability to grow on methanol, and linkage analysis suggested that several of the affected genes were clustered on the chromosome. These data were used to propose a model for the genetic organization of dissimilatory and assimilatory methylotrophic functions in which the methanol dehydrogenase gene and several genes of the assimilation pathway (the serine pathway) were suggested to be in an operon, while the genes for the C-1–specific cytochrome *c*, the formate dehydrogenase, and a regulatory gene were not linked to each other or to the other C-1 genes (O'Connor and Hanson, 1978; O'Connor, 1981a). Unfortunately, it was not possible to further characterize this system without more mutants or a high-frequency gene transfer system.

More recently, the conjugative plasmid R68.45 has been used to mobilize chromosomal DNA and complement a methanol dehydrogenase mutant of *M. organophilum* XX (Allen and Hanson, 1982). Efforts are now under way to determine the location of the methanol dehydrogenase and other C-1 genes using R68.45 (R. S. Hanson, personal communication).

In *Pseudomonas* AM1, a facultative methylotroph, induction data have suggested that methanol dehydrogenase activity is regulated separately from other C-1 functions (McNerney and O'Connor, 1980). A piece of *Pseudomonas* AM1 DNA containing sequences necessary for methanol dehydrogenase activity has been cloned by Gautier and Bonewald (1980) using the Inc Q vector R1162. This piece of cloned DNA complemented the mutant M15A, which lacks methanol dehydrogenase activity. It is not known whether the cloned DNA contained any of the other C-1 genes, or whether it contained any methanol-inducible promotors.

In our laboratory we have been studying an autotrophic methanol-utilizer, *Xanthobacter* H4.14. This bacterium is capable of growing heterotrophically on a wide variety of organic substrates or autotrophically on H_2/CO_2, formate, or methanol. *Xanthobacter* H4.14 appears to have a methanol dehydrogenase, a formaldehyde dehydrogenase, and an NAD-linked formate dehydrogenase, all of which are induced during growth on methanol (C. Weaver and M. Lidstrom-O'Connor, unpublished). We have successfully used pJB4J1, the IncP-mu "suicide" plasmid containing the kanamycin transposon, Tn5, (Beringer et al., 1978) to generate transposon mutants in *Xanthobacter* H4.14. Several amino acid auxotrophs and mutants deficient in their ability to grow on methanol have been isolated (D. Anderson and M. Lidstrom-O'Connor, unpublished). The methanol mutants are all capable of growing on H_2/CO_2, indicating that their carbon assimilatory pathways are intact. The abilities of whole cells of these mutants to oxidize various C-1 compounds as determined using a Rank oxygen electrode are shown in table 3. It should be noted that we have evidence suggesting that the methanol dehydrogenase in this strain is capable of oxidizing formaldehyde at high rates, as in other methylotrophs. Therefore, the whole cell oxidation of formaldehyde in some cases represents a combination of activities of both formaldehyde and methanol dehydrogenases. Transposon insertion mutations tend to be polar within an operon, and, therefore, the preliminary data presented in table 3 are consistent with a model suggesting that some of the C-1 oxidation functions may be present in an operon in *Xanthobacter* H4.14. Further studies are currently under way to test this model, which include complementation of the mutants with a library of chromosomal DNA cloned into an IncP vector.

TABLE 3

THE ABILITY OF WHOLE CELLS OF METHANOL MUTANTS IN XANTHOBACTER H4.14 TO OXIDIZE C-1 COMPOUNDS

Strain	Oxidation Capabilities[a]		
	Methanol	Formaldehyde	Formate
Wild type	+ + +	+ + +	+ + +
M1	+ + +	+ + +[b]	−
M2	+ + +	+ + +	+ + +
M3	−	−	−
M4	+	+ + +	+ + +
M5	+	+	+

[a]+ + +, strong oxidation; +, weak oxidation; −, no detectable oxidation.

[b]Enzyme assays in extracts suggest that formaldehyde dehydrogenase activity is missing.

GENETICS OF METHANOTROPHY

Even less is known concerning the genetics of methanotrophy than the genetics of methylotrophy. No single mutants deficient in methanotrophic functions have been reported. However, growth on methane may be a plasmid-encoded function in some facultative methanotrophs. In both *M. organophilum* XX and *M. organophilum* CRL26-SB1, the ability to grow on methane is an unstable property, and may correlate with the presence of plasmids (Hanson, 1980; Haber and Hanson, 1982; Haber et al., 1983). Efforts are being made to determine whether methanotrophic genes are indeed present on these plasmids (R. S. Hanson, personal communication).

Initially our laboratory had observed a similar instability and plasmid correlation in another facultative methanotroph, *Methylobacterium ethanolicum,* but we have since shown this culture to consist of two bacteria in a stable syntrophism: an obligate methanotroph that contains three plasmids, and a facultative autotrophic methylotroph that contains no detectable plasmids. Our interest in plasmid-encoded functions in the methanotrophs prompted us to carry out a survey of plasmids in obligate methanotrophs. Of ten different strains screened, nine were found to contain detectable plasmid DNA (A. Wopat and M. E. Lidstrom-O'Connor, unpublished). Cross-hybridization experiments using Southern blots (Southern, 1975) and ^{32}P-labeled plasmid DNA have shown that regions of homology exist between plasmids from obligate methanotrophs of diverse types and origins. Further experiments are under way to delineate these regions of homology and determine whether any C-1 genes are plasmid-encoded in obligate methanotrophs.

ACKNOWLEDGEMENTS

The new data that I have described are the result of excellent work done by Craig Weaver, Gail Fulton, David Nunn, Ann Wopat, and Dawn Anderson. This work was supported by grants from the Department of Energy (DE AM06076RL02225) and the National Institutes of Health (2 RO1 GM 27766-04). I thank R. S. Hanson for providing a preprint of unpublished data.

LITERATURE CITED

Allen, L. N., and R. S. Hanson. 1982. Abstracts of the XIIIth International Congress for Microbiology, p. 60.

Anthony, C. 1975. The microbial metabolism of C_1 compounds: the cytochromes of *Pseudomonas* AM1. Biochem. J. 146:289–98.

Bamforth, C. W., and M. L.-O'Connor. 1979. The isolation of pleiotropic mutants of *Pseudomonas aminovorans* deficient in their ability to grow on methylamine and an examination of their enzymic constitution. J. Gen. Microbiol. 110:143–49.

Beringer, J. E., J. L. Beyon, A. V. Buchanan-Wallaston, and A. W. G. Johnson. 1978. Transfer of the drug-resistance transposon Tn5 to *Rhizobium*. Nature (London) 276:633–34.

Colby, J., H. Dalton, and R. Whittenbury. 1979. Biological and biochemical aspects of microbial growth on C-1 compounds. Ann. Rev. Microbiol. 33:481–517.

Ditta, G. S., Stanfield, D. Corbin, and D. R. Helinski. 1980. Broad host range DNA cloning system for Gram-negative bacteria: construction of gene bank of *Rhizobium meliloti*. Proc. Natl. Acad. Sci. USA 77:7347–51.

Ehhalt, D. H., and U. Schmidt. 1978. Sources and sinks of atmospheric methane. Pure Appl. Geophys. 116:452–64.

Gautier, F., and R. Bonewald. 1980. The use of plasmid R1162 and derivatives for gene cloning in the methanol-utilizing *Pseudomonas* AM1. Molec. Gen. Genet. 178:375–80.

Haber, C. L., L. N. Allen, and R. S. Hanson. 1983. Methylotrophic bacteria: biochemical diversity and genetics. Science 221:1147–52.

Haber, C. L., and R. S. Hanson. 1982. Abstracts of the XIIIth International Congress for Microbiology, p. 60.

Hanson, R. S. 1980. Ecology and diversity of methylotrophic organisms. Adv. Appl. Microbiol. 26:3–39.

Harder, W., and J. R. Quayle. 1971. Aspects of glycine and serine biosynthesis during growth of *Pseudomonas* AM1 on C_1 compounds. Biochem. J. 121:763–69.

Hennam, J. F., A. E. Cunningham, G. S. Sharp, and K. T. Atherton. 1982. Expression of eukaryotic coding sequences in *Methylophilus methylotrophus*. Nature (London) 297:80–82.

Heptinstall, J., and J. R. Quayle. 1970. Pathways leading to and from serine during growth of *Pseudomonas* AM1 on C_1 compounds or succinate. Biochem. J. 117:563–72.

Higgins, I. J., D. J. Best, and R. C. Hammond. 1980. New findings in methane-utilizing bacteria highlight their importance in the biosphere and their commercial potential. Nature (London) 286:561–64.

Higgins, I. J., D. J. Best, R. C. Hammond, and D. C. Scott. 1981. Methane-oxidizing microorganisms. Microbiol. Rev. 45:556–90.

Holloway, B. W. 1981. The application of *Pseudomonas*-based genetics to methylotrophs. *In* H. Dalton (ed.), Microbial growth on C-1 compounds, pp. 317–24. Heyden, London.

Hou, C. T., R. N. Patel, and A. I. Laskin. 1980. Epoxidation and ketone formation by C-1 utilizing microbes. Adv. Appl. Microbiol. 26:41–69.

Jeyaseelan, K., and J. R. Guest. 1979. Transfer of antibiotic resistance to facultative methylotrophs with plasmid R68.45. FEMS Microbiol. Lett. 6:87–90.

Marison, I. W., and M. M. Attwood. 1982. A possible alternative mechanism for the oxidation of formaldehyde to formate. J. Gen. Microbiol. 128:1441–46.

McNerney, T., and M. L. O'Connor. 1980. Regulation of the enzymes associated with C-1 metabolism in three facultative methylotrophs. Appl. Environ. Microbiol. 40:370–75.

O'Connor, M. L. 1981a. Extension of the model concerning linkage of genes coding for C-1 related functions in *Methylobacterium organophilum*. Appl. Environ. Microbiol. 41:437–41.

O'Connor, M. L. 1981b. Regulation and genetics in facultative methylotrophic bacteria. *In* H. Dalton (ed.), Microbial growth on C-1 compounds, pp. 294–300. Heyden, London.

O'Connor, M. L., and R. S. Hanson. 1977. Regulation of enzymes in *Methylobacterium organophilum*. J. Gen. Microbiol. 101:327–32.

O'Connor, M. L., and R. S. Hanson. 1978. Linkage relationships between mutants in *Methylobacterium organophilum* impaired in their ability to grow on one-carbon compounds. J. Gen. Microbiol. 104:105–11.

O'Connor, M. L., A. E. Wopat, and R. S. Hanson. 1977. Genetic transformation in *Methylobacterium organophilum*. J. Gen. Microbiol. 98:265–72.

Quayle, J. R. 1972. The metabolism of one-carbon compounds by microorganisms. Adv. Microb. Physiol. 7:119–203.

Reeburgh, W. S. 1982. A major sink and flux control for methane in marine sediments: anaerobic consumption. *In* K. A. Fanning and F. T. Manheim (eds.), The dynamic environment of the ocean floor, pp. 203–17. Heath, Lexington, Massachusetts.

Rudd, J. W. M., and C. D. Taylor. 1980. Methane cycling in the aquatic environment. Adv. Aquat. Microbiol. 2:77–150.

Salem, A. R., A. J. Hacking, and J. R. Quayle. 1974. Lack of malyl-CoA lyase in a mutant of *Pseudomonas* AM1. J. Gen. Microbiol. 81:525–27.

Southern, E. M. 1975. Detection of specific sequences among DNA fragments separated by gel electrophoresis. J. Molec. Biol. 98:503–17.

Warner, P. J., I. J. Higgins, and J. W. Drozd. 1980. Conjugative transfer of antibiotic resistance to methylotrophic bacteria. FEMS Microbiol. Lett. 7:181–85.

Williams, E., and B. W. Bainbridge. 1971. Genetic transformation in *Methylococcus capsulatus*. J. Appl. Bacteriol. 34:683–87.

Williams, E., A. Shimmin, and B. W. Bainbridge. 1977. Mutation in the obligate methylotrophs *Methylococcus capsulatus* and *Methylomonas albus*. FEMS Microbiol. Lett. 2:293–96.

Windass, J. D., M. J. Worsey, E. M. Pioloi, D. Pioloi, P. T. Barth, K. T. Atherton, E. C. Dart, D. Byrom, K. Powell, and P. J. Senior. 1980. Improved conversion of methanol to single-cell protein by *Methylophilus methylotrophus*. Nature (London) 287:396–400.

JAMES I. FREA

Methanogenesis: Its Role in the Carbon Cycle

14

INTRODUCTION

Methane-producing microorganisms are ubiquitous and functional in a wide variety of anaerobic environments, including the rumen of ruminant animals, anaerobic sewage-digesters, and in freshwater and marine sediments. They are present to a lesser extent in the digestive tract of a variety of invertebrate and vertebrate animals (including humans) and the wetwood of some trees. They are usually in close association with actively decomposing organic matter.

The magnitude of methanogen activity is not generally fully appreciated. About half of the organic carbon degraded by anaerobic microbes is eventually converted to methane (Higgins et al., 1981). Biogenic sources of methane are major contributors to atmospheric methane; they release at least as much to the atmosphere as is produced from natural gas wells. It is important to recognize that biogenic methane released to the atmosphere is only a small fraction of that actually produced because the methanotrophic bacteria oxidize most of it before it ever reaches the atmosphere. Nonetheless, the methane that does reach the atmosphere is equivalent to 0.5% of the annual biological production of dry organic matter. Methane is thus one of the most abundant organic compounds on earth, and the microbes that produce it play a major role in the cycling of carbon. Not only is their role quantitatively significant, but they are key factors in organic turnover in many environments. Without methanogens, the decomposition of organic carbon would cease with the accumulation of small organic molecules such as organic acids and neutral compounds.

GENERAL OVERVIEW OF ORGANIC MATTER DECOMPOSITION

Methanogenesis is largely a process mediated by autotrophic microbes that oxidize H_2 with concomitant reduction of CO_2 to CH_4. In fact, all but one of the known methanogens are capable of this process (Mah et al., 1977). While generating energy and reducing equivalents from the oxidation of H_2, some of the methanogens derive essentially all of their carbon for growth from CO_2 (Zeikus, 1977). Methanogens, although requiring some organic carbon for growth (usually acetate), acquire a large fraction of their cellular carbon from CO_2. Thus, the methanogenic microbial population has a large component of autotrophic members.

METHANOGENS IN THE CARBON CYCLE

Growth Substrates

Methanogens have a limited substrate range. They have been shown to obtain their carbon primarily from acetate, formate, methanol, methylamine, and CO_2. Assimilation of carbon by methanogens growing on one-carbon units (CO_2, methanol) apparently occurs by unique pathways. They are not known to employ the Calvin-Benson cycle or the serine or hexulose pathway of the methylotrophs.

When methanogens assimilate one carbon units, participation of the CH_4 production system appears to be obligatory because the methyl unit of the intermediate acetate arises through it by reduction of CO_2 (fig. 1). Growth of methanogens on H_2/CO_2 apparently requires B_{12} in the synthesis of acetate from $-CH_3$ and CO_2 (fig. 1; Kenealy et al., 1980). Acetate is the common starting point for synthesis of cellular materials.

The metabolism of acetate (exogenous or generated from CO_2 reduction) is accomplished via different but related systems as exemplified by *Methanobacterium thermoautotrophicum* (which does not require acetate for growth but derives as much as 60% of its cellular carbon from it if acetate is available) and by *Methanosarcina barkeri* (which will grow on acetate as its sole carbon source).

Autotrophic assimilation of CO_2 by *M. thermoautotrophicum* occurs via pyruvate synthase, phosphoenolpyruvate carboxylase, and α-ketoglutarate synthase reactions (Delwiche and Bryan, 1976; Zeikus et al., 1977; Daniels and Zeikus, 1978; Fuchs and Stupperich, 1978; Fuchs et al., 1978). Pyruvate synthase is a key enzyme because it catalyzes the reductive carboxylation of acetyl CoA to pyruvate. Hexoses and pentoses, as well as dicarboxylic acids, arise through pyruvate from acetyl CoA (Fuchs and Stupperich, 1980). Short-

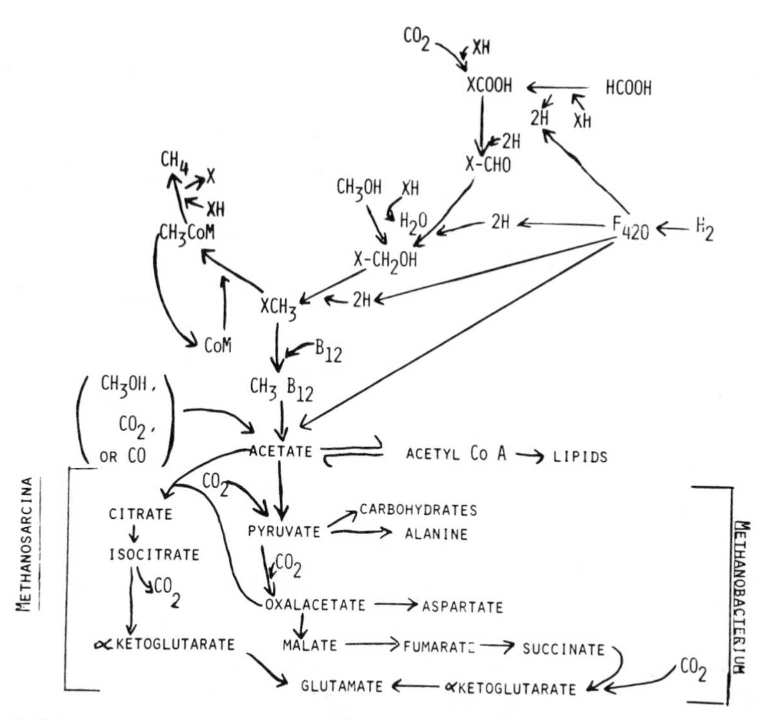

Fig. 1. Scheme for the reduction of CO_2 to CH_4 and cellular constituents by the methanogens. X represents a carrier molecule that has not been identified.

term labeling studies have demonstrated that the carbon from CO_2 appears predominantly in alanine, a finding consistent with the proposal that acetyl CoA is converted to pyruvate by carboxylation (Stupperich and Fuchs, 1981). Pyruvate is phosphorylated to phosphoenolpyruvate, which is then carboxylated to oxalacetate by PEP carboxylase (Stupperich and Funchs, 1981). Dicarboxylic acids (oxalacetate to α-ketoglutarate) are generated via the reductive reactions in the reverse TCA cycle.

The CO_2 metabolism by *M. barkeri* occurs using a different route for the synthesis of α-ketoglutarate (Zeikus et al., 1977; Weimer and Zeikus, 1979). Pyruvate synthase activity is also a key function in this pathway. In this synthetic system, α-ketoglutarate is synthesized by the oxidative reactions of the citric acid cycle.

Most methanogens can derive energy for growth and reducing equivalents for synthesis from the oxidation of hydrogen. Excess electrons are disposed of in the reduction of CO_2 to methane (Zeikus, 1977). Some methanogens are capable of acetate degradation (in the presence or absence of H_2) with the production of CH_4 and growth (Winfrey and Zeikus, 1979a; Boone, 1982; Lovley and Klug, 1982). Formate and methanol (and methylamine) can serve also as sources of energy and reducing equivalents for growth. Formate is actually split into H_2 and CO_2 by formate dehydrogenase, so its use is essentially similar to the use of H_2/CO_2. Methanol is oxidized to CO_2 in the process of energy conservation and the production of reducing equivalents. A recent report (Oremland et al., 1982) shows that trimethylamine and methionine may be precursors of methane. A small but growing list of energy and reducing equivalent sources is developing, with H_2 and acetate predominating in the majority of systems studied.

Temperature and pH

Methane production in natural environments is influenced by temperature. Several studies have shown that decreased temperature leads to decreased methane production (Mallard and Frea, 1972; Zeikus, 1977). Methanogens have been found to grow at temperatures as low as 4°C and as high as 70°C. Most have temperature optima between 30° and 45°C with the exception of *M. thermoautotrophicum,* which has a temperature optimum of 65°C.

Generally, methanogens require pH in the range of 6.5–8.0 for growth and activity. They are particularly sensitive to acidic environments. The optimum pH for methanogens from the alkaline Big Soda Lake was found to be 9.7. Methane production by methanogens from Big Soda Lake was decreased by a factor of more than 100 at pH 7.7 (Oremland et al., 1982).

Nutritional Requirements

Typically methanogens fulfill their nitrogen requirements through the assimilation of ammonia (Bryant et al., 1971). Assimilation of ammonia present in the environment at low levels occurs through the activity of glutamate synthase and glutamine synthetase. When excess ammonia is present, *M. thermoautotrophicum* incorporates it directly into alanine through the action of alanine dehydrogenase. Glutamate dehydrogenase activity is absent (Kenealy et al., 1982).

The methanogens fulfill their sulfur requirements from sulfate, sulfide, cysteine, and methionine. Thus, sulfate (0.16–0.52 mM) is required by some of the methanobacteria and *Methanospirillum* when grown in cysteine-sulfide–containing media (Patel et al., 1978). Sulfate is normally a constituent of media recommended for the growth of methanogens. Growth and methane production can usually be optimized only if sulfides are present in the medium; the soluble inorganic sulfides seem to be the most effective (Wellinger and Wuhrmann, 1977; Scherer and Sahm, 1981b). Cysteine stimulates growth of *M. barkeri* but cannot replace inorganic sulfide as a sulfur source (Scherer and Sahm, 1981b). *M. thermoautotrophicum,* however, is able to use either inorganic sulfide or cysteine as sulfur source (Kenealy et al., 1982).

Nickel, cobalt, molybdenum, and selenium are required by the methanogens studied (Jones and Stadtman, 1977; Schönheit et al., 1979; Scherer and Sahm, 1981a). Concentrations in the nanomolar range generally suffice.

Patel and colleagues (1978) found that the methanogens have a requirement for ferrous iron with the optimum concentration in the 0.3–0.9 mM range. Corder (1982) determined, however, that ferrous ion at 36 μM levels in a defined medium supported growth of fourteen different methanogenic strains. The medium could be used successfully for the enrichment and isolation of a wide variety of methanogens from diverse environments.

Historically methanogens have been grown in complex media containing yeast extract, a protein digest, rumen fluid, anaerobic digester extract, or a mixture of amino acids and vitamins. Several studies have shown that these supplements provide growth factors for some methanogens but are unnecessary for others. *Methanobrevibacter ruminantium* M-1 requires coenzyme M (2-mercaptoethanesulfonic acid) and α-methylbutyrate (Bryant et al., 1971), which is supplied by rumen fluid. *Methanococcus voltae* requires leucine, isoleucine, and pantothenic acid (Whitman et al., 1982). *M. barkeri* displays a requirement for riboflavin (Scherer and Sahm, 1981a). *M. bryantii* M.o.H. requires only thiamine in a chemically defined mineral salts medium. In the same medium, *M. smithii* requires thiamine and biotin for growth. Although *Methanospirillum hungatei* JF grew through multiple transfers in the defined medium without vitamins, the addition of thiamine, biotin, and lipoic acid slightly accelerated growth (Corder, 1982).

IMPACT ON THE CARBON CYCLE

Flow of Carbon

Organic matter is characteristically complex and polymeric, as found in nature. The first step in decomposition must be its degradation to simple soluble monomers. This initial step is accomplished by a heterogeneous group of biological entities, both aerobic and anaerobic. Methanogens are not directly involved in this step.

The release of simple soluble monomers (e.g., long-chain fatty acids, amino acids, simple carbohydrates) by nonmethanogenic organisms provides substrates for the growth of the nonmethanogenic organisms. Products of their metabolic activity include alcohols, neutral compounds, short-chain (volatile) fatty acids, H_2, CO_2, and NH_3. These products serve as substrates for a variety of organisms, some of which are methanogens and some are non-methanogens. The nonmethanogens further degrade monomers to products that support methanogenesis. The methanogens terminate the decomposition process with the production of CH_4 (and CO_2).

The decomposition process can be characterized also on the respiratory category of the actively decomposing population. In this discussion the population that functions aerobically will not be considered because methanogenesis is not functional under such conditions. Aerobic conditions inhibit methanogens and the products of aerobic metabolism generally are not significant substrates for methane production. The population that is active under anaerobic conditions is the consortium that can provide the conditions and substrates for the growth and activity of methanogens. Three general classes of microbes, based on respiratory function, may be present in anaerobic environments.

One of these populations is the denitrifiers, a diverse group containing mostly heterotrophic bacteria that prefer to function aerobically. If anaerobic conditions exist and nitrate is available, these bacteria metabolize much as they do aerobically; thus, their products are not useful to the methanogens, and the methanogens are generally inactive. In addition, nitrate and nitrite are inhibitory to the methanogens.

Another group functioning under anaerobic conditions is the group of fermenting bacteria. These microbes are critical because they provide as products of their activity both the substrates for growth and the low redox conditions (< -200 mV) necessary for survival and activity of the methanogens (Cappenberg, 1974).

A third group of bacteria, the sulfate-reducers, is functional in anaerobic environments. These bacteria have been shown to effectively outcompete the methanogens for fermentative products and may produce toxic levels of S^{2-} if sulfate is present in sufficient quantities (Macgregor and Keeney, 1973; Cappenberg, 1974; Winfrey and Zeikus, 1977; Lovley et al., 1982; Senior et al., 1982).

Flow of electrons

Oxygen depletion through consumption by the aerobic population leads to a series of changes including a decrease in Eh, a decrease in oxidized species, an increase in reduced species of compounds, and a concurrent change in the nature of the metabolic activity of the microbial population. The O_2 level decreases and finally disappears with the redox potential falling to about +350 mV as a consequence of the activity of the aerobic population. Nitrate is the next to disappear, and by the time the potential reaches about +100 mV, it is essentially gone; denitrifiers are thus rendered inactive. The decrease to about +100 mV results from the combined activity of the denitrifiers and facultative anaerobes (fermenters).

As the potential decreases, the activity of the sulfate-reducers increases; the potential drops below −150 mV as sulfide production increases. Methane producers begin to be active as the potential falls below −200 mV and in the absence of sulfate become a dominant population below −150 mV (Connell and Patrick, 1968; Bell, 1969; Connell and Patrick, 1969; Meek et al., 1969). It is generally held that methanogenesis is the terminal process in anaerobic decomposition of organic matter and becomes significant only if nitrate and sulfate are absent or are depleted through denitrification and sulfate reduction (Patrick and DeLuane, 1972; Cappenberg, 1974; Kelly and Chynoweth, 1979; Lovley et al., 1982).

Thermodynamic considerations

The sequence described for the decrease in redox potential is repeated when the basis of analysis is decreasing thermodynamic energy yields. For purposes of comparison, only acetate as the model energy source is considered. Acetate is degradable by denitrifiers, sulfate-reducers, and methanogens.

Denitrifiers can oxidize acetate to CO_2 and water with concurrent reduction of nitrate to nitrogen, as follows:

$$5\ CH_3COO^- + 8\ NO_3^- + 3\ H^+ \rightarrow 10\ HCO_3^- + 4\ N_2 + 4\ H_2O\ . \qquad (1)$$

This reaction has a ΔG^o of -792 kJ mol^{-1} of acetate. Thus even at low acetate concentrations considerable energy is available for growth of the denitrifying population.

Acetate-decomposing sulfate-reducing bacteria (equation 2) have been described (Widdel and Pfennig, 1977), and their potential role in sulfate-containing environments has been recently documented (Sørensen et al., 1981).

$$CH_3COO^- + SO_4^{2-} \rightarrow HS^- + 2\ HCO_3^-\ . \qquad (2)$$

The ΔG^o for this reaction is -47 kJ mol^{-1} of acetate.

Some methanogens are also capable of decomposing acetate (equation 3) and they can derive their carbon and energy from the decomposition.

$$CH_3COO^- + H^+ \rightarrow CH_4 + CO_2 \ . \quad (3)$$

The ΔG^o is -8.6 kJ mol^{-1}.

With the common substrate acetate, the sequence of populations on a thermodynamic basis would thus be denitrifiers, sulfate-reducers, and then methanogens.

Hydrogen is generated in the anaerobic (fermentative) decomposition of organic matter. It would therefore be available as a substrate for the growth of microbes. In the case of anaerobic respiratory activity, hydrogen is not a product of organic matter decomposition so that hydrogen-utilizing populations would be deprived of growth substrate. With sufficient nitrate available, denitrifiers would dominate fermenters, and the fermentative production of hydrogen (and other reduced species besides N_2) would be minimal.

In the absence of nitrate, fermentative organic matter decomposition would lead to the production of hydrogen in addition to acetate. The hydrogen would be the subject of competition by sulfate-reducers and methanogens, since both groups are capable of hydrogen utilization.

Sulfate-reducers employ the reaction:

$$SO_4^{2-} + H^+ + 4\ H_2 \rightarrow HS^- + 4\ H_2O \ . \quad (4)$$

It has a ΔG^o of -152.2 kJ mol^{-1} of hydrogen.

Methanogens reduce CO_2 at the expense of H_2 by the reaction:

$$HCO_3^- + 4\ H_2 + H^+ \rightarrow CH_4 + 3\ H_2O \ . \quad (5)$$

The ΔG^o of this reaction is -131 kJ mol^{-1} of hydrogen.

On a strictly thermodynamic basis sulfate-reducers would have the advantage over methanogens.

Thermodynamic considerations would argue that methanogens would not be competitive and probably not even functional where either denitrifiers or sulfate-reducers are active. It has been suggested that a more thermodynamically favorable reaction will not necessarily exclude another thermodynamically favorable reaction (McCarty, 1972). Recent studies of methanogen activity on sediments where sulfate reduction is occurring supports that position (Lovley et al., 1982; Senior et al., 1982).

Kinetic Conditions

Where two populations compete for the same substrate, one will generally dominate, and this dominance may lead to the near exclusion of the activity of the second. Sulfate-reducers and methanogens compete for both hydrogen and acetate. The primary competition in environments where sufficient sulfate is available seems to be for hydrogen (Hungate et al., 1970; Senior et al., 1982). Acetate-utilizing methanogens seem to be generally absent from sediments where sulfate-reducers are active.

Competition can be assessed on a kinetic basis. That population which has the higher affinity for a given commonly utilizable substrate should outcompete others that have a lower affinity.

Hungate and colleagues (1970) found that methanogens in the rumen, as well as some in pure cultures, were quite capable of hydrogen utilization. They determined the K_m of hydrogen utilization to be about 120 Pa.

Strayer and Tiedje (1978b) analyzed hydrogen conversion to methane in lake sediments. Their experimental value for the K_m of hydrogen uptake was about 240 Pa. In later similar studies, Lovley and colleagues (1982) determined the K_m for hydrogen utilization by lake sediment methanogens to be about 600 Pa. In the same study, they also determined the K_m for hydrogen utilization by sulfate-reducers to be about 140 Pa. Thus, although there is some variability in hydrogen affinity as shown by different experiments, the affinity of the methanogens for H_2 is less than that of the sulfate-reducers. On the basis of kinetics alone then, it would be reasonable to expect that in the presence of adequate sulfate, the sulfate-reducers would lower the hydrogen concentration sufficiently to inhibit the methanogens.

Capacity Considerations

For a given population to compete for a substrate, that population must be of adequate size with its members generally capable of utilizing the substrate or carrying out the reaction. Capacity refers to the size of the functional population.

Although denitrifiers and sulfate-reducers are not significant in some environments where methanogens function (for example, the rumen, intestinal tract, and sludge-digesters), they are potentially important in freshwater and marine sediments. Thus, a brief analysis of the capacity of all three groups would be useful.

Denitrification is a characteristic of a relatively few species of microbes, but, nonetheless, the number of potential denitrifiers in a wide range of environments is significant. It is reasonable to expect that the size of the denitrifying population (particularly in sediments) will be quite adequate to dominate under anaerobic conditions. It is somewhat less likely that the denitrifying capacity will be dominant since denitrification is repressed under aerobic conditions and becomes active only after a lag when anaerobic conditions are imposed. Thus, it requires an induction period (40 min to 3 h) for the population to develop functional denitrifying ability. Nitrate is required as an inducer for some bacteria.

Reducing equivalents from organic carbon compounds (particularly simple water-extractable compounds) play an important role in the activity of denitrifiers. The presence of abundant organic carbon and anaerobic conditions shift the reduction toward ammonia and away from the nitrogen oxides (Delwiche and Bryan, 1976; Payne, 1981; Knowles, 1982).

Sulfate-reducing bacteria also are widely distributed. They are active in marine and freshwater sediments, sewage, and soils provided that there is an adequate supply of sulfate and simple organic acids or hydrogen. Sewage may have 10^7 sulfate-reducers ml^{-1}, and marine sediments have shown similar counts (Postgate, 1965). Where decomposing organic matter has exhausted oxygen and adequate sulfate is available, active populations of sulfate-reducers can be expected to develop. Lovley and colleagues (1982) found that the addition of sulfate to lake sediments (naturally low in sulfate) led to only a slow inhibition of methanogenesis because of H_2S produced by sulfate-reducers. Less than 10% of the hydrogen turnover could be attributed to sulfate-reducers. The sulfate-reducing capacity of the system is maintained at a low level by the limiting sulfate availability. The addition of saturating concentrations of sulfate does not initially lead to increased sulfate reduction if the capacity has been exceeded.

The methanogenic capacity also varies. Population sizes as high as 10^9 per g of dry sediment in aquatic sediments have been shown (Zeikus and Winfrey, 1976; Strayer and Tiedje, 1978a; Ward and Frea, 1980). Similar population sizes are present in the rumen and in anaerobic sludge-digesters. Methanogen capacity varies and is related to the characteristics of the environment (availability of H_2/CO_2 and/or acetate, low redox potential, presence or absence of an active competing population such as sulfate-reducers).

Contributions of Methanogens

Methanogens accomplish two main functions in the maintenance of the carbon cycle: (1) conversion of organic carbon to inorganic gaseous carbon and (2) provision of an outlet for the reducing equivalents generated by the decomposition of organic matter in anaerobic environments. In most environments where methane production occurs, acetate is a key product of the decomposition of organic matter. In the rumen the acetate generated from carbohydrate decomposition is used by the animal itself. In other environments acetate does not accumulate but is converted to CH_4 and CO_2, which are then removed by ebullition when their concentration exceeds their solubility. About two-thirds of the CH_4 produced is estimated to arise from acetate (Kenealy et al., 1982).

Pure cultures of methanogens *(Methanosarcina)* have been shown to metabolize acetate to produce CH_4 and CO_2, with the CH_4 coming from the methyl carbon of the acetate (Buswell and Solo, 1948; Stadtman and Barker, 1949; Stadtman and Barker, 1951; Zeikus et al., 1975; Weimer and Zeikus, 1978; Zehnder et al., 1980). Tracer studies have long indicated that acetate decomposition in natural environments occurs by this route (Pine and Barker, 1956; Winfrey and Zeikus, 1979a, b).

A second path for the production of methane was suggested many years ago. It was proposed that the methyl group of acetate was converted to CO_2 and that the CO_2 was subsequently reduced to methane. This route implies the cooperation of an acetate oxidizing, H_2-liberating organism and an H_2-oxidizing, CO_2-reducing methanogen. Some bacteria (particularly *Desulfuromonas acetoxidans*) are capable of the acetate oxidation, but pure culture studies have not documented the coupled production of hydrogen as the reduced species (Pfennig and Biebl, 1976). The anaerobic oxidation of the methyl group of acetate to CO_2 has been observed in studies of lake sediments (Cappenberg and Prins, 1974; Winfrey et al., 1977; Winfrey and Zeikus, 1979a). Some oxidation may be the result of methanogens (Winfrey and Zeikus, 1979b) since *Methanosarcina* has been shown to convert the methyl of acetate to CO_2 in the presence of methanol or methylamine and less in the presence of H_2/CO_2 (Zeikus et al., 1975; Weimer and Zeikus, 1978). A recent description (Zehnder et al., 1980) of a methanogenic acetate-oxidizing isolate that cannot use H_2 suggests that methanogens may be responsible directly for the complete conversion of acetate to CO_2 as well as to CH_4. This organism is present in sewage sludge; because of its very high affinity for acetate ($K_m \approx$ 0.46 mM compared with $K_m \approx$ 5 mM for *Methanosarcina* [Winfrey and Zeikus, 1979b], which is also present in sewage sludge), this organism may outcompete the sarcina, which otherwise grows faster and is more efficient.

A more general and widely distributed role of the methanogens in the carbon cycle is the removal of the H_2 generated in the anaerobic decomposition of organic matter. Proton reduction is a major mechanism for the disposal of electrons generated in the oxidation of organic matter. Thermodynamically, the reaction producing H_2 from protons becomes more favorable as the concentration of H_2 decreases (Smith and Mah, 1978). Thus, any activity that removes H_2 or maintains low H_2 concentration enhances anaerobic organic matter decomposition.

Where oxygen, nitrate, and sulfate are limiting, methanogens play a unique role. They are able to capitalize on an energy-producing reaction for which there is little competition, i.e., the oxidation of H_2 coupled to the reduction of CO_2 to produce CH_4. By oxidizing environmental H_2 (and thus decreasing its concentration), the methanogens provide the environment in which mixed populations of anaerobic organic compound-decomposers are active.

Investigations conducted over more than two decades have established that the activities of a variety of bacteria in anaerobic ecosystems are directly affected by the utilization of H_2 by methanogenesis. Hungates's early studies (1966) of rumen microflora led to the suggestion that electrons produced in the rumen fermentation were diverted from the production of reduced products such as alcohol to the generation of H_2, which was used by the methanogenic population. Hungate found that pure cultures of nonmethanogenic rumen bacteria formed reduced products not found in the normal rumen fermentation. Instead, methane was a product of the rumen fermentation. Similar phenomena have been observed in other rich anaerobic environments.

There is considerable interaction between nonmethanogenic and methanogenic organisms, in particular in the coupled redox reactions in single substrate fermentations. The term "interspecies H_2 transfer" has been used to describe this coupled redox interaction (Iannotti et al., 1973). Methanogens are prominent, although not the exclusive, H_2-oxidizers in this coupled oxidation-reduction interaction. The methanogens provide a mechanism that permits the following actions to occur in anaerobic environments:

1. fermentation of reduced low molecular weight molecules that would normally be (unfermentable) terminal products;
2. more complete oxidations of carbohydrates;
3. acquisition of H_2 and CO_2 by methanogens from otherwise nonutilizable substrates;
4. shift of electron flow from reduction of organic compounds to proton reduction.

The importance of interspecies H_2 transfer is clear; the part played by methanogens in methane-producing environments warrants further analysis.

Methanobacterium omelianskii, first described by Barker (1940), was originally considered to be capable of CH_4 production from the oxidation of ethanol (with concurrent production of acetate). Bryant and colleagues (1967) successfully separated the "*M. omelianskii*" into two organisms. One, the S organism, oxidized ethanol to acetate and H_2, and the other, *M. bryantii* converted H_2 and CO_2 to CH_4. The S organism grew well on ethanol only in the presence of the methanogen because the S organism was inhibited by the H_2 produced in the ethanol oxidation. Maintenance of low H_2 partial pressures by the methanogen permitted good growth. It was later shown (Reddy et al.,

1972) that S organism grew well by itself on pyruvate where electrons were disposed of by the production of ethanol.

Considerable evidence has accumulated supporting the role of methanogens in scavenging H_2 in rumen fermentations. Much of the experimentation has been done with pure cultures of various rumen isolates; thus, the function of the microbes and their interactions is inferred and not directly determined.

Latham and Wolin (1977) found that the anaerobic cellulolytic rumen bacterium *Ruminococcus flavefaciens* produces primarily succinate and some acetate, formate, H_2, and CO_2 from cellulose in monoculture. However, in co-culture with the rumen methanogen *M. ruminantium,* acetate was the major product, succinate was present in small amounts, formate was absent, H_2 did not accumulate, and much CH_4 was formed. They concluded that this type of interaction may determine the flow of cellulose carbon in the rumen fermentation.

McInerney and colleagues (1981a, b) reported the isolation of an anaerobic butyrate-degrading bacterium (similar to *Syntrophomonas wolfei*) from a bovine rumen fluid enrichment. When grown in co-culture with H_2-using organisms such as *Methanospirillum, Methanosarcina,* or *Desulfovibrio, Syntrophomonas* degraded saturated short-chain fatty acids (butyrate through octanoate) to acetate or acetate and propionate. Growth and fatty acid degradation occurred only in the presence of the H_2-using microbe. Addition of H_2 to the medium prevented the growth and butyrate degradation by the *Syntrophomonas*.

An oxalate-degrading enrichment of rumen bacteria (Dawson et al., 1980) produced 1 mol CH_4 per 3.8 mol of oxalate degraded. The addition of H_2 or formate inhibited oxalate degradation (but not methanogenesis), and the addition of chloroform inhibited both oxalate degradation and methane production. The oxalate degradation could be uncoupled from methane production if the enrichments were grown in continuous cultures at dilution rates of 0.078 h^{-1}. The uncoupled degradation was not inhibited by H_2 or formate, and formate was the primary product. Other hydrogen-utilizing bacteria (*Vibrio succinogenes* and *Desulfovibrio* G-11) did not compete for hydrogen with the methanogens in the enrichments. Increased cell yields in the presence of the methanogens was a primary effect of the microbial interactions, since oxalate degradation was not obligately coupled to H_2 removal by the methanogens.

A triculture of an anaerobic rumen fungus with *Methanobrevibacter* and *Methanosarcina* converted cellulose (from a variety of natural sources) to CH_4 and CO_2. A hexose equivalent was converted to 2 mol each of CH_4 and CO_2. The fungus in co-culture with either of the methanogens degraded cellulose, but the final products included, in addition to CH_4 some acetate, lactate, formate, and ethanol. Only the triculture functioned to provide complete conversion to CH_4 and CO_2. In the triculture, acetate temporarily accumulated but disappeared as the fermentation progressed (Mountfort et al., 1982).

Vogels and colleagues (1980) found that nine genera of rumen ciliates formed ectosymbioses with methanogenic bacteria. The ciliates are anaerobic and produce H_2 as a result of metabolic activity. The metabolism of the ciliates is inhibited by H_2; the attachment of methanogens would be of selective value to both organisms.

Methanogens also play a critical role in the fermentation characteristic of anaerobic sludge digestion. In sludge digestion the fermentative bacteria hydrolyze polymers and ferment the soluble compounds to H_2, CO_2, and volatile acids. The volatile acids (primarily propionate, butyrate, and longer carbon chain fatty acids) are converted to H_2, CO_2, and acetate by the H_2-producing acetogenic population. The acetate and H_2 serve as substrates for CH_4 production by the methanogens. It has been noted frequently that the fatty acids accumulate in digesters where either retention times are too short or where high rates of organic matter loading occur. The rate-limiting step in digestion with methane production is the degradation of fatty acids. The key is the efficiency of H_2 conversion to methane through interspecies hydrogen transfer (Mackie and Bryant, 1981). A variety of studies have shown that about 70% of the methane produced arises from the methyl group of acetate (Jeris and McCarty, 1965; Smith and Mah, 1966; Mountfort and Asher, 1978) in the mesophilic fermentation. The remainder of the methane is produced from CO_2 reduction by H_2. In the digestion at 60°C, the proportion of CH_4 from acetate rises to about 80% with a concurrent decrease in that which comes from H_2 oxidation (Mackie and Bryant, 1981). Of the methane produced, about 23% can be traced back to propionate and butyrate (13–17% and 7–9%, respectively) via acetate and H_2 (Kaspar and Wuhrman, 1978; Mackie and Bryant, 1981). CH_4 production from H_2 decreases in significance after fresh substrate is added to the digester and increases as the time after addition lengthens (Mountfort and Asher, 1978; Mackie and Bryant, 1981). Early stages of the fermentation after feeding degrade compounds yielding high acetate-to-hydrogen ratios. A continuously fed reactor would probably not show as clear a trend toward acetate production but would be more likely to have characteristics similar to the aged reactor.

Studies of "artificial consortia" (known mixed cultures of methanogens and other organisms) have provided evidence that supports the conclusions drawn from the mixed cultures involving natural populations in digesters or fermenters started (and maintained) with digester sludge. The first such study reported (Winter and Wolfe, 1979) found that a co-culture of an adapted "*Methanosarcina barkeri*" strain MS and *Acetobacterium woodii* was successful in the complete conversion of carbohydrates (fructose or glucose) to methane and CO_2. The *Acetobacterium* converted fructose to 3 mol of acetate, and only when the acetate level rose above 10 mM did the *Methanosarcina* convert appreciable quantities of the acetate to methane. When *Acetobacterium woodii* was co-cultured with *Methanobacterium* strain AZ, the fructose in the growth medium was converted to 2 mol of acetate and 1 mol each of CO_2 and CH_4. Similar results were obtained when each of four other H_2-oxidizing strains of methanogenic bacteria were grown with the *Acetobacterium* in fructose medium (Winter and Wolfe, 1980).

Weimer and Zeikus (1977) observed that the rate of decomposition of cellulose (and, to a much lesser extent, cellobiose) by a co-culture of *Clostridium thermocellum* and *M. thermoautotrophicum* was approximately the same as that with the monoculture of the *Clostridium*. The H_2 produced by the cellulose-decomposer was converted to methane, but essentially none of the acetate was converted to methane. However, this cellulose-decomposer is not inhibited by H_2; thus, the presence of the methanogen shifts the products away from H_2 and ethanol to acetic acid and CH_4. The main advantage of the co-culture was the increased growth rate of the cellulolytic organism and the earlier appearance of cellulolytic activity.

Sulfate-reducers *(Desulfovibrio desulfuricans* or *Desulfovibrio vulgaris)* degrade lactate to acetate, H_2, and CO_2 in the absence of sulfate when co-cultured with methanogens. Co-culture with methanogens that oxidize H_2 only leads to the conversion of lactate to acetate and CH_4. Co-culture with *Methanosarcina,* which oxidizes H_2 and degrades acetate, leads to the conversion of lactate ultimately to CH_4 and CO_2. Until all of the lactate disappears, there is stoichiometric production of acetate and CH_4. After all of the lactate is degraded, the acetate is converted to CH_4 and CO_2 (McInerney and Bryant, 1981).

Decomposition of organic matter in sediments is often dependent on the activity of methanogens. In sediments, methanogens are frequently the key microbial group that permits other decomposers to continue to function, again because of their role in H_2 removal. However, in sediments their role is not as exclusive as in the rumen and the sludge-digester because other microbial groups (denitrifiers and sulfate-reducers) also remove H_2. When nitrate and sulfate are available, these groups have been shown to outcompete the methanogens. Nonetheless, the methanogens are generally present and active in sediments and are often important in the accomplishment of the complete decomposition of organic matter. As in the rumen and the sludge-digester, there is a close interaction of the fermentative bacteria, the acetogenic hydrogen-producing bacteria, and methanogens in sediments. Chloroform inhibition of sediment methanogenesis results in an immediate accumulation of H_2 and volatile fatty acids, with acetate and propionate constituting 82% and 13% of the accumulated volatile fatty acids (Boone, 1982).

In a study to determine the role of immediate precursors to methane, Winfrey and Zeikus (1979b) found that about 40% of the methane produced in sediment arose from CO_2 reduction. The reduction of CO_2 and conversion of acetate to CH_4 occurred simultaneously, and the addition of H_2 did not decrease CH_4 production from acetate, nor did the addition of acetate decrease production from the reduction of CO_2 by H_2. In these sediments methane production was limited by the availability of acetate and H_2 because of the competition for these substrates by nonmethanogens.

Considerable evidence has accumulated to suggest that major competitors for H_2 in sediments are sulfate-reducers if adequate sulfate is present (which is usually true in marine environments, and less often in freshwater sediments) (Martens and Berner, 1974; Abram and Nedwell, 1978a, b; Mountfort et al., 1980). Similarly acetate conversion to methane occurs primarily in the absence of sulfate-reducers (Sansone and Martens, 1981).

Shlomi and colleagues (1978) found that benzoate degradation by an enrichment from black mud was inhibited by inhibitors of methane production. Acetate and butyrate rapidly accumulated, since acetate conversion to methane was blocked and the conversion of butyrate to acetate was inhibited by the lack of removal of H_2 through CH_4 production.

A report by Oremland and colleagues (1982) describes a study of methanogenesis in the sediments of an alkaline hypersaline desert lake. They found that methane production was stimulated by methanol, trimethylamine, and methionine. H_2, acetate, and formate addition had little stimulatory effect.

A role for methanogens in the carbon cycle is the production of vitamins, amino acids, and accessory growth factors that are necessary for other members of the community. They are capable of converting simple substrates (H_2, CO_2, acetate, minerals) to cellular material. Certainly at their death their cellular contents become available to other cells. Additionally, it is possible that some metabolites reach the environment even while the cells are alive and functional.

The question of direct contribution of nutrients to other cells by methanogens has received little attention. Ward and colleagues (1978) reported on the results of a methanogenic acetate enrichment from a sludge-digester. After initial enrichment the culture was maintained for 2 years by weekly transfer. This stable enrichment contained *Methanosarcina* and several associated nonmethanogenic bacteria at concentration of 10^8 ml^{-1}. Some of the nonmethanogens (particularly their ''satellite'' bacteria) neither assimilated nor metabolized acetate. Since acetate was the only organic carbon and energy source, the nutritional requirements of these nonmethanogens was linked to substrates and growth factors produced from acetate by the acetate-utilizing members of the population. Only methanogens were capable of acetate metabolism and thus provided nutrients for the nonmethanogens. A wide range of dependency was shown. Some of the dependents required specific growth factors, whereas others required only an organic carbon and energy source.

Methanogens are rich in some vitamins and in fact can produce them (for example, B_{12}) (Ward et al., 1978). Some methanogens have been shown to excrete amino acids (Zehnder and Wuhrmann, 1977).

Methanogens thus may well make a significant contribution to the carbon cycle through the less obvious role of supplier of nutrients and metabolites to other microbes in the community.

Directions for the Future

The role of the methanogens in the carbon cycle has been clarified considerably in the last two decades. Several areas require attention in order to answer questions that have arisen or to resolve paradoxes that are now apparent.

Discovery of the methanogens was followed by several decades where study was focused on isolation, enumeration, and evaluation of the growth requirements of the methanogens. The last decade has seen a shift toward an emphasis on the physiology of methane production. Left unexplained are phenomena that are encountered routinely in cultivation of the methanogens and are recorded anecdotally in the literature.

One frequently cited observation is the rather distinct differences in response to redox potential among the methanogens. For example, *Methanobrevibacter* survives exposure to relatively elevated redox potentials and will initiate growth in much less reduced media than will *Methanospirillum*. Why is oxygen so toxic to methanogens, and why is there a differential response to somewhat elevated redox potentials? How do the methanogens survive the inevitable apparent exposure to elevated redox potentials in natural environments? Methanogens can be isolated from environments with high redox potential such as dry soil, the water column of freshwater lakes, dried manure, and other similar sites.

Another area where further clarification is necessary involves the general nutritional needs of the methanogens. The isolation and growth of methanogens has generally been accomplished in complex media, media that frequently contain ill-defined ingredients (for example, rumen fluid, yeast extract, sludge amendments, and so forth). Even the chemically defined media contain a complex array of ingredients many of which may have no substantial function. It has been found that most of the methanogens will grow satisfactorily on a defined medium containing no amino acids, no or few (1–3) vitamins, and a limited variety of mineral components (Corder, 1982). This medium (methanogen minimal medium) serves as a satisfactory maintenance vehicle, gives growth and methanogenesis comparable to more complex media, and functions extremely well as an enrichment and isolation medium for methanogens from a wide variety of sources. It is useful in genetic studies and studies involving the physiological and nutritional character of the methanogens.

Another facet of the differences in the understanding of nutrition of the methanogens is the question of energy and growth substrates. Well documented are the functions of acetate, H_2/CO_2 and formate, and to a lesser extent methanol and methylamine. There is reason to question whether there may be other substrates for the methanogens. The report by Oremland and colleagues (1982) that H_2/CO_2, acetate, and formate do not stimulate methanogenesis in sediments from an alkaline-hypersaline lake but that instead stimulation by methanol, trimethylamine, and methionine occurred suggests that further investigation is necessary.

Another question that deserves investigation has to do with the contributions of the methanogens in mixed cultures in ways that are more subtle than relieving the accumulation of H_2 and acetate in natural environments. Many of the methanogens grow quite satisfactorily without organic compounds being available, and, if they need organic matter, that need is frequently satisfied by acetate. Many of them are self-sufficient in terms of vitamins, amino acids, and other similar growth factors. Thus they can be a source of amino acids, vitamins, and nucleic acid precursors for other organisms. Little is known about this role for the methanogens, but what is known suggests that at least under some circumstances they function to provide growth factors for other organisms. It is not unlikely that in so doing they help themselves in that the organisms they provide with growth factors may be part of the community that provides them with H_2, CO_2, and acetate. Investigation of this role may help explain some of the observations that methanogen populations fluctuate and vary in composition in relation to season (in sediments), in relation to diet composition (in ruminants), and in relation to sludge composition (in anaerobic digestion). Work in this area might be particularly fruitful.

A further need in the study of the role of the methanogens in the carbon cycle has to do with how methanogens interact with other organisms and each other. There are two observations that exemplify the questions that exist. The triculture of an anaerobic rumen fungus, *Methanobrevibacter*, and *Methanosarcina* is not stable and cannot be routinely transferred (Mountfort et al., 1982). It was speculated that stability could not be attained because of the slow growth of the *Methanosarcina* and the loss of the viability of the fungus. It was observed that the fungus disappeared after seven days of incubation with the *Methanobrevibacter*. No clearcut experimental evidence is available to explain the lack of stability of the triculture.

Mixed cultures of methanogens and other organisms that are stable over many transfers have been reported (for example, the culture [Ward et al., 1978] of *Methanosarcina* and several obligately anaerobic nonmethanogens enriched from sludge and maintained for two years by weekly transfers). Enrichment of stable mixed cultures is successful, but the establishment of stable mixed cultures by combining pure cultures is frequently not sucessful; too little is known about the needs of the individual members and their interactions.

In work with methanogenic enrichments from a variety of environments using the methanogen minimal medium, it was observed that the enrichment from a given source frequently demonstrates a sequence of methanogens. Different sources may show different sequences. There does not seem to be a clear relationship between the size of the population of the various methanogens and the sequence of enriched populations. That is, the most numerous methanogen is not necessarily the first to appear nor is it necessarily that which will finally come to dominate. What is clear is that the methanogens each respond differently, and two H_2-oxidizers may not complete equally well in enrichment (or in a change in their native environment).

It has also been observed that different methanogens dominate different sites in a large continuous ecosystem, which was shown in a study of Lake Erie and Cuyahoga River sediments where *M. ruminantium, M. bryantii, barkeri,* and *M. hungatei* dominate in different areas. Their distribution seemed to relate to the sediment character and composition (Ward and Frea, 1978). A better understanding of these interactions must await further study.

One of the most perplexing questions concerning the methanogens revolves around the role of acetate and the acetate-splitting methanogens. It seems to be generally well documented and accepted that acetate is the substrate for 60–90% of the CH_4 produced in the rumen, sludge digestion, and sediments. What is not well explained is the apparent paucity of acetate-splitting methanogens in these environments; only two have been described, *M. barkeri* and *M. soehngenii* (Smith and Mah, 1966). Evidence from enrichments for acetate-splitting methanogens seldom support the observed acetate-splitting activity in the environment from which the enrichments were made. Analyses for specific acetate-splitters (for example by fluorescent antibody techniques) similarly do not usually document the presence of known acetate-utilizers in populations large enough to explain the activity. Because of the dominance of the activity, the acetate-splitters should be the dominant population.

Two possible explanations exist for these observations. Either the known splitters are present in larger numbers and our analytical techniques are so insensitive we cannot document their presence; or, another population(s) is responsible for the activity and we have not developed techniques for its detection and isolation. Clearly, more work is necessary to explain how acetate is being converted in anaerobic habitats.

LITERATURE CITED

Abram, J. W., and D. B. Nedwell. 1978a. Inhibition of methanogenesis by sulphate reducing bacteria competing for transferred hydrogen. Arch. Microbiol. 117:89–92.

Abram, J. W., and D. B. Nedwell. 1978b. Hydrogen as a substrate for methanogenesis and sulphate reduction in anaerobic saltmarsh sediment. Arch. Microbiol. 117:93–97.

Balderston, W. L., and W. J. Payne. 1976. Inhibition of methanogenesis in salt-marsh sediments and whole-cell suspensions of methanogenic bacteria by nitrogen oxides. Appl. Environ. Microbiol. 32:264–69.

Barker, H. A. 1940. Studies upon the methane fermentation. IV. The isolation and culture of *Methanobacillus omelianskii*. Antonie van Leeuwenhoek 6:201–20.

Bell, R. G. 1969. Studies on the decomposition of organic matter in flooded soil. Soil Biol. Biochem. 1:105–16.

Boone, D. R. 1982. Terminal reactions in the anaerobic digestion of animal waste. Appl. Environ. Microbiol. 43:57–64.

Bryant, M. P., S. F. Tzeng, I. M. Robinson, and A. E. Joyner. 1971. Nutrient requirements of methanogenic bacteria. Adv. Chem. Ser. 105:23–40.

Bryant, M. P., E. A. Wolin, M. J. Wolin, and R. S. Wolfe. 1967. *Methanobacillus omelianskii:* a symbiotic association of two species of bacteria. Arch. Mikrobiol. 59:20–31.

Buswell, A. M., and F. W. Sollo, Jr. 1948. The mechanism of methane fermentation. J. Amer. Chem. Soc. 70:1778–80.

Cappenberg, T. E. 1974. Interrelations between sulfate-reducing and methane-producing bacteria in bottom deposits of a fresh-water lake. I. Field observations. Antonie van Leeuwenhoek 40:285–95.

Cappenberg, T., and H. Prins. 1974. Interrelations between sulfate-reducing and methane-producing bacteria in bottom deposits of a fresh-water lake. III. Experiments with ^{14}C-labeled substrates. Antonie van Leeuwenhoek 40:457–69.

Connell, W. E., and W. H. Patrick, Jr. 1968. Sulfate reduction in soil: effect of redox potential and pH. Science 159:86–87.

Connell, W. E., and W. H. Patrick, Jr. 1969. Reduction of sulfate to sulfide in a water logged soil. Soil Sci. Soc. Amer. Proc. 33:711–15.

Corder, R. E. 1982. The development of an integrated anaerobic system for the cultivation and characterization of methanogenic bacteria. Ph.D. dissertation, Ohio State University, Columbus, Ohio.

Daniels, L., and J. G. Zeikus. 1978. One-carbon metabolism in methanogenic bacteria: analysis of short-term fixation products of $^{14}CO_2$ and $^{14}CH_3OH$ incorporated into whole cells. J. Bacteriol. 136:75–84.

Dawson, K. A., M. J. Allison, and P. A. Hartman. 1980. Characteristics of anaerobic oxalate-degrading enrichment cultures from the rumen. Appl. Environ. Microbiol. 40:840–46.

Delwiche, C. C., and B. A. Bryan. 1976. Denitrification. Ann. Rev. Microbiol. 30:241–62.

Fuchs, G., and E. Stupperich. 1978. Evidence for an incomplete reductive carboxylic acid cycle in *Methanobacterium thermoautotrophicum*. Arch. Microbiol. 118:121–25.

Fuchs, G., and E. Stupperich. 1980. Acetyl CoA, a central intermediate of autotrophic CO_2 fixation in *Methanobacterium thermoautotrophicum*. Arch. Microbiol. 127:267–72.

Fuchs, G., E. Stupperich, and R. K. Thauer. 1978. Function of fumarate reductase in methanogenic bacteria *(Methanobacterium)*. Arch. Microbiol. 119:215–18.

Higgins, I. J., D. J. Best, R. C. Hammond, and D. Scott. 1981. Methane-oxidizing microorganisms. Microbiol. Rev. 45:556–90.

Hungate, R. E. 1966. The rumen and its microbes. Academic Press, New York.

Hungate, R. E., W. Smith, T. Bauchop, I. Yu, and J. C. Rabinowitz. 1970. Formate as an intermediate in the bovine rumen fermentation. J. Bacteriol. 102:389–97.

Iannotti, E. L., D. Kafkewitz, M. J. Wolin, and M. P. Bryant. 1973. Glucose fermentation products of *Ruminococcus albus* grown in continuous culture with *Vibrio succinogenes*: changes caused by interspecies transfer of H_2. J. Bacteriol. 114:1231–40.

Jeris, J. S., and P. L. McCarty. 1965. The biochemistry of methane fermentation using ^{14}C tracers. J. Water. Poll. Cont. Fed. 37:178–92.

Jones, J. B., and T. C. Stadtman. 1977. *Methanococcus vannielii:* culture and effects of selenium and tungsten on growth. J. Bacteriol. 130:1404–6.

Kaspar, H. F., and K. Wuhrmann. 1978. Kinetic parameters and relative turnovers of some important catabolic reactions in digesting sludge. Appl. Environ. Microbiol. 36:1–7.

Kelly, C. A., and D. P. Chynoweth. 1979. Methanogenesis: a measure of chemoorganotrophic (heterotrophic) activity in anaerobic lake sediments. *In* J. W. Costerton and R. R. Colwell (eds.), Native aquatic bacteria: enumeration, activity, and ecology, pp. 164–79. American Society for Testing and Materials, Philadelphia.

Kenealy, W. R., T. E. Thompson, K. R. Schubert, and J. G. Zeikus. 1982. Ammonia assimilation and synthesis of alanine, aspartate, and glutamate in *Methanosarcina barkeri* and *Methanobacterium thermoautotrophicum*. J. Bacteriol. 150:1357–65.

Knowles, R. 1982. Denitrification. Microbiol. Rev. 46:43–70.

Latham, M. J., and M. J. Wolin. 1977. Fermentation of cellulose by *Ruminococcus flavefaciens* in the presence and absence of *Methanobacterium ruminantium*. Appl. Environ. Microbiol. 34:297–301.

Lovley, D. R., D. F. Dwyer, and M. J. Klug. 1982. Kinetic analysis of competition between sulfate reducers and methanogens for hydrogen in sediments. Appl. Environ. Microbiol. 43:1373–79.

Lovley, D. R., and M. J. Klug. 1982. Intermediary metabolism of organic matter in the sediments of a eutrophic lake. Appl. Environ. Microbiol. 43:552–60.

Macgregor, A. N., and D. R. Keeney. 1973. Methane formation by lake sediments during *in vitro* incubation. Water Resour. Bull. 9:1153–58.

Mackie, R. I., and M. P. Bryant. 1981. Metabolic activity of fatty acid-oxidizing bacteria and the contribution of acetate, propionate, butyrate, and CO_2 to methanogenesis in cattle waste at 40° and 60°C. Appl. Environ. Microbiol. 41:1363–73.

Mah, R. A., D. M. Ward, L. Baresi, and T. L. Glass. 1977. Biogenesis of methane. Ann. Rev. Microbiol. 31:309–41.

Mallard, G. E., and J. I. Frea. 1972. Methane production in Lake Erie sediments: temperature and substrate effects. Proc. 15th Conf. Great Lakes Research 1972:87–93.

Martens, C. S., and R. A. Berner. 1974. Methane production in the interstitial waters of sulfate-depleted marine sediments. Science 185:1167–69.

McCarty, P. L. 1972. Energetics of organic matter degradation. *In* R. Mitchell (ed.), Water pollution microbiology, pp. 91–118. John Wiley and Sons, New York.

McInerney, M. J., and M. P. Bryant. 1981. Anaerobic degradation of lactate by syntrophic associations of *Methanosarcina barkeri* and *Desulfovibrio* species and effect of H_2 on acetate degradation. Appl. Environ. Microbiol. 41:346–54.

McInerney, M. J., M. P. Bryant, R. B. Hespell, and J. W. Costerton. 1981a. *Syntrophomonas wolfei* gen. nov. sp. nov., an anaerobic, syntrophic, fatty acid-oxidizing bacterium. Appl. Environ. Microbiol. 41:1029–39.

McInerney, M. J., R. I. Mackie, and M. P. Bryant. 1981b. Syntrophic association of a butyrate-degrading bacterium and *Methanosarcina* enriched from bovine rumen fluid. Appl. Environ. Microbiol. 41:826–28.

Meek, B. D., L. B. Grass, and A. J. MacKenzie. 1969. Applied nitrogen losses in relation to oxygen status of soils. Soil Sci. Soc. Amer. Proc. 33:575–78.

Mountfort, D. O., and R. A. Asher. 1978. Changes in proportions of acetate and carbon dioxide used as methane precursors during the anaerobic digestion of bovine waste. Appl. Environ. Microbiol. 35:648–54.

Mountfort, D. O., R. A. Asher, and T. Bauchop. 1982. Fermentation of cellulose to methane and carbon dioxide by a rumen anaerobic fungus in a triculture with *Methanobrevibacter* sp. strain RA1 and *Methanosarcina barkeri*. Appl. Environ. Microbiol. 44:128–34.

Mountfort, D. O., R. A. Asher, E. L. Mays, and J. M. Tiedje. 1980. Carbon and electron flow in mud and sandflat intertidal sediments at Delaware Inlet, Nelson, New Zealand. Appl. Environ. Microbiol. 39:686–94.

Oremland, R. S., L. Marsh, and D. J. DesMarais. 1982. Methanogenesis in Big Soda Lake, Nevada: an alkaline, moderately hypersaline desert lake. Appl. Environ. Microbiol. 43:462–68.

Patel, G. B., A. W. Khan, and L. A. Roth. 1978. Optimum levels of sulphate and iron for the cultivation of pure cultures of methanogens in synthetic media. J. Appl. Bacteriol. 45:347–56.

Patrick, W. H., and R. D. DeLuane. 1972. Characterization of the oxidized and reduced zones in a flooded soil. Soil Sci. Soc. Amer. Proc. 36:573–76.

Payne, W. J. 1981. Denitrification. John Wiley and Sons, New York.

Pfennig, N., and H. Biebl. 1976. *Desulfuromonas acetoxidans* gen. nov. and sp. nov., a new anaerobic, sulfur-reducing, acetate-oxidizing bacterium. Arch. Microbiol. 110:3–12.

Pine, M. J., and H. A. Barker. 1956. Studies on the methane fermentation XII. The pathway of hydrogen in the acetate fermentation. J. Bacteriol. 71:644–48.

Postgate, J. R. 1965. Recent advances in the study of the sulfate-reducing bacteria. Bacteriol. Rev. 29:425–41.

Reddy, C. A., M. P. Bryant, and M. J. Wolin. 1972. Characteristics of S organism isolated from *Methanobacillus omelianskii*. J. Bacteriol. 109:539–45.

Sansone, F. J., and C. S. Martens. 1981. Methane production from acetate and associated methane fluxes from anoxic coastal sediments. Science 211:707–9.

Scherer, P., and H. Sahm. 1981a. Effect of trace elements and vitamins on the growth of *Methanosarcina barkeri*. Acta Biotechnol. 1:57–65.

Scherer, P., and H. Sahm. 1981b. Influence of sulphur-containing compounds on the growth of *Methanosarcina barkeri* in a defined medium. Eur. J. Appl. Microbiol. Biotechnol. 12:28–35.

Schönheit, P., J. Moll, and R. K. Thauer. 1979. Nickel, cobalt, and molybdenum requirement for growth of *Methanobacterium thermoautotrophicum*. Arch. Microbiol. 123:105–7.

Senior, E., E. B. Lindström, I. M. Banat, and D. B. Nedwell. 1982. Sulfate reduction and methanogenesis in the sediment of a saltmarsh on the east coast of the United Kingdom. Appl. Environ. Microbiol. 43:987–96.

Shlomi, E. R., A. Lankhorst, and R. A. Prins. 1978. Methanogenic fermentation of benzoate in an enrichment culture. Microb. Ecol. 4:249–61.

Sørensen, J., D. Christensen, and B. B. Jørgensen. 1981. Volatile fatty acids and hydrogen as substrates for sulfate-reducing bacteria in anaerobic marine sediment. Appl. Environ. Microbiol. 42:5–11.

Smith, P. H., and R. A. Mah. 1966. Kinetics of acetate metabolism during sludge digestion. Appl. Microbiol. 14:368–71.

Smith, M. R., and R. A. Mah. 1978. Growth and methanogenesis by *Methanosarcina* strain 227 on acetate and methanol. Appl. Environ. Microbiol. 36:870–79.

Stadtman, T. C., and H. A. Barker. 1949. Studies on the methane fermentation. VII. Tracer experiments on mechanism of methane fermentation. Arch. Biochem. 21:256–64.

Stadtman, T. C., and H. A. Baker. 1951. Studies on the methane fermentation. IX. The origin of methane in the acetate and methanol fermentations by *Methanosarcina*. J. Bacteriol. 61:81–86.

Strayer, R. F., and J. M. Tiedje. 1978a. Application of the fluorescent-antibody technique to the study of a methanogenic bacterium in lake sediments. Appl. Environ. Microbiol. 35:192–98.

Strayer, R. F., and J. M. Tiedje. 1978b. Kinetic parameters of the conversion of methane precursors to methane in a hypereutrophic lake sediment. Appl. Environ. Microbiol. 36:330–40.

Stupperich, E., and G. Fuchs. 1981. Products of CO_2 fixation and ^{14}C labelling pattern of alanine in *Methanobacterium thermoautotrophicum* pulse-labelled with $^{14}CO_2$. Arch. Microbiol. 130:294–300.

Vogels, G. D., W. F. Hoppe, and C. K. Stumm. 1980. Association of methanogenic bacteria with rumen ciliates. Appl. Environ. Microbiol. 40:608–12.

Ward, D. M., R. A. Mah, and I. R. Kaplan. 1978. Methanogenesis from acetate: a nonmethanogenic bacterium from an anaerobic acetate enrichment. Appl. Environ. Microbiol. 35:1185–92.

Ward, T. E., and J. I. Frea. 1980. Sediment distribution of methanogenic bacteria in Lake Erie and Cleveland Harbor. Appl. Environ. Microbiol. 39:597–603.

Wellinger, A., and K. Wuhrmann. 1977. Influence of sulfide compounds on the metabolism of *Methanobacterium* strain AZ. Arch. Microbiol. 115:13–17.

Weimer, P. J., and J. G. Zeikus. 1977. Fermentation of cellulose and cellobiose by *Clostridium thermocellum* in the absence and presence of *Methanobacterium thermoautotrophicum*. Appl. Environ. Microbiol. 33:289–97.

Weimer, P. J., and J. G. Zeikus. 1978. Acetate metabolism in *Methanosarcina barkeri*. Arch. Microbiol. 119:175–82.

Weimer, P. J., and J. G. Zeikus. 1979. Acetate assimilation pathway of *Methanosarcina barkeri*. J. Bacteriol. 137:332–39.

Whitman, W. B., E. Ankwanda, and R. S. Wolfe. 1982. Nutrition and carbon metabolism of *Methanococcus voltae*. J. Bacteriol. 149:852–63.

Widdel, F., and N. Pfennig. 1977. A new anaerobic, sporing, acetate-oxidizing, sulfate-reducing bacterium, *Desulfotomaculum* (emend.) *acetoxidans*. Arch. Microbiol. 112:119–22.

Winfrey, M. R., D. R. Nelson, S. C. Klevickis, and J. G. Zeikus. 1977. Association of hydrogen metabolism with methanogenesis in Lake Mendota sediments. Appl. Environ. Microbiol. 33:312–18.

Winfrey, M. R., and J. G. Zeikus. 1977. Effect of sulfate on carbon and electron flow during microbial methanogenesis in freshwater sediments. Appl. Environ. Microbiol. 33:275–81.

Winfrey, M. R., and J. G. Zeikus. 1979a. Microbial methanogenesis and acetate metabolism in a meromictic lake. Appl. Environ. Microbiol. 37:213–21.

Winfrey, M. R., and J. G. Zeikus. 1979b. Anaerobic metabolism of immediate methane precursors in Lake Mendota. Appl. Environ. Microbiol. 37:244–53.

Winter, J. U., and R. S. Wolfe. 1979. Complete degradation of carbohydrates to carbon dioxide and methane by syntrophic cultures of *Acetobacterium woodii* and *Methanosarcina barkeri*. Arch. Microbiol. 121:97–102.

Winter, J. U., and R. S. Wolfe. 1980. Methane formation from fructose by syntrophic associations of *Acetobacterium woodii* and different strains of methanogens. Arch. Microbiol. 124:73–79.

Wolin, M. J. 1981. Fermentation in the rumen and human large intestine. Science 213:1463–68.

Zehnder, A. J. B., B. A. Huser, T. D. Brock, and K. Wuhrmann. 1980. Characterization of an acetate-decarboxylating, non-hydrogen-oxidizing methane bacterium. Arch. Microbiol. 124: 1–11.

Zehnder, A. J. B., and K. Wuhrmann. 1977. Physiology of a *Methanobacterium* strain AZ. Arch. Microbiol. 111:199–205.

Zeikus, J. G. 1977. The biology of methanogenic bacteria. Bacteriol. Rev. 41:514–41.

Zeikus, J. G., G. Fuchs, W. Kenealy, and R. K. Thauer. 1977. Oxidoreductases involved in cell carbon synthesis of *Methanobacterium thermoautotrophicum*. J. Bacteriol. 132:604–13.

Zeikus, J. G., P. J. Weimer, D. R. Nelson, and L. Daniels. 1975. Bacterial methanogenesis: acetate as a methane precursor in pure culture. Arch. Microbiol. 104:129–34.

Zeikus, J. G., and M. R. Winfrey. 1976. Temperature limitation of methanogenesis in aquatic sediments. Appl. Environ. Microbiol. 31:99–107.

G. DENNIS SPROTT AND KEN F. JARRELL

Electrochemical Potential and Membrane Properties of Methanogenic Bacteria

15

Bacteria can maintain an electrochemical potential of protons[1] ($\Delta\bar{\mu}_{H^+}$) composed of electrical and chemical components according to the relationship:

$$\Delta\bar{\mu}_{H^+} = \Delta\psi - \frac{2.3\ RT}{F}\ \Delta pH \qquad (1)$$

where $\Delta\psi$ is the electrical potential and ΔpH the pH difference across the membrane (Mitchell, 1966). Recently these components were demonstrated and measured in cells of *Methanospirillum hungatei* and *Methanobacterium thermoautotrophicum* (Jarrell and Sprott, 1981). Conflicting results of Sauer and colleagues (1981), where the absence of an electrical potential in *M. thermoautotrophicum* was claimed, appear to be explained by the omission of a hydrophobic counterion necessary to obtain uptake of triphenylmethylphosphonium cation ($TPMP^+$). An internal pH of 6.6 to 6.8 occurs in *M. thermoautotrophicum* (Jarrell and Sprott, 1981; Sauer et al., 1981), *M. hungatei* (Jarrell and Sprott, 1981), and *M. bryantii* (Jarrell and Sprott, 1983a) with the transmembrane pH difference being partially maintained at pH values above or below the cytoplasmic pH.

Reduction of CO_2 to methane requires 4 pairs of electrons from H_2 according to the reaction:

$$CO_2 + 4\ H_2 \rightarrow CH_4 + 2\ H_2O \qquad \Delta G^{\circ\prime} = -131\ kJ\ mol^{-1}. \qquad (2)$$

A scheme based on the concept of C-carriers during reduction of CO_2 (Barker, 1956) and incorporating the discovery of coenzyme M (McBride and

Wolfe, 1971) and methanopterin has been proposed (Vogels et al., 1982). Energy input is required at the step where formate-dihydromethanopterin is converted to formaldehydemethanopterin (Vogels et al., 1982). The energy may be derived from a transmembrane electrochemical potential (Kell et al., 1981), since methanogenesis from CO_2 requires closed-membrane vesicles (Sauer et al., 1980) and is sensitive to carbonylcyanide-m-chlorophenylhydrazone (CCCP) (Roberton and Wolfe, 1970; Sauer et al., 1981).

Membrane studies with methanogenic bacteria are often complicated by the ineffectiveness exhibited by classical reagents such as carbonyl cyanide-p-trifluoromethoxyphenylhydrazone (FCCP), or N,N′-dicyclohexylcarbodiimide (DCCD) (Jarrell and Sprott, 1982a; Sprott and Jarrell, 1982). In some cases the unexpected resistance may reflect an inability of the reagent to penetrate the cell wall. A spheroplast preparation sensitive to various ionophores and uncouplers would be useful in studies of the $\Delta\bar{\mu}_{H^+}$ and its relationship to methanogenesis. In the present report, we demonstrate the formation of spheroplasts of *M. hungatei* capable of active methanogenesis, measure both components of $\Delta\bar{\mu}_{H^+}$ as a function of external pH, and compare ionophore effects in cells and spheroplasts.

MATERIALS AND METHODS

Organism and Growth Conditions

M. hungatei GP1 was grown at 35°C (150 rpm) under an atmosphere of CO_2/H_2 (1:4, v/v) in a chemically defined medium containing acetate. The composition of medium JMA (K^+ content 47 mM, Na^+ 13 mM) and its reduction by the Hungate procedure have been described (Jarrell and Sprott, 1982a).

Spheroplast Formation

Ten-ml aliquots of logarithmic phase cells (2.5–4.0 mg dry weight) were harvested under CO_2/H_2 in modified Corex glass centrifuge tubes (Sprott and Jarrell, 1981). Supernatants were removed under N_2 pressure by inverting the tubes and inserting a needle through the butyl septum for fluid outflow. Each cell pellet was resuspended in 5 ml of spheroplast solution and 10 μl 3 N NaOH (stored under N_2) injected both immediately and again after 15 min incubation. The spheroplast solution was prepared by heating 0.5 M sucrose dissolved in 50 mM sodium carbonate (pH 9.5) to boiling under N_2, and by flushing the hot solution with a stream of N_2 (200 ml min^{-1}) for 15 min. The solution was dispensed in a 100 ml volume by syringe into each 1 *l* growth bottle containing 200 mg of dithiothreitol (DTT) under N_2. The bottles were sealed with butyl stoppers and stored overnight before use. Following an exposure time in spheroplast solution of 30 min at 35°C (150 rpm) the cells were centrifuged and resuspended into 5 ml of JMA medium or various buffers (containing 1 mM $MgCl_2$). The buffers were reduced as for growth medium preparation, except N_2 replaced the CO_2/H_2 during reduction and dispensing. To obtain some extreme pH values it was necessary to adjust the pH of the spheroplast suspensions with small volumes of 2 N HCl or 2 N NaOH (stored under N_2).

Chemical and Electrical Potentials

Details have been described for measuring the penetrations into cell pellets of ^{14}C-inulin, 5,5-[2-^{14}C] dimethyl-2,4-oxazolidinedione ([2-^{14}C] DMO), and ^{14}C-methylamine (Sprott and Jarrell, 1981; Jarrell and Sprott, 1981). ^{14}C-urea was used at a final concentration of 0.1 mM, and 0.15 μCi μmol^{-1}. Briefly, labels were added to the cells or spheroplasts (5 ml) in sealed centrifuge tubes under CO_2/H_2. After 15 min incubation at 35°C (150 rpm, tubes horizontal), the samples were centrifuged at ambient temperature (9700 *g*, 10 min). Radioactivity was measured in the supernatant and pellets (extracted with 1 M $HClO_4$). For the determination of ΔpH, the pH of the supernatant was measured with an Orion digital pH meter. The electrical potential was determined using a filtration technique as before (Jarrell and Sprott, 1981) except that the filters were presoaked in a solution of 10 μM $TPMP^+$ before use (Sprott et al., 1983). Binding of ^3H-$TPMP^+$ was corrected for by subtracting uptake occurring in preparations of cells or spheroplasts heat-treated for 15 min at 90°C. Care was taken to allow at least a 12 h solubilization time in Aquasol before counting the filters.

Ionophore Effects

Ionophores and DCCD were added as ethanolic solutions. Concentrations used are shown in tables 3 and 4 (p. 265). Pretreatment with inhibitors was for 15 min (C_2H_2, 1.5 h) unless stated otherwise.

Dry Weights

All activities are expressed on the basis of cell dry weights, determined by drying cell pellets at 62°C to a constant weight. About 6 to 8% of the dry weight is composed of ash.

CH_4 Synthesis

CH_4 was analyzed by gas-chromatography as before (Jarrell et al., 1982).

Materials

Labeled compounds were purchased from New England Nuclear, Lachine, Quebec, except ^{14}C-methylamine from Amersham, Oakville, Ont. The sources of ionophores and inhibitors is described elsewhere (Jarrell and Sprott, 1981).

RESULTS

Preparation and Activity of Spheroplasts

Osmotically fragile spheroplasts of *M. hungatei* can be formed by treatment with DTT at pH values above 8 (Sprott et al., 1979). Furthermore, it was noted previously that spheroplast formation was prevented by concentrations of 0.5 M sucrose, which corresponded to the point where cell plasmolysis began. Cells exposed to excess DTT at pH 8.8 to 8.9 for 30 min in various sucrose concentrations are shown in fig. 1 (top panels). In 0.5 M sucrose the cells maintained their normal shape with no spheroplasts being formed. However, a decrease in concentration to 0.4 or 0.3 M sucrose resulted in a rounding-up of the cytoplasmic contents within the cell sheath and in a progressive increase in the release of spheroplasts. Upon centrifugation and resuspension in JMA growth medium, the treated cells formed spheroplasts immediately, with a consistent yield of 95 to 99% (fig. 1, bottom panel). Remnants of the empty sheath can also be seen.

To obtain consistently high yields of spheroplasts, it was necessary to conduct the DTT reaction under a 100% N_2 atmosphere, since an alkaline pH could not be maintained reproducibly if CO_2 was present. Control cells, incubated at pH 8.8 under N_2 for 30 min, recovered most of their CH_4 synthesis activity when placed in a CO_2/H_2 atmosphere (fig. 2). The activity exhibited by spheroplasts depended on the method of preparation. Retention of the cytoplasmic contents in the sheath during the DTT reaction by employing 0.5 M sucrose gave a spheroplast preparation with up to 70% of the activity found in cells. Because 0.3 M and 0.5 M sucrose appear equally effective in preventing lysis of the spheroplasts, it is likely that the sheath protected the CH_4 synthesis reactions during exposure to DTT and alkaline pH. The cell-sheath can protect also against the lytic action of detergents (Sprott et al., 1979). As a consequence of these findings, 0.5 M sucrose was used during the preparation of spheroplasts. Such preparations produced CH_4

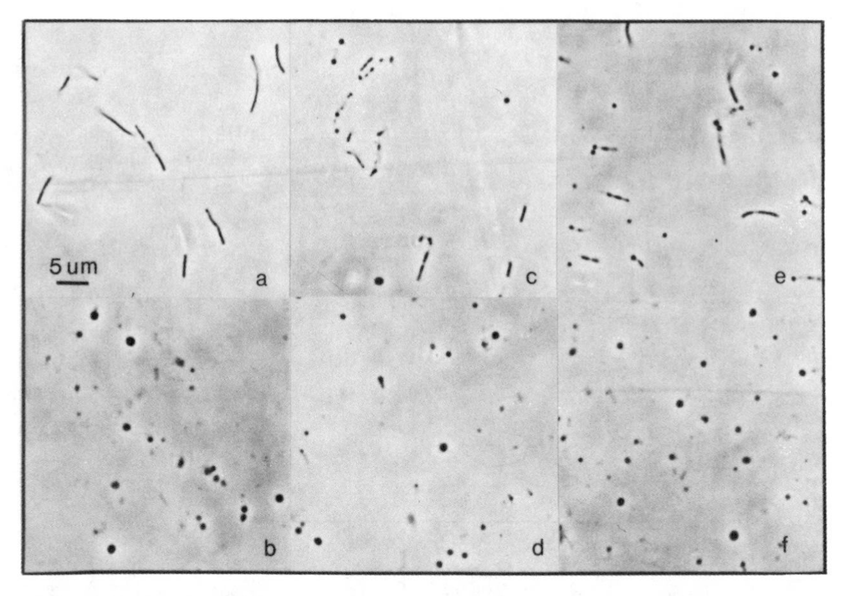

Fig. 1. Phase contrast photographs of *M. hungatei* during spheroplast formation. In the top panels, cell-forms are shown following incubation for 30 min in 50 mM sodium carbonate, DTT, and sucrose either 0.5 M (a), 0.4 M (c) or 0.3 M (e). Just following sampling for photography, the preparations (a, c, and e) were centrifuged and resuspended in JMA. Photographs of the resulting spheroplasts (b, d, and f) were taken within 5 to 10 min following resuspension (b prepared from a, d from c, and f from e).

at a constant rate for at least 3 h; incubation for 16 h resulted in a dramatic decline in activity, without lysis.

To allow a direct comparison between cells and spheroplasts, all activities are reported on the basis of *cell* dry weight. Spheroplast preparations contain the cell-sheath, which can be isolated from osmotic lysates of spheroplasts with yields of 13% of the cell dry weight (Sprott et al., 1983). Attempts to separate the sheath from the spheroplasts have been unsuccessful using anaerobic Percoll gradients.

Chemical and Electrical Potentials

Measurements of $\Delta\bar{\mu}_{H^+}$ and intracellular ion concentrations require an accurate measure of the cytoplasmic volume. In cells of *M. hungatei,* inulin, sucrose and glucose all penetrate the sheath but not the cytoplasmic membrane (Sprott and Jarrell, 1981). Because the spheroplasts appear to retain

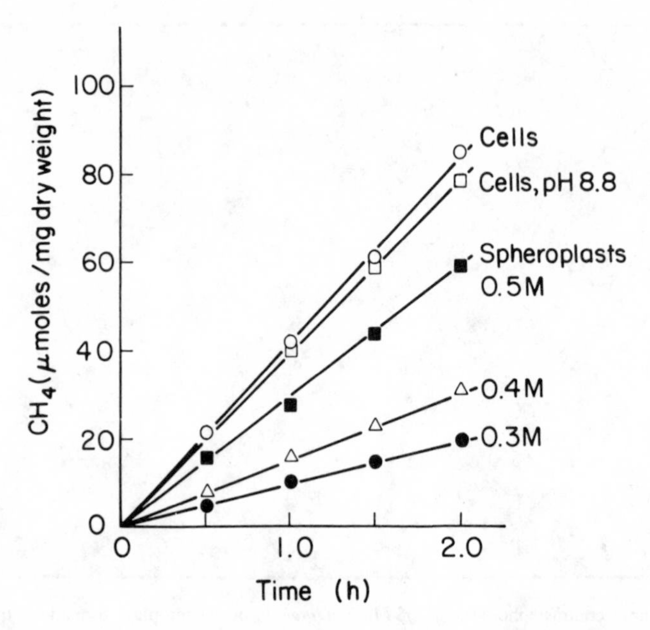

Fig. 2. Rates of CH_4 synthesis in cells and spheroplasts. Cells were centrifuged anaerobically (2.85 mg per tube) and treated for 30 min as follows: in JMA (○); in alkaline buffer (50 mM sodium carbonate, final pH 8.8), sucrose 0.3 M, but no DTT (□); in alkaline buffer, DTT included and sucrose either 0.5 M (■), 0.4 M (△) or 0.3 M (●). The treated cell preparations were again centrifuged and pellets resuspended in 5 ml JMA (pH 6.9). Each 5 ml preparation was placed in a sealed 60 ml serum bottle containing an atmosphere of CO_2/H_2 (1:4, v/v) and incubated at 35°C, 150 rpm. Methane was analyzed in 0.2 ml volumes of headspace gas by gas-chromatography.

only fragments of the unstudied inner wall (Sprott et al., 1979), we used inulin as the nonpenetrating marker. The total space (extracellular and cytoplasmic) in each pellet of spheroplasts was routinely measured with urea, which compared favorably with gravimetric values (table 1). The internal space of the spheroplast preparation was the same as found for cells of *M. hungatei* (Sprott and Jarrell, 1981).

After treatment of the cells with spheroplast solution, the spheroplasts were usually formed by resuspension of the treated-cell pellets in JMA medium. However, when the effects of external pH were measured, buffers reduced with cysteine-sodium sulfide replaced the JMA medium. Buffers of high concentration (100 mM) or containing greater than 10 mM $MgCl_2$ resulted in

TABLE 1

DETERMINATION OF INTERNAL SPACE IN SPHEROPLASTS OF *M. HUNGATEI*

Test Compound	Penetration (%)[a]	Internal Space ($\mu l\ mg^{-1}$)[b]
None (gravimetric)	103 ± 3.5	
^{14}C-urea	100 ± 3.5	
^{14}C-inulin	81 ± 1.5	2.04 ± 0.13

[a]The volume of liquid in a pellet of spheroplasts that was penetrated by urea was taken as 100 percent. The total weight of fluid in each pellet was determined gravimetrically after drying to a constant weight.

[b]The mean and standard error for 2 independent experiments are presented as $\mu l\ mg^{-1}$ dry weight of starting cell material.

lower release of spheroplasts. Consequently, the treated-cell pellets were resuspended in the desired 50 mM buffer with 1 mM $MgCl_2$. This produced consistently high yields (95 to 99%) of intact spheroplasts.

Accumulation of a weak acid (DMO) or weak base (methylamine) by spheroplasts occurred over a wide range of external pH (fig. 3). Neither compound is metabolized significantly by *M. hungatei* (Jarrell and Sprott, 1981). The magnitude of the pH gradients (ΔpH) compared well to those found in cells of *M. hungatei* (Jarrell and Sprott, 1981). The spheroplast preparation lysed at acidic pH (< 4.8), and showed an abrupt lysis point at pH 10 where the turbidity of the suspension cleared immediately. Binding of the labels was not pronounced, since little penetration above 100% (equilibration point) was observed for DMO at pH 7.3 or for methylamine at pH 6.1.

The distributions of hydrophobic phosphonium cations have been used to measure electrical potentials in a wide variety of cells (Rottenberg, 1979). Electrical potentials measured using $TPMP^+$, with tetraphenylboron as counter anion, were 79 mV for *M. hungatei,* 119 mV for *M. thermoautotrophicum,* and 127 mV for *M. bryantii* (Jarrell and Sprott, 1981, 1983a). The relatively small gradient of $TPMP^+$ established in cells of *M. hungatei* appears related to interference by the sheath or inner wall, since in spheroplasts a much larger electrical potential was calculated (table 2). An optimum probe concentration of 10 μM is observed. Similar results were obtained if the filters containing the spheroplasts were washed with 0.1 M LiCl or 0.1 M $MgCl_2$. Furthermore, the distribution of $TPMP^+$ was similar in cells of *M. bryantii* assayed either by the filter method or by rapid anaerobic centrifugation through oil (Jarrell and Sprott, 1982a).

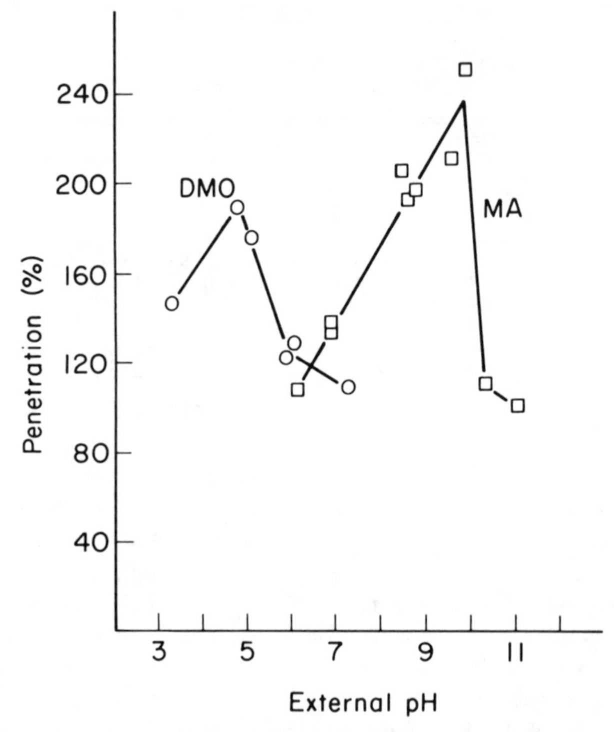

Fig. 3. Accumulation of DMO and methylamine (MA) by spheroplasts as a function of external pH. Cells (2.9 mg) were treated with spheroplasting solution (see ''Methods''), centrifuged, and resuspended in anaerobic 50 mM buffers (sodium phosphate, citrate-sodium phosphate, and MES) to which 1 mM $MgCl_2$ was added. The centrifuge tubes containing spheroplasts were incubated horizontally at 35°C (100 rpm) and ^{14}C-methylamine added. After 15 min the tubes were centrifuged at ambient temperature. Radioactivity in the supernatants and pellets was extracted with 1 M $HC10_4$ (Jarrell and Sprott, 1981). Penetrations of 100% represent equilibration of the label across the cytoplasmic membrane.

A dramatic decrease in $\Delta\tilde{\mu}_{H^+}$ of spheroplasts occurred as the external pH was increased above pH 5 (fig. 4). This occurred because the electrical potential did not increase sufficiently to compensate for the disappearance of the chemical potential. Above an external pH of 6.5, the pH gradient was reversed (inside acid). Also, the internal pH remained very constant between pH 6.5–6.7 with an external pH of 4.8–7.0, but increased gradually as the external pH was adjusted more alkaline.

TABLE 2

MEASUREMENT OF THE ELECTRICAL POTENTIAL IN CELLS AND SPHEROPLASTS OF *M. HUNGATEI* AND EFFECTS OF PROBE CONCENTRATION[a]

Preparation	$TPMP^+$ (μM)	$TPMP^+$ (In/Out)	Δψ (mV)
Cells	10	16.2	74
Spheroplasts	10	49.6	103
Spheroplasts	20	28.5	89

[a]The uptake of ^{3}H-TPMP was measured in the presence of 2 μM TPB^-. Each reaction mixture (5.0 ml) contained 4.7 mg dry weight of cell material. To correct for nonspecific uptake of $TPMP^+$ a heat-treated control (90°C, 15 min) was subtracted for each treatment.

Reagent Effects on Electrical Potential and Internal K^+ Concentrations

The electrical potential declined partially upon treatment with either valinomycin or 0.2 M KCl, and dramatically if both were present together (table 3). Dramatic inhibitions by higher concentrations of KCl did not require valinomycin. A partial collapse of the electrical potential was caused by monensin, nigericin, acetylene, and DCCD (500 μM). Other agents stimulated the uptake of $TPMP^+$. Gramicidin and especially TCS fell into this latter category. Also observed was the requirement for TPB^- to measure appreciable uptake of $TPMP^+$.

An increase in the electrical potential (inside negative) could be explained by the loss of internal K^+. Treated cells (table 4) were either centrifuged anaerobically from JMA medium (K^+ 47 mM) or centrifuged and washed with K^+-free buffer (0.1 M TES, neutralized with NaOH, and reduced with cysteine-sodium sulfide as for growth medium preparation). In either instance nigericin, C_2H_2, 500 μM DCCD, or Triton X-100 were the most effective in promoting K^+ efflux. In contrast to the results with cells, all the treatments caused marked K^+ loss from spheroplasts. Valinomycin was particularly effective in spheroplasts, but caused little K^+ outflow from the cells.

Treated preparations of cells and spheroplasts retained their normal phase-dark appearance with the following exceptions. Both cell and spheroplast preparations appeared severely plasmolyzed when suspended in the presence of 0.55 M KCl. Lower concentrations (0.15 M) of K^+ had no effect. Triton X-100 treatment lysed the spheroplast preparation completely (pH 6.9), but

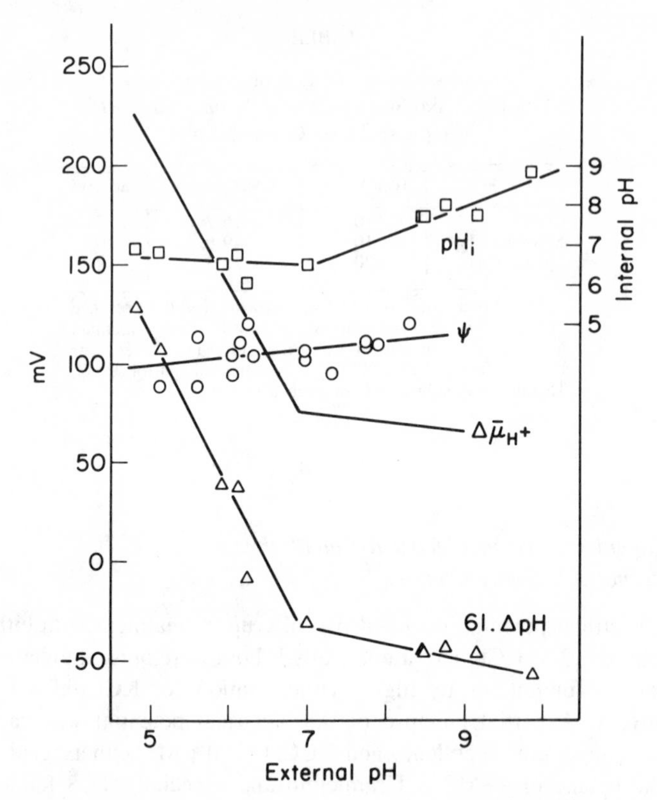

Fig. 4. The electrochemical potential in spheroplasts of *M. hungatei*. Spheroplasts were incubated in anaerobic buffers (see legend of fig. 3) under a CO_2/H_2 atmosphere in the presence of labeled compounds. Internal pH values and chemical potentials (61·ΔpH) were calculated from DMO distributions according to Mitchell et al. (1979), and from methylamine distributions as described by Rottenberg (1979). Electrical potentials were calculated from the distribution of ^{3}H-TPMP$^+$ with 2 μM tetraphenylboron present (0.46 mg of starting cell material per filter).

had no visible effect on intact cells. The lytic effect of Triton X-100 was much less pronounced at more acidic pH. Severely plasmolyzed cells and clumps of lysed vesicles remained after heating at 90°C for 15 min.

Maintenance of the electrical potential for a 2 h period did not require CO_2 or H_2 (table 5). Considerable inhibition occurred when CO_2/N_2 replaced the normal CO_2/H_2 gas phase. Also, increases in pH occurred when CO_2 was absent, but this cannot completely explain the unexpected lack of inhibition noted for H_2, N_2, and, especially, air atmospheres.

TABLE 3

EFFECTS OF IONOPHORES AND OTHER TREATMENTS ON THE ELECTRICAL POTENTIAL OF SPHEROPLASTS[a]

Treatment	$TPMP^+$ (In/Out)	mV
None	46	101
None, less TPB^-	17	75
Valinomycin (20 μM)	30	90
KCl (0.2 M)	34	93
KCl (0.55 M)	15	72
Valinomycin (20 μM) + KCl (0.2 M)	17	75
Valinomycin (20 μM) + KCl (0.55 M)	(<1)	(+ inside)
Monensin (20 μM)	39	97
Nigericin (30 μM)	21	81
Gramicidin (10 $\mu g/ml^{-1}$)	57	107
TCS (20 μM)	122	127
DNP (1 mM)	47	102
C_2H_2 (290 μM)[b]	35	94
DCCD (20 μM)	58	108
DCCD (500 μM)	32	92

[a]Spheroplasts (4.65 mg) were suspended in 5 ml JMA medium contained in 60 ml serum bottles, final pH 6.8. The gas phase was CO_2/H_2 (1:4, v/v). Uptake of $TPMP^+$ was measured after incubation for 5 min (TPB^- 2μM) by filtering 0.465 mg aliquots in triplicate.

[b]Concentration of dissolved C_2H_2.

TABLE 4

EFFECTS OF IONOPHORES AND OTHER TREATMENTS ON THE INTRACYTOPLASMIC K^+ CONCENTRATIONS (mM) IN SPHEROPLASTS AND CELLS[a]

Treatment	CELLS		SPHEROPLASTS
	JMA	K^+-free Buffer	K^+-free Buffer
	Exp. 1	Exp. 2	Exp. 3
None (ethanol)	154	191	138.0
Valinomycin	145	173	0.6
Nigericin	106	89	22.0
Monensin	130	149	25.0
Gramicidin	138	214	20.0
TCS	ND[b]	144	53.0
DCCD, 20 μM	ND	214	68.0
DCCD, 500 μM	ND	76	32.0
C_2H_2	114	103	69.0
Triton X-100 (0.05%, w/v)	ND	21	12.0

[a]Cells (1.7 mg dry weight) or spheroplasts (prepared from 5.8 mg cells) were incubated with the inhibitor in JMA medium in 60 ml serum bottles under CO_2/H_2 for 1.5 h. The suspensions were centrifuged anaerobically (9700 *g*, 10 min) and the pellets washed with JMA medium or a K^+-free buffer. K^+ was released from the cell or spheroplast pellets by treatment with n-butanol at 85°C (Sprott and Jarrell, 1981). Results were obtained from three separate experiments.

[b]ND, not determined.

TABLE 5

EFFECTS OF HEAD-SPACE GAS ON THE ELECTRICAL POTENTIAL IN SPHEROPLASTS[a]

Head-Space Gas	$TPMP^+$ (In/Out)	mV
CO_2/H_2 (20:80, v/v)	52	105
CO_2/N_2 (20:80, v/v)	24	84
H_2 (100%)	66	111
N_2 (100%)	72	113
Air (20% O_2)	78	115

[a]Spheroplasts prepared under a N_2 atmosphere from 4.2 mg of cells, and suspended in 5 ml JMA medium, were placed in 60 ml serum bottles containing the indicated gas mixtures. The reaction mixes were incubated at 35°C for 1.5 h. Electrical potential was measured as in Materials and Methods. After incubation, the pH had shifted from 6.55 (CO_2/H_2 and CO_2/N_2) to 7.65 (H_2), 7.72 (N_2) and 7.87 (air).

Ionophores and CH_4 Synthesis

As noted for the retention of internal K^+ concentrations (table 4), valinomycin, nigericin, and gramicidin were effective in inhibiting CH_4 synthesis in spheroplast preparations but not in cell suspensions (table 6). Particularly effective inhibitors for both preparations were valinomycin with 0.25 M KCl, 0.5 M KCl, DNP, C_2H_2, 500 μM DCCD, air, Triton X-100 (in spite of the absence of cell lysis), and heat treatment (table 6).

Electrical Potential and CH_4 Synthesis

Sufficiently high concentrations of $TPMP^+$ will neutralize the inside negative charge (Michel and Oesterhelt, 1976), and provide a means of determining the effect of the electrical potential on CH_4 synthesis. CH_4 synthesis was completely inhibited within the range of $TPMP^+$ concentrations tested (0.02–1.0 mM) (fig. 5), as previously found in membrane extracts of *M. ruminantium* (Sauer et al., 1979). The K_i for inhibition in cells was 145 μM and in spheroplasts 63 μM. Because the bromide salt of $TPMP^+$ was used, possible effects of the bromide ion were evaluated and found to be insignificant (fig. 5).

TABLE 6

EFFECTS OF IONOPHORES AND OTHER TREATMENTS ON THE RATE OF CH_4 SYNTHESIS IN CELLS AND SPHEROPLASTS[a]

Treatment	Activity Remaining (%)	
	Cells	Spheroplasts
Ethanol (0.4%, v/v)	100	114
Valinomycin	99	9
KCl (0.20 M)	89	100
KCl (0.25 M)	79	88
KCl (0.55 M)	2	3
Valinomycin + KCl (0.25 M)	10	3
Valinomycin + KCl (0.55 M)	5	8
Monensin	105	92
Nigericin	98	36
Gramicidin	114	25
TCS	32	22
DNP	0	0
C_2H_2	4	13
DCCD (20 μM)	86	23
DCCD (500 μM)	9	7
O_2 (20%, v/v)	5	0
Triton X-100	2	0
Heat	0	0

[a]The rates of CH_4 synthesis in JMA medium (pH 6.9) without any treatment were 42 μmol h^{-1} mg^{-1} for cells and 25 μmol h^{-1} mg^{-1} for spheroplasts.

DISCUSSION

Intact spheroplasts of *M. hungatei* were prepared that retained considerable methanogenic activity. Since most of the intracellular K^+ was retained, only a small percentage could have lysed during preparation. The usefulness of spheroplasts in these studies resides primarily in their increased sensitivity, compared with cells, to certain of the ionophores. Although nigericin and gramicidin inhibited methanogenesis only in spheroplasts, the most pronounced difference was with valinomycin. This reagent abolished both the intracellular K^+ pool and CH_4 synthesis in spheroplasts without significantly affecting either function in cells. It was previously concluded that valinomycin was ineffective in *M. hungatei* cells because in the presence of valinomycin ^{86}Rb-uptake did not occur, while an electrical potential was measured using $TPMP^+$ in the presence of TPB^- (Jarrell and Sprott, 1981). However, under certain conditions valinomycin can have an effect on intact

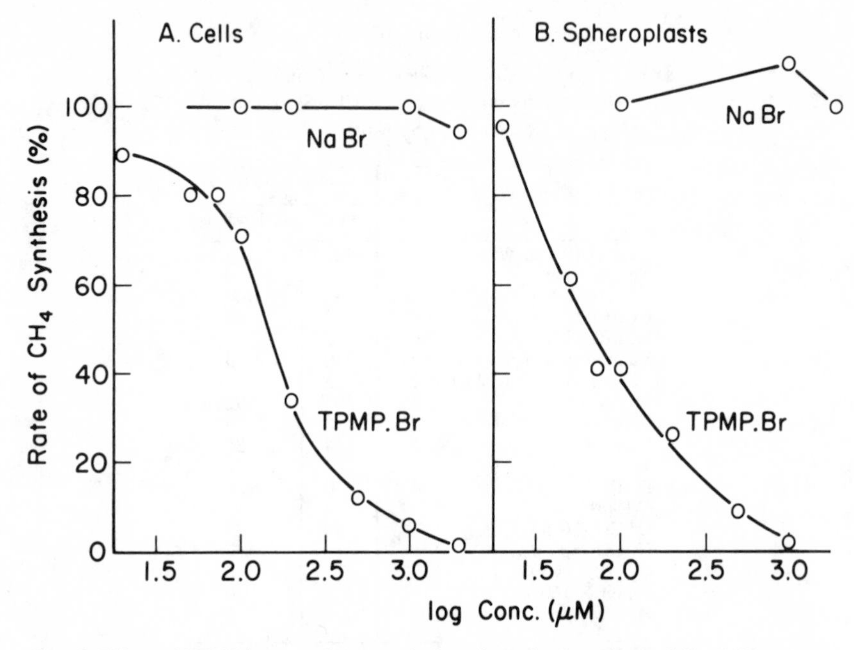

Fig. 5. Effect of $TPMP^+$ on methanogenesis in cells and spheroplasts. Cells (1.79 mg) or spheroplasts prepared from 3.58 mg of cells were incubated in 60 ml serum vials in 5 ml JMA medium under an atmosphere of CO_2/H_2. $TPMP^+Br^-$ was added as an ethanolic solution. The rate of CH_4 synthesis determined over a 90 min period was linear and is expressed as per cent of control receiving an equivalent amount of ethanol (0.2%, v/v). Effects of bromide anion were determined using sodium bromide.

cells, as shown by the valinomycin-dependent inhibition of CH_4-synthesis in the presence of 0.25 M KCl.

The cytoplasmic membrane of *M. hungatei* has been isolated and shown to consist of 45 to 50% protein, 35 to 37% lipid, and 10 to 12% carbohydrate (Sprott et al., 1983). Of particular interest is the lipid fraction, which consists primarily (94%) of C_{40}-tetraether and C_{20}-diether lipids. The completed structures of these ether-linked lipids have been determined, and on a dry weight basis 64% of the total membrane lipid is present as two dibiphytanyldiglycerol tetraether phosphoglycolipids (Kushwaha et al., 1981). Since the dimensions of the non-cyclic C_{40} chains are sufficient to span the cytoplasmic membrane (Langworthy et al., 1982), unusual physicochemical properties are predicted. Thus, the penetration and mode of action of ionophores, as well as permeabilities to ions and other compounds, may be different from those found for the more extensively studied eubacterial membranes.

The $\Delta\bar{\mu}_{H^+}$ at neutral pH would be approximately 130 mV for *M. thermoautotrophicum* (Jarrell and Sprott, 1981), 127 mV for *M. bryantii* (Jarrell and Sprott, 1983a), 80 mV for *M. hungatei* (Jarrell and Sprott, 1981), and 100 mV for spheroplasts of *M. hungatei* (present study). In all cases the most striking feature is the absence of a ΔpH at normal growth pH values of 6.6–6.8 (Jarrell and Sprott, 1981; Sauer et al., 1981). An intracellular production of H^+ in *M. thermoautotrophicum* has been shown (Spencer et al., 1980) and interpreted to indicate that an appreciable ΔpH could exist between an intravesicular space and the cytosol (Spencer et al. 1980; Sauer et al., 1981). However, thin sections of *M. hungatei* under a variety of conditions have never revealed significant amounts of internal membrane structures (G. D. Sprott, K. F. Jarrell, L. Sowden, and J. R. Colvin, unpublished).

The existence of an electrical potential across the cytoplasmic membrane of methanogenic bacteria seems certain. For some methanogens the uptake of ^{86}Rb in the presence of valinomycin or the uptake of hydrophobic phosphonium cations has been demonstrated (Jarrell and Sprott, 1981; Butsch et al., 1981). A transcytoplasmic membrane potential can be inferred from the high concentrations of intracellular K^+ found in methanogens, ranging from 150 mM in *M. hungatei* (Sprott and Jarrell, 1981) to nearly 0.9 M in *Methanobrevibacter smithii* (Jarrell and Sprott, 1982b) since the accumulation of K^+ is expected to depend at least in part on $\Delta\psi$ (Bakker and Harold, 1980). Furthermore, using spheroplasts sensitive to valinomycin, a valinomycin-dependent collapse of $\Delta\psi$ is observed when the external K^+ concentration is 0.2 M. In the absence of valinomycin, increases in the concentration of KCl to levels higher than in the cytoplasm progressively collapsed the electrical potential. A similar decline in $\Delta\psi$ of *Streptococcus faecalis* and *Escherichia coli*, as external KCl was added, indicated that the influx of K^+ was electrogenic (Bakker and Mangerich, 1981). Other reagents causing a partial decline in $\Delta\psi$ were nigericin, monensin, C_2H_2, and high concentrations of DCCD. In the case of nigericin and monensin this was unexpected, since nigericin usually mediates an electroneutral K^+/H^+ exchange and monensin an electroneutral Na^+/H^+ exchange (Bakker, 1979). An effect of C_2H_2 on $\Delta\psi$ was predicted, since this reagent results in the loss of ATP in methanogens and causes an imposed ΔpH to collapse, suggesting an increase in H^+ permeability (Sprott et al., 1982). Similarly, high concentrations of DCCD can affect $\Delta\psi$ by nonspecific membrane damage, as shown by the loss of K^+ from both cells and spheroplasts (table 4), and low concentrations can increase $\Delta\psi$ by blocking H^+ movement through the F_o-portion of the ATPase (Linnett and

Beechey, 1979). The protonophore TCS as well as gramicidin, which normally provides a membrane channel through which Na^+, K^+, and H^+ can pass (Bakker, 1979), had the unexpected result of increasing $\Delta\psi$ in spheroplasts. If the distribution of $TPMP^+$ really gives a quantitative measure of the membrane potential, then these reagents must be acting not simply as protonophores or would have collapsed as H^+ influx occurred. Recently we have found that TCS and gramicidin give the expected collapse of $\Delta\psi$ in cells of *M. bryantii* (Jarrell and Sprott, 1983b). This result was obtained only when the phosphonium assay for $\Delta\psi$ was replaced by determining the distribution of ^{86}Rb in the presence of valinomycin. These discrepancies have yet to be resolved.

An involvement of $\Delta\bar{\mu}_{H^+}$ in CO_2 reduction to CH_4 based on inhibitions of methanogenesis by protonophores seems likely (Roberton and Wolfe, 1970; Sauer et al., 1979, 1980). This is supported here by the total inhibition of methanogenesis by $TPMP^+$ (K_i of 145 and 63 μM in cells and spheroplasts, respectively) and K^+ at concentrations sufficient to discharge $\Delta\psi$. Furthermore, in the presence of 0.2 M KCl a valinomycin-dependent potent inhibition of methanogenesis corresponded to a similar loss of $\Delta\psi$.

The conservation of energy as ion gradients could occur by at least three different models. In the first two models an outwardly directed H^+ pump could be accomplished by hydrogenase or ATPase reactions. It is unlikely that *M. hungatei* uses ATP hydrolysis to establish $\Delta\bar{\mu}_{H^+}$, since 20 μM DCCD results in a dramatic decline in ATP content (Sprott and Jarrell, 1982) without any appreciable effect on $\Delta\psi$ (table 3). This supports other evidence that methanogens require ATPase activity for ATP synthesis (Mountfort, 1978; Doddema et al., 1979). The energy available as $\Delta\bar{\mu}_{H^+}$ may be used to establish an electrochemical Na^+ gradient ($\Delta\bar{\mu}_{H^+}$) through Na^+/H^+ antiport, but this activity has yet to be reported in a methanogen. In a third possible model, Na^+ efflux as the primary event to establish a $\Delta\mu_{Na^+}$ seems attractive. *M. hungatei* has an outwardly directed Na^+ pump (Sprott and Jarrell, 1981), and methanogens require Na^+ for growth and CH_4 synthesis (Perski et al., 1982). In all of the studies reported here, Na^+ added as Na sulfide was present. The lack of monensin effects at pH 6.8 should not be considered as evidence against Na^+ pumping because this ionophore is much more effective at acid pH in *M. bryantii* (Jarrell and Sprott, 1983a) and *M. thermoautotrophicum* (Perski et al., 1982). Regardless of whether a Na^+ or H^+ pump is considered as the primary event, oxygen was expected to serve as an inhibitor of $\Delta\psi$. The absence of such an inhibition (table 5) is difficult to explain at present. A point to consider is that these studies always measured the maintenance of $\Delta\psi$ rather than its formation.

ACKNOWLEDGMENTS

Kathleen Shaw is gratefully acknowledged for excellent technical support. The photographic work was done by H. Turner.

1. The abbreviations used are as follows: electrochemical potential of protons, $\Delta\bar{\mu}_{H^+}$; electrical potential, $\Delta\psi$; chemical potential, $61 \cdot \Delta pH$; triphenylmethylphosphonium cation, $TPMP^+$; carbonylcyanide-m-chlorophenylhydrazone, CCCP; carbonylcyanide-p-trifluoromethoxyphenylhydrazone, FCCP; N,N′-dicyclohexylcarbodiimide, DCCD; dithiothreitol, DTT; 5,5-dimethyl-2,4-oxazolidinedione, DMO; intracytoplasmic pH, pH_i; tetraphenylboron, TPB^-; 3,3′,4′,5-tetrachlorosalicylanilide, TCS; 2,4-dinitrophenol, DNP; N,N-bis [2-hydroxyethyl]-2-aminoethane sulfonic acid, TES; 2[N-morpholino] ethanesulfonic acid, MES.

LITERATURE CITED

Bakker, E. P. 1979. Ionophore antibiotics. *In* F. E. Hahn (ed.), Antibiotics, 5:67–97. Springer-Verlag, New York.

Bakker, E. P., and F. M. Harold. 1980. Energy coupling to potassium transport in *Streptococcus faecalis*: interplay of ATP and the proton motive force. J. Biol. Chem. 255:433–40.

Bakker, E. P., and W. E. Mangerich. 1981. Interconversion of components of the bacterial proton motive force by electrogenic potassium transport. J. Bacteriol. 147:820–26.

Barker, H. A. 1956. Biological formation of methane. *In* Bacterial fermentations, pp. 1–27. John Wiley & Sons, New York.

Doddema, H. J., C. van der Drift, G. D. Vogels, and M. Veenhuis. 1979. Chemiosmotic coupling in *Methanobacterium thermoautotrophicum:* hydrogen-dependent adenosine 5′-triphosphate synthesis by subcellular particles. J. Bacteriol. 140:1081–89.

Butsch, B. M., K. W. Hanselmann, B. A. Melandri, and R. Bachofen. 1981. Measurement of the membrane potential in whole cells of *Methanobacterium thermoautotrophicum* and other bacteria by use of an electrode sensitive to the tetraphenylphosphonium cation (TPP^+). Experientia 37:1225.

Jarrell, K. F., J. R. Colvin, and G. D. Sprott. 1982. Spontaneous protoplast formation in *Methanobacterium bryantii*. J. Bacteriol. 149:346–53.

Jarrell, K. F., and G. D. Sprott. 1981. The transmembrane electrical potential and intracellular pH in methanogenic bacteria. Can. J. Microbiol. 27:720–28.

Jarrell, K. F., and G. D. Sprott. 1982a. Nickel transport in *Methanobacterium bryantii*. J. Bacteriol. 151:1195–1203.

Jarrell, K. F., and G. D. Sprott. 1982b. High intracellular potassium concentrations in methanogens most closely related to extreme halophiles. 13th Int. Congr. Microbiol., abstract P5:1, p. 27, Boston, Massachusetts.

Jarrell, K. F., and G. D. Sprott. 1983a. The effects of ionophores and metabolic inhibitors on methanogenesis and energy-related properties of *Methanobacterium bryantii*. Arch. Biochem. Biophys. 225:33–41.

Jarrell, K. F., and G. D. Sprott. 1983b. Measurement and significance of the membrane potential in *Methanobacterium bryantii*. Biochim. Biophys. Acta 725:280–88.

Kell, D. B., H. J. Doddema, J. G. Morris, and G. D. Vogels. 1981. Energy coupling in methanogens. *In* H. Dalton (ed.), Microbial growth on C-1 compounds, pp. 159–70. Heyden & Sons, London.

Kushwaha, S. C., M. Kates, G. D. Sprott, and I. C. P. Smith. 1981. Novel polar lipids from the methanogen *Methanospirillum hungatei* GPI. Biochim. Biophys. Acta 664:156–73.

Langworthy, T. A., T. G. Tornabene, and G. Holzer. 1982. Lipids of archaebacteria. Zbl. Bakt. Hyg. I. Abt. Orig. C 3:228–44.

Linnett, P. E. and R. B. Beechey. 1979. Inhibitors of the ATP synthetase system. *In* S. Fleischer and L. Packer (eds.), Methods in enzymology, 55:472–518. Academic Press, New York.

McBride, B. C., and R. S. Wolfe. 1971. A new coenzyme of methyl transfer, coenzyme M. Biochemistry 10:2317–24.

Michel, H., and D. Oesterhelt. 1976. Light-induced changes of the pH gradient and the membrane potential in *Halobacterium halobium*. FEBS Lett. 65:175–78.

Mitchell, P. 1966. Chemiosmotic coupling in oxidative and photosynthetic phosphorylation. Biol. Rev. 41:445–502.

Mitchell, W. J., I. R. Booth, and W. A. Hamilton. 1979. Quantitative analysis of proton-linked transport systems. Glutamate transport in *Staphylococcus aureus*. Biochem. J. 184:441–49.

Mountfort, D. O. 1978. Evidence for ATP synthesis driven by a proton gradient in *Methanosarcina barkeri*. Biochem. Biophys. Res. Commun. 85:1346–51.

Perski, H. J., P. Schönheit, and R. K. Thauer. 1982. Sodium dependence of methane formation in methanogenic bacteria. FEBS Lett. 143:323–26.

Roberton, A. M., and R. S. Wolfe. 1970. Adenosine triphosphate pools in *Methanobacterium*. J. Bacteriol. 102, 43–51.

Rottenberg, H. (1979). The measurement of membrane potential and ΔpH in cells, organelles, and vesicles. *In* S. Fleischer and L. Packer (eds.), Methods in enzymology, 55:547–69. Academic Press, New York.

Sauer, F. D., J. D. Erfle, and S. Mahadevan. 1979. Methane synthesis without the addition of adenosine triphosphate by cell membranes isolated from *Methanobacterium ruminantium*. Biochem. J. 178:165–72.

Sauer, F. D., J. D. Erfle, and S. Mahadevan. 1980. Methane production by the membranous fraction of *Methanobacterium thermoautotrophicum*. Biochem. J. 190:177–82.

Sauer, F. D., J. D. Erfle, and S. Mahadevan. 1981. Evidence for an internal electrochemical proton gradient in *Methanobacterium thermoautotrophicum*. J. Biol. Chem. 256:9843–48.

Spencer, R. W., L. Daniels, G. Fulton, and W. H. Orme-Johnson. 1980. Product isotope effects on in vivo methanogenesis in *Methanobacterium thermoautotrophicum*. Biochemistry 19:3678–83.

Sprott, G. D., J. R. Colvin, and R. C. McKellar. 1979. Spheroplasts of *Methanospirillum hungatei* formed upon treatment with dithiothreitol. Can. J. Microbiol. 25:730–38.

Sprott, G. D., and K. F. Jarrell. 1981. K^+, Na^+, and Mg^{2+} content and permeability of *Methanospirillum hungatei* and *Methanobacterium thermoautotrophicum*. Can. J. Microbiol. 27:444–51.

Sprott, G. D., and K. F. Jarrell. 1982. Sensitivity of methanogenic bacteria to dicyclohexylcarbodiimide. Can. J. Microbiol. 28:982–86.

Sprott, G. D., K. F. Jarrell, K. M. Shaw, and R. Knowles. 1982. Acetylene as an inhibitor of methanogenic bacteria. J. Gen. Microbiol. 128:2453–62.

Sprott, G. D., K. M. Shaw, and K. F. Jarrell. 1983. Isolation and chemical composition of the cytoplasmic membrane of the archaebacterium *Methanospirillum hungatei*. J. Biol. Chem. 258:4026–31.

Vogels, G. D., J. T. Keltjens, T. J. Hutten, and C. van der Drift. 1982. Coenzymes of methanogenic bacteria. Zbl. Bakt. Hyg. I. Abt. Orig. C 3:258–64.

LEONARD A. HOOK, ROBERT E. CORDER, PAUL T. HAMILTON,
JAMES I. FREA, AND JOHN N. REEVE

Development of a Plating System for Genetic Exchange Studies in Methanogens Using a Modified Ultra-Low Oxygen Chamber

16

INTRODUCTION

Methanogens constitute a morphologically diverse group of archaebacteria whose salient physiological characteristic is the production of methane gas. They are strict anaerobes generally capable of metabolizing a limited number of carbon sources (CO_2, formate, acetate, methanol, and methylamine) only if the Eh of their growth milieu is poised below −330 mV (Smith and Hungate, 1958). Use of pre-reduced growth media, prepared by the Hungate technique (Hungate, 1969), was an early milestone in the development of anaerobic culture techniques needed to study the biology of methanogens. More recent innovations include the use of syringes for culture transfer (Macy et al., 1972), cultivation of methanogens in sealed serum bottles (Miller and Wolin, 1974), and pressurization of the headspace of culture vessels up to 202.6 kPa using a mixture of 20% CO_2:80% H_2 (v/v) (Balch and Wolfe, 1976). Obtaining growth of methanogens on solid media is complicated by their extreme intolerance of oxygen and has generally been accomplished either by using anaerobic roll tubes (Hungate, 1969; Holdeman and Moore, 1972) or by using Petri plates incubated inside a Freter-type anaerobic glove box under an oxygen-free atmosphere (Aranki and Freter, 1972). Rapid growth has been obtained by incubating culture plates inside a plastic, ultra-low oxygen chamber (Edwards and McBride, 1975) or inside a pressurized, stainless-steel cylinder (Balch and Wolfe, 1976). In both methods the atmosphere inside the inner incubation vessel consisted of 20% CO_2:80% H_2 (v/v), which is the optimum stoichiometric ratio for methanogenesis.

For ease of manipulation, the method of choice for handling and growth of methanogens on solid media is the pressurized, ultra-low oxygen chamber housed in an anaerobic glove box. It has, however, been necessary for individual investigators to construct the ultra-low oxygen chamber, as this system was not commercially available. In this report we describe the rapid growth of methanogens on an agar-solidified, minimal medium in Petri plates incubated in pressurized Oxoid model HP 11 anaerobe jars. The jars function as ultra-low oxygen chambers (Edwards and McBride, 1975) when incubated within a Forma Scientific model 1024/1030 anaerobic glove box equipped with incubators to house the anaerobe jars. The completely assembled system consists of readily available manufactured items and requires only a minimum amount of in-house construction. The system described has proved to be extremely useful in the plating of mixed cultures for isolation of new methanogens and in genetic studies requiring mutant selection and replica plating.

MATERIALS AND METHODS

Assembly of Components

The Forma Scientific model 1024 Freter-type anaerobe chamber (Forma Scientific, Marietta, Ohio) consisted of (1) an automatic sequence airlock, (2) a stainless steel 109 × 74 × 76 cm enclosure with a vinyl front window, and (3) one pair of gloves. It was factory-equipped with a 0.065 m^3 incubator (20–60°C) and two stainless-steel utility shelves. We added a Forma Scientific model 1030 anaerobic cabinet, which doubled the working area and provided an additional pair of gloves. The model 1030 cabinet also housed a 0.128 m^3 incubator that, combined with the incubator space in the basic unit, provided sufficient room for simultaneous incubation of ten Oxoid anaerobe jars.

Each Oxoid Model HP 11 anaerobe jar (Oxoid USA, Columbia, Maryland) comprised a clear, polycarbonate jar, a rubber-coated wire Petri dish holder (14–plate capacity), and a steel lid fitted with a pressure/vacuum gauge, two Shraeder valves, and a pressure relief valve. The knurled-knob bolt on each jar, normally used to secure the lid, was replaced with a 10 × 60 mm, stainless-steel, metric hex-head bolt (Metro Industries, Columbus, Ohio). The jars withstood extended incubations at 37°C while pressurized to the maximum recommended pressure of 82.7 kPa. The two Shraeder valves allowed jars either to be evacuated to a vacuum of 760 mm Hg, or allowed gas exchange by pressurizing through one valve while venting the exiting gas through the other valve.

Anaerobic Gas Supply

The Forma chamber was fitted with a gas-inlet valve (Whitey Valve, no. B1-RS4, Scioto Valve, Westerville, Ohio) and a toggle-switch vent valve (Whitey no. B-1GS4) to facilitate pressurization of the Oxoid anaerobe jars inside the anaerobe glove box. The valves were mounted in the ceiling of the chamber (Swagelok bulkhead union no. B-400-61), near the junction of the two units so that anaerobe jars incubated in either half of the chamber could be gassed from the same lines. Pressurizing gases (usually a mixture of 20% CO_2:80% H_2, v/v) were delivered to the anaerobe jars, inside the chamber, using 6 mm Tygon tubing fitted with a Milton no. S-698 female air chuck. The female air chuck was the type commonly used to fill automobile tires and is available in automotive parts stores.

The gases supplied to the Oxoid anaerobe jars (20% CO_2:80% H_2, v/v) and to the working space of the anaerobe chamber (90% N_2:10% H_2, v/v) were mixed with a Matheson model 7481T four-channel gas mixer (Matheson Gas Products, Dayton, Ohio). Gas flows to the gas mixer were controlled by four toggle-switch on/off gas valves (Whitey no. B-1GS4). To provide optimum gas flows at the correct pressure to the anaerobe chamber (approximately 40 $l\ min^{-1}$ at 68.9 kPa) and to the gassing manifold (see below) (approximately 6 $l\ min^{-1}$ at 275.6 kPa), the gas mixer was equipped with three Model 603 and one Model 605 rotameters (Matheson Gas Products, Dayton, Ohio). Gas pressure was regulated with a Matheson Model 3476 step-down regulator. A 3-way ball valve (Whitey no. B-42XS4) directed the gas mixtures either to the anaerobe chamber or to the gassing manifold via an all copper oxygen scrubber (Balch and Wolfe, 1976). The design of the gassing manifold was a modification of the system described by Balch and Wolfe (1976). The gassing manifold was constructed of a 41-cm-long piece of 2.54-cm-diameter copper pipe fitted with nine Nupro no. B-4J valves (Nupro Co., Willoughby, Ohio), a 3-way ball valve (Whitey no. 1-RS-4), and a pressure/vacuum gauge. The manifold was used to out-gas open vessels, according to the Hungate technique (Hungate, 1969), and to exchange gases in the headspace of sealed anaerobe tubes (Bellco no. 2048-00150, Bellco Glass, Vineland, New Jersey) or serum bottles by using a vacuum pump connected to the gassing manifold. Gases were delivered, via 6 mm Tygon tubing, to culture vessels through 2 ml cotton-filled, syringe barrels (Balch and Wolfe, 1976) connected to the tubing by Becton-Dickinson (B-D) stainless steel Luer-Lok tips (part nos. B-D 3081 and B-D 3096).

Media

Agar-based growth media were prepared anaerobically according to the Hungate technique (Hungate, 1969); autoclaved in modified, 1 l Wheaton bottles (Balch and Wolfe, 1976), and dispensed into plastic, disposable, 100 × 15 mm Petri dishes inside the anaerobe chamber. The methanogen growth medium (MMM) was Medium I of Balch et al. (1979) with the following modifications (g l^{-1}): NH_4Cl, 2.7; $FeSO_4 \cdot 7H_2O$, 0.01; $MgCl_2 \cdot 6H_2O$, 0.1; L-cysteine·HCl·H_2O, 0.1; $Na_2S \cdot 9H_2O$, 0.1; $NiCl_2$, 0.13. Sodium formate, yeast extract, trypticase, and the following trace elements were omitted: Zn, Cu, B, Ca, Mn, and Al. Sterile cysteine-sulfide reducing agent (final concentration of 0.025%, w/v) was added aseptically after the growth medium had been autoclaved. To ensure full reduction before inoculation, solidified plates were exposed to H_2S for 24 h in anaerobe jars that contained open, 16 × 150 mm test tubes half-filled with a solution of 2.5% (w/v) $Na_2S \cdot 9H_2O$. Inoculated plates were incubated in the Oxoid anaerobe jars at 37°C with the headspace gas (20% CO_2:80% H_2, v/v) maintained at 82.7 kPa by repressurization at approximately 24 h intervals.

Organisms

Methanogens used in this study are listed in table 1. Novel methanogen isolates are described in table 2.

Determination of Plating Efficiencies

Organisms growing in MMM medium were diluted in the same medium. Viable counts were obtained by spreading 0.1 ml aliquots, from appropriate

TABLE 1

METHANOGEN STRAINS USED IN THIS STUDY

Organism	Cell Morphology	Source or Reference
Methanococcus vannielii	coccus	R. S. Wolfe
Methanococcus maripaludis	coccus	Jones et al., 1983
Methanospirillum hungatei JFl	spirillum	M. P. Bryant
Methanosarcina barkeri MS	sarcina	M. P. Bryant
Methanobrevibacter smithii PS	rod	M. P. Bryant
Methanobacterium formicicum MFl	rod	J. P. Robinson
Methanobacterium bryantii M.o.H.	rod	M. P. Bryant
Methanococcus deltae ΔRC	coccus	Corder et al., 1983
Methanogenium olentangyi ER/RC	coccus	Corder et al., 1983

TABLE 2

NEW METHANOGEN ISOLATES OBTAINED FROM NATURAL SAMPLES USING THE FORMA SCIENTIFIC 1024/1030 ANAEROBE CHAMBER, OXOID HP 11 ANAEROBE JARS, AND METHANOGEN MINIMAL MEDIUM PLATES[a]

Organism	Cell Morphology	Source	Isolated By	O.S.U. No.
Methanococcus ΔRC	Coccus	Miss. R. Delta	R. Corder	981
Methanococcus ΔLH	Coccus	Miss. R. Delta	L. Hook	962
Methanogenium ER/RC	Coccus	Olentangy R.	R. Corder	961
Cuyahoga Isolate PC[2]	Coccus	Cuyahoga R.	R. Corder	934
Maine Isolate MMC	Coccus	Penobscot Bay	R. Corder	952
Methanobacterium formicicum RC	Rod	Cuyahoga R.	R. Corder	969
M. formicicum LH2	Rod	Cuyahoga R.	L. Hook	951
M. formicicum LH4	Rod	Cuyahoga R.	L. Hook	956
Methanobacterium MMY	Rod	Penobscot Bay	R. Corder	971
Methanobacterium CS-8	Rod	Sewage digester	L. Hook	984
Methanobacterium OLM	Rod	Olentangy R.	T. Rosvanis	903
Methanobacterium Tenn. R.	Rod	Tennessee R.	P. Hamilton	725
Methanobacterium Ohio R.	Rod	Ohio R.	P. Hamilton	683
Methanobacterium BR-10	Rod	Miss. R.	L. Hook	831
M. thermoautotrophicum RC	Rod	Cuyahoga R.	R. Corder	967
M. thermoautotrophicum LH	Rod	Cuyahoga R.	L. Hook	904
Methanobacterium JP 1B	Rod	Sewage digester	L. Hook	657
Methanobacterium Rum 5	Rod	Bovine rumen	L. Hook	664
Methanobacterium JP 6	Rod	Sewage digester	L. Hook	668
Methanobacterium GC-3A	Rod	Alaska oil well	R. Fleming	671
Methanobacterium GC-2B	Rod	Alaska oil well	R. Fleming	672
Methanobacterium GC-3B	Rod	Alaska oil well	R. Fleming	673
M. formicicum DJF-1	Rod	Cuyahoga R.	D. Feldhake	674
Methanobacterium sp. DJF-2	Rod	Cuyahoga R.	D. Feldhake	675
Methanobacterium "Brevi" DJF-3	Rod	Cuyahoga R.	D. Feldhake	900
Methanosarcina barkeri	Sarcina	Penobscot Bay	T. Rosvanis	922
Methanosarcina Rumen Isolate	Sarcina	Bovine rumen	R. Corder	669
Methanobrevibacter JP3	Rod	Sewage digester	L. Hook	665
Methanobrevibacter JP4	Rod	Sewage digester	L. Hook	666
Methanobrevibacter JP2A	Rod	Sewage digester	L. Hook	667

[a]The new methanogenic isolates listed here were obtained from various anaerobic samples. An inoculum was introduced into 20 ml of methanogen minimal medium in 150 ml serum bottles containing a headspace gas of 20% $CO_2 - 80\%$ H_2 v/v at 275 kPa. These enrichment cultures were incubated at 37°C with shaking. Gas utilization was monitored. Positive enrichments were streaked onto plates of MMM and the plates incubated in Oxoid anaerobe jars at 82.7 kPa. Isolated colonies were picked and transferred to broth, regrown, and tested for purity by microscopic observation and by transfer to MMM supplemented with 0.5% w/v glucose, 0.5% w/v yeast extract, and 0.5% w/v trypticase (under N_2 headspace). All isolates have been stored as frozen suspensions in MMM plus 10% v/v glycerol under liquid nitrogen. The designations given to these novel isolates reflect the morphology of the organism as determined by light microscopy. Permanent taxonomic assignments must await detailed characterization of each isolate.

dilutions, onto plates containing MMM solidified by addition of 1.5% agar. Microscopic cell counts were obtained by use of a Petroff-Hausser counting chamber.

RESULTS

The modified Forma Scientific model 1024/1030 anaerobe chamber is shown in figure 1. Copper gassing lines can be seen connected to the chamber ceiling. Two gas valves, mounted in the rear wall (factory-installed), were used for connecting vacuum lines to suction flasks inside the chamber. A modified, Oxoid model HP 11 anaerobe jar is shown in figure 2. Milton no. S-698 female air chucks are shown fitted with retainer clips, which enabled the user to clip the devices onto the Shraeder valves so that, if necessary, manual adjustments in the gas mixture or flow rates could be made during gassing (fig. 2A).

Addition of the Matheson model 7481T four-channel gas mixer to the system (fig. 3) allowed a greater flexibility in the proportioning of gas mixtures and also simplified gas cylinder purchases by removing the need for special (and expensive) premixed gases. In routine use, after presetting the four individual gas-flow needle valves on the gas mixer, it was possible to switch from one gas mixture to another in a matter of seconds by using toggle-switch on/off valves and adjusting the Matheson 3476 step-down regulator (fig. 3).

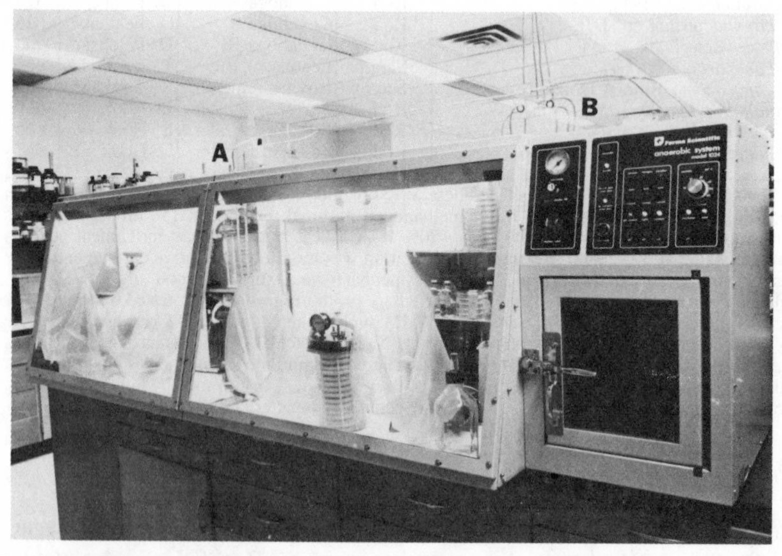

Fig. 1. Forma Scientific model 1024/1030 anaerobic glove box. Copper gas lines feed either the anaerobe jars (A) or the working space of the chamber (B).

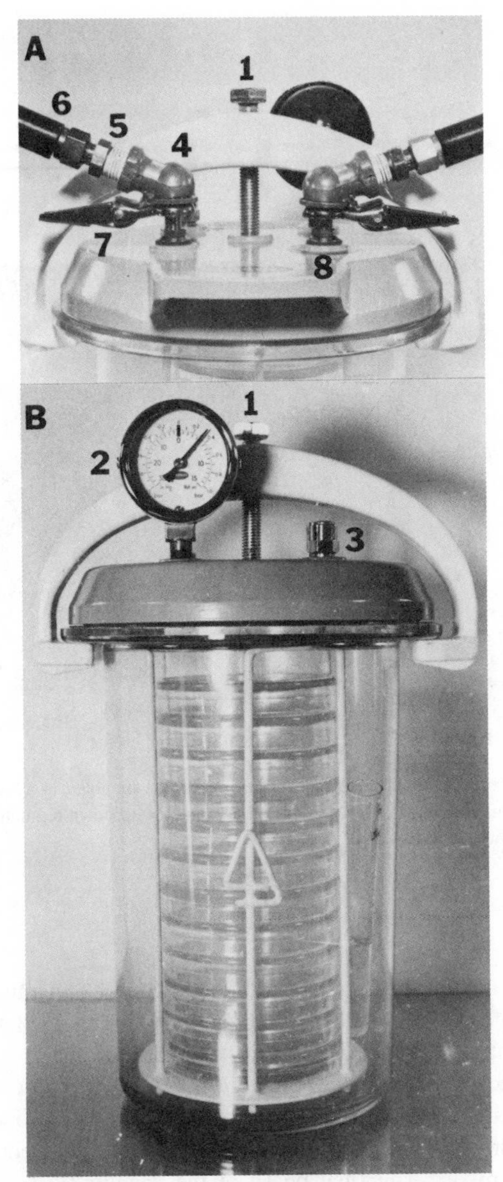

Fig. 2. Oxoid model HP 11 anaerobe jar. (1) Replacement 10 × 60 mm, stainless steel, metric, hex-head bolt; (2) pressure/vacuum gauge; (3) pressure relief valve; (4) Milton no. S698 female air chuck; (5) Cajon no. B-4-TA-1-4 adapter; (6) 0.64 cm copper tube fitted with Swagelok connector and rubber hose handle; (7) retainer clip; (8) Shraeder valve.

Fig. 3. Matheson model 7481T four-channel gas mixer (A) mounted in a wooden cabinet; (B) toggle switch on/off gas valves; (C) Matheson model 3476 step-down regulator; (D), 3-way ball valve; (E) gas flow needle valves.

The gassing manifold (fig. 4), a modification of the apparatus described by Balch and Wolfe (1976), could be used for anaerobic manipulations of culture vessels (Hungate technique) at the same time that the anaerobe chamber was in use, even though both devices drew gas mixtures from a common source. During the vacuum/flush cycling of the airlock of the anaerobe chamber, N_2 fill gas was drawn directly from a supply cylinder not connected to the Matheson gas mixer. The final phase of the vacuum/flush cycle filled the airlock with the gas mixture inside the working space of the anaerobe chamber. This gas transfer caused a slight pressure drop within the working space that was replenished at the convenience of the user. The working space gas

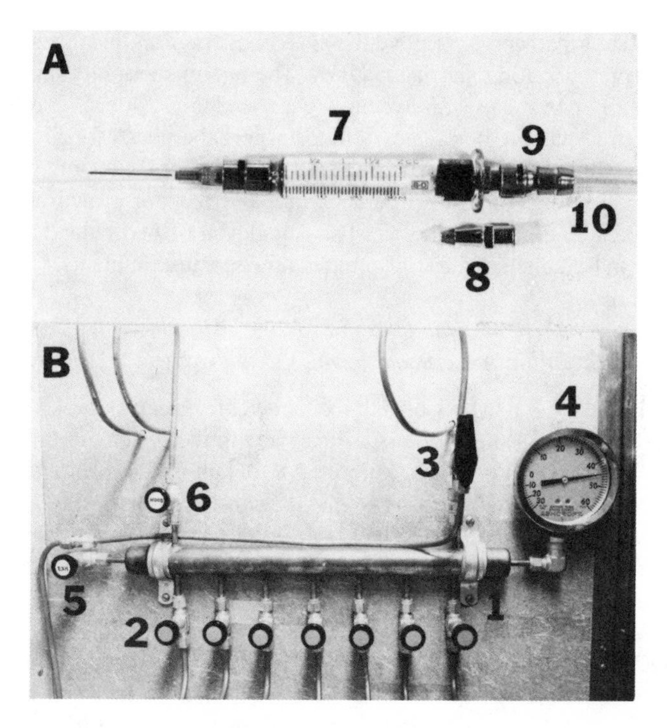

Fig. 4. Modified gassing manifold of Balch and Wolfe (1976). (1) 2.54 cm capped, copper pipe, (2) Nupro valves; (3) 3-way ball valve; (4) pressure/vacuum gauge; (5) exhaust port; (6) gas feed to anaerobe chamber (for pressurizing anaerobe jars); (7) cotton-filled, 2 ml syringe barrel; (8) B-D no. 3081 Luer-Lok tip; (9) B-D no. 3096 Luer-Lok tip; (10) 6 mm diameter Tygon tubing.

mixture (90% N_2:10% H_2, v/v) was directed from the Matheson gas mixer to the anaerobe chamber when the mixer was available.

The minor modifications made to the gassing manifold design of Balch and Wolfe (1976) were intended to simplify construction as well as improve durability. Tygon-tubing gas lines fitted with Luer-Lok tips (fig. 4A) allowed higher gas flow rates and were more flexible and durable than polyethylene tubes connected with Vacutainer needle-holders as described previously (Balch and Wolfe, 1976). The 2.54 × 41 cm copper pipe (fig. 4B), when pressurized, acted as a gas reservoir that delivered uniform flow rates to each of the seven gassing lines. The Forma Scientific anaerobe chamber/Oxoid anaerobe jar system had certain advantages over previously described systems. The system was capable of coincidentally housing and incubating 140 plates without the need to remove the anaerobe jars from the anaerobe chamber for repressurization and incubations. The jars were rapidly repressurized *in situ* based upon the pressure/vacuum gauge readings for each jar. Since the jars were transparent, it was possible to inspect the plates for growth while they were still inside the closed jars. Two independent incubators were available within the anaerobic environment so that effects of growth at different temperatures could be monitored. This should be of particular value in the isolation and characterization of temperature sensitive mutants.

Growth of Methanogens on Solid Media Using the Forma Scientific Anaerobe Chamber Oxoid Anaerobe Jar System

Established and newly isolated methanogenic species (tables 1 and 2) generally grew luxuriously on plates incubated in the system described. Standard loop-streaking procedures were used to obtain isolated colonies (fig. 5). Viable counts were determined by plating aliquots from serial dilutions of suspensions of methanococci. The viable counts were routinely in excess of 50% of the total counts (table 3). The plating efficiencies were not unreasonable considering that methanococci are fragile and prone to lysis during manipulation.

Spreading 0.1 ml of a stationary phase culture of a methanogen onto a plate gave confluent growth in 48 h at 37°C. Antibiotic sensitivities were determined by placing antibiotic-impregnated disks, or spotting 5 μl of an antibiotic solution, onto the lawns after 24 h of incubation (fig. 5). In agreement with previous reports (Pecher and Böck, 1981; Smith and Mah, 1981; Weisburg and Tanner, 1982), we found that methanogens were resistant to most commonly used antibiotics but were sensitive to bromoethanesulfonate,

TABLE 3

PLATING EFFICIENCIES OF METHANOCOCCAL SPECIES[a]

Organism	A_{600}[b]	Plate Count (cfu ml^{-1})	Petroff-Hausser Count (cells ml^{-1})	Plating Efficiency (%)
Methanococcus vannielii	0.37[c]	1.18×10^8	2.10×10^8	56.2
Methanococcus voltae	0.78	2.81×10^8	3.63×10^8	77.4
Methanococcus deltae	0.80	2.79×10^8	5.46×10^8	50.4

[a]Averages of three replicates.

[b]Optical densities of cultures were determined in optically matched anaerobic culture tubes (Bellco Glass, Vineland, N.J.) in a Bausch and Lomb Spectronic 20 spectrophotometer.

[c]The culture was a 1:1 dilution of a late-log phage culture.

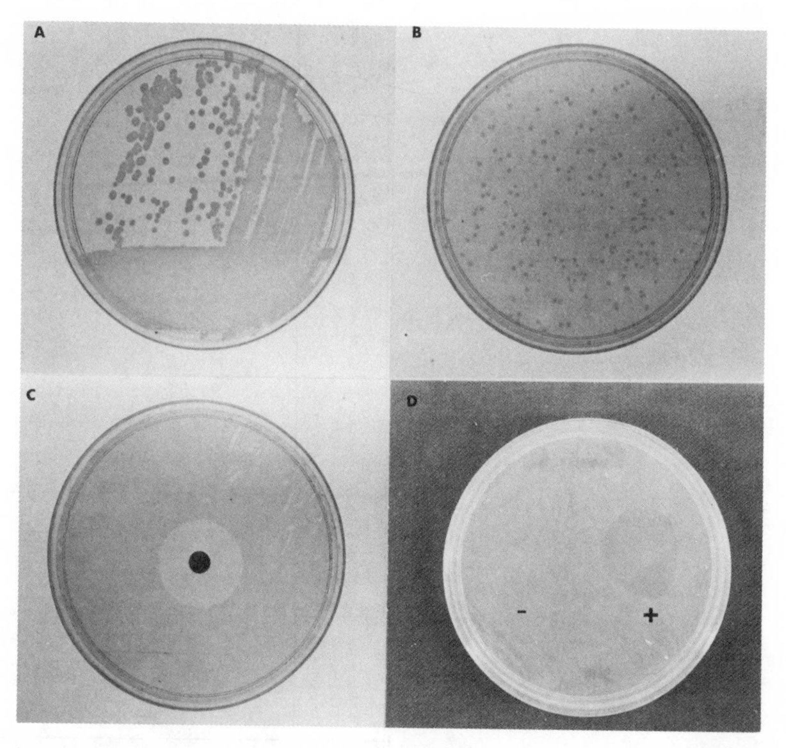

Fig. 5. Demonstration of four common bacteriological plating procedures for methanogens using the Forma Scientific anaerobe chamber and Oxoid anaerobe jars. (A) Streak plate: new methanogen, sewage sludge isolate *Methanobacterium* JP12A, streaked onto MMM agar and incubated at 37°C for 7 days; (B) spread plate: 0.1 ml of a dilution of a *Methanococcus vannielii* culture spread on MMM agar and incubated for 5 days at 37°C; (C) antibiotic disc sensitivity test: *Methanobacterium* JP12A lawn grown on MMM agar in the presence of a disc impregnated with 1 μg leucinostatin; (D) antibiotic spot sensitivity test: *Methanococcus vannielii* was grown as a lawn on MMM agar. A 5 μl volume of pyrrolnitrin (1 mg ml^{-1} in dimethylsulfoxide [DMSO]) was spotted on the growing lawn as indicated by +. An identical volume of DMSO alone had no effect on methanogen growth (−). This system could also be used to expose growing methanogens to mutagenic agent or to test for the presence of viruses.

monensin, chloramphenicol, and, in some cases, to bacitracin. All methanogens tested were found to be sensitive to leucinostatin, metronidazole, and pyrrolnitrin (table 4). Sensitivity to other antibiotics was species-dependent. Spotting of 5 μl of an antibiotic solution onto a growing lawn of a sensitive species resulted in an area of growth inhibition. Colonies isolated from the

TABLE 4

ANTIBIOTIC SENSITIVITIES OF METHANOGENS[a]

Antibiotic	*Methano-genium olentangyi*	*Methano-coccus deltae*	*Methano-coccus vannielii*	*Methano-coccus maripaludis*	*Methano-spirillum hungatei* JF	*Methano-sarcina barkeri* MS	*Methanobre-vibacter smithii* PS	*Methano-bacterium formicicum* MF1	*Methano-bacterium bryantii* M.o.H.
Actinomycin	—	100	—	—	—	—	—		
Adriamycin	10	10	10	10	—	—	1	—	—
Anthelmycin	—	—	—	—	—	—		—	—
Anthelvencin	—	100	—	—	—	100	—	—	—
Bleomycin sulfate	—	—	—	—	—	—	—	—	—
Cefotaxime	—	—	—	—	—	—	—	—	—
Cefoxitin	—	—	—	—	—	—	—	—	—
Cephaloridine	—	—	—	—	—	—	—	—	—
Efrapeptin	10	10	100	100		100	—		—
Echinocandin	—	100	—	—	—	—	—		—
Hygromycin	—	—	—	—	—	—	—	—	—
Leucinostatin	1	1	1	10	1	1	10		10
Metronidazole	1	1	1	1	1	1	1	1	10
Monensin	1	1	1	1	1	1		1	100
Mycophenolic acid	—	—	100	—	—	—	—	—	—

Neomycin sulfate	—	—	—	—	—	100	—	—	—
Pleuromutilin	—	10	—	10	—	—			10
Pyrrolnitrin	100	10	100	100	10	10	10	100	100
Rutamycin	—	10	—	—	—	—		—	—
Sinefungin	—	—	—	—	—		—		—
Tobramycin	—	—	—	—	—	—	—	—	—
Tunicamycin	—	10	—	—	—	100	—		—
A201A	—	10	100	—	—	—	—	—	—
A2315	—	—	1	1		—	—	—	—
A7413	100	1	—	10	100		100	—	—
A21978C	—	—	1	—		100	—	—	—
A23187	—	—	—	—	—	1	—	10	
A41030A	100	100	—	—	—	—	—	10	—
G418	—	—	—	—	—	—	—	—	—
Trimethoprim	—	—	—	—	—	—	—	—	—

[a]Difco absorbent discs were impregnated with 1, 10, or 100 μg of the test antibiotic and placed on lawns of the methanogens growing on the defined minimal medium. Antibiotic sensitivities are indicated as the least quantity of antibiotic that caused a measurable zone of growth inhibition. The symbol (—) indicates no zone of inhibition. A blank indicates that test was not completed. Adriamycin was purchased from Adria Laboratories, Wilmington, DE 19899; monesin and neomycin sulfate were purchased from Sigma, St. Louis, MO. Metronidazole was a gift from J. B. Cornett, (Sterling-Winthrop Research Institute); G418 was a gift from J. Davis (Biogen, Geneva); and the remaining antibiotics were generously provided by R. L. Hamill (Lilly Research Laboratories, Indianapolis, Ind.). The unnamed antibiotics (excepting G418) are listed by their Eli Lilly numerical designation.

periphery of these areas are currently being screened to derive drug-resistant mutants of the sensitive species.

DISCUSSION

Facilities are now available that allow the use of standard procedures for isolating and characterizing drug-resistant mutants, auxotrophic mutants, and/or temperature-sensitive mutants of methanogens. Previous observations were confirmed that mutants of methanogenic species can be obtained that have an increased resistance to bromoethanesulfonate (Smith and Mah, 1981). Isolates of *Methanococcus vannielii* resistant to some of the antibiotics, which normally inhibit the growth of this species, were also obtained. The availability of drug-resistance-conferring mutations and plating techniques that allow the selection of recombinants has provided the bases for screening protocols for a genetic exchange system in the methanogens. Furthermore, our recent studies indicate that *M. vannielii* may be particularly applicable for further characterization of a genetic exchange system in methanogens.

ACKNOWLEDGMENTS

This study was supported by grant DE-AC02-81ER10945 from the Department of Energy. J. N. R. is the recipient of Research Career Development Award 5K04AG00108 from the National Institute of Aging.

LITERATURE CITED

Aranki, A., and R. Freter. 1972. Use of anaerobic glove boxes for the cultivation of strictly anaerobic bacteria. Amer. J. Clin. Nutr. 25:1329–34.

Balch, W. E., and R. S. Wolfe. 1976. New approach to the cultivation of methanogenic bacteria: 2-mercaptoethane-sulfonic acid (HS-CoM)-dependent growth of *Methanobacterium ruminantium* in a pressurized atmosphere. Appl. Environ. Microbiol. 32:781–91.

Corder, R. E., L. A. Hook, J. M. Larkin, and J. I. Frea. 1983. Isolation and characterization of two new methane-producing cocci: *Methanogenium olentangyi,* sp. nov., and *Methanococcus deltae,* sp. nov. Arch. Microbiol. 134:28–32.

Edwards, T., and B. C. McBride. 1975. New method for the isolation and identification of methanogenic bacteria. Appl. Microbiol. 29:540–45.

Holdeman, L. V., and W. E. C. Moore. 1972. Roll-tube technique for anaerobic bacteria. Amer. J. Clin. Nutr. 25:1314–17.

Hungate, R. E. 1969. A roll tube method for cultivation of strict anaerobes. *In* J. R. Norris and D. W. Ribbons (eds.), Methods in microbiology, 3B:117–32. Academic Press, New York.

Jones, W. J., M. J. B. Paynter, and R. Gupta. 1983. Characterization of *Methanoccocus maripaludis* sp. nov., a new methanogen isolated from salt marsh sediment. Arch. Microbiol.134:91–97.

Macy, J. M., T. E. Snellen, and R. E. Hungate. 1972. Use of syringe methods for anaerobiosis. Amer. J. Clin. Nutr. 25:1318–23.

Miller, T. L., and M. J. Wolin. 1974. A serum bottle modification of the Hungate technique for cultivating anaerobes. Appl. Microbiol. 27:985–87.

Pecher, T., and A. Böck. 1981. *In vivo* susceptibility of halophilic and methanogenic organisms to protein synthesis inhibitors. FEMS Microbiol. Lett. 10:295–97.

Smith, P. H., and R. E. Hungate. 1958. Isolation and characterization of *Methanobacterium ruminantium* n. sp. J. Bacteriol. 75:713–18.

Smith, M. R., and R. A. Mah. 1981. 2-Bromoethanesulfonate: a selective agent for isolating resistant *Methanosarcina* mutants. Curr. Microbiol. 6:321–26.

Weisburg, W. G., and R. S. Tanner. 1982. Aminoglycoside sensitivity of archaebacteria. FEMS Microbiol. Lett. 14:307–10.

PAUL T. HAMILTON AND JOHN N. REEVE

Cloning and Expression of Archaebacterial DNA from Methanogens in *Escherichia coli*

17

INTRODUCTION

Methane biogenesis ranks with processes such as photosynthesis and nitrogen fixation as a major factor in the global cycling of nutrients. Methane-generating microorganisms (methanogens) play a vital role in the environment by catalyzing the final stage in the decomposition of waste biomass to methane. As the opportunities to use methane as a fuel increase, it is to be expected that methanogens may also soon play a much larger role in the world's economy. The study of methanogens themselves has been impeded by their extreme sensitivity to oxygen; they require redox potentials below -330mV for growth, and only recently have techniques been developed by which methanogens can easily be cultivated as pure cultures in the laboratory (Balch et al., 1979; Hook et al., 1984). The advent of these techniques has facilitated a rapid expansion in the study of the biology of methanogens and the biochemistry of methane biogenesis. Results obtained to date indicate that methanogens form an extremely diverse group (Balch et al., 1979; Prévot, 1980; Wolfe, 1979; Zeikus, 1977). They span the full range of prokaryotic morphological types and have DNA contents that range from 27.5 to 52 mol% G+C. It has, in fact, been proposed that methanogens are not true prokaryotes but representatives of a third biological kingdom called the Archaebacteria (Balch et al., 1977; Woese and Fox, 1977; Woese et al., 1978). This kingdom, which includes extreme halophilic and acidothermophilic microorganisms in addition to the methanogens, was proposed because its members have several properties radically different from both prokaryotic and eukaryotic species. Archaebacteria have unique structural subunits in their cell envelopes and lipids (Kandler and König, 1978). Their

RNA and protein-synthesizing systems are resistant to most of the antibiotics that inhibit these processes in prokaryotes and eukaryotes (Pecher and Böck, 1981; Weisburg and Tanner, 1982; Hook et al., 1984). Their DNA-dependent RNA-polymerases are radically different from those of classical prokaryotes (Stetter et al., 1980), and comparative analyses of the sequences of tRNAs and rRNAs have provided convincing evidence that, whereas the different archaebacterial types are related to each other, they are only very distantly related in evolutionary terms to current prokaryotic and eukaryotic species (Balch et al., 1979).

Although these biochemical properties suggest that the Archaebacteria, including methanogens, are fundamentally different from classical prokaryotes, we recently demonstrated that DNA isolated from methanogens could be transcribed and translated into polypeptides when cloned in *Escherichia coli* (Reeve et al., 1982). This result has been confirmed by other workers (Bollschweiler and Klein, 1982). Our interpretation was that Archaebacteria must therefore employ the standard genetic code. This has now been directly substantiated by the cloning and sequencing of a bacteriorhodopsin gene from the archaebacterium *Halobacterium halobium* (Dunn et al., 1981). We have continued to investigate the expression of methanogen-derived DNA in *E. coli* and here report the cloning and initial characterization of expression of DNA sequences that function to complement auxotrophic mutations in *E. coli*. A preliminary report of these results has been presented (Hamilton and Reeve, 1982).

MATERIALS AND METHODS

Bacteria

The organisms used in this study are listed in table 1.

Growth media

Initially methanogens were grown under two atmospheres pressure of H_2:CO_2 (80:20) in a medium containing (per liter) 4 g tryptone, 2 g Na_2CO_3, 2 g CH_3COONa, 2 g yeast extract, 0.3 g K_2HPO_4, 0.3 g KH_2PO_4, 0.6 g $(NH_4)_2SO_4$, 0.6 g NaCl, 0.13 g $MgSO_4 \cdot 6H_2O$, 8 mg $CaCl_2 \cdot H_2O$, 0.125 g cysteine, and 0.125 g $Na_2S \cdot 9H_2O$. In later experiments the growth medium was changed to that described by Hook and colleagues (1984). Cultures were routinely checked for contamination by microscopic examination and by plating samples on Brain-Heart Infusion plates (Difco Laboratories, Detroit, Mich.) that were incubated both aerobically and anaerobically.

Escherichia coli strains used for preparation of plasmid DNAs and minicells were grown in L-broth (1% w/v Bacto-tryptone, 0.5% w/v yeast extract, 0.5% w/v NaCl) supplemented with either ampicillin (100 $\mu g\ ml^{-1}$) or tetracycline (5 $\mu g\ ml^{-1}$) as appropriate. Selection and screening of transformants was on L-broth plates (L-broth containing 15 g agar l^{-1}) containing 100 μg ampicillin ml^{-1} or 5 μg tetracycline ml^{-1}. Complementation of *E. coli* auxotrophic requirements was tested on plates containing M9-minimal salts (Miller, 1972) supplemented with 40 μg of the appropriate amino acids ml^{-1}; 50 μg adenosine ml^{-1} and 1 μg of appropriate vitamins ml^{-1}. The plates were incubated aerobically at 37°C.

DNA Isolation from Methanogens

Methanococci do not have rigid, peptidoglycan-like cell envelopes (Jones et al., 1977). Therefore, cells washed and resuspended in TE (10 mM Tris-HCl; 1 mM EDTA) were lysed by addition of sodium dodecylsulfate (SDS) (final concentration, 1% w/v). Lysis of the other methanogenic species (table 1) was obtained by subjecting frozen, concentrated cell suspensions to cryoimpacting as described by Smucker and Pfister (1975). DNA was purified from the cell lysates by phenol extraction as described by Marmur (1961).

The simple lysis procedure described above for methanococci can be accomplished in the loading wells of agarose gels and electrophoresis then used to screen for the presence of plasmids. To date, plasmids have not been detected by this technique in the methanococcal species investigated.

Cloning in Plasmid pBR322

*Hind*III-digested, methanogen-derived DNA was ligated to *Hind*III-digested pBR322 using T4 DNA ligase. The reaction mixture was then used to transform competent cells of *E. coli* 5K (Lederberg and Cohen, 1974). Ampicillin-resistant, tetracycline-sensitive transformants were selected using the D-cycloserine enrichment technique for recombinants (Bolivar and Backman, 1979). Plasmid DNA was prepared from mixtures of many different transformed clones and used to transform *E. coli* χ760 or *E. coli* χ985. Clones were selected for their ability to grow in the absence of one of the auxotrophic requirements of *E. coli* χ760 or *E. coli* χ985 (table 1).

Agarose Gel Electrophoresis, Plasmid DNA Preparation, and Nick Translation

Restriction enzymes were purchased from Bethesda Research Laboratories (Gaithersburg, Md.) and used as recommended by the supplier. Preparation of

TABLE 1

BACTERIAL STRAINS USED IN THIS STUDY

Name	Relevant Genetic Characteristics	Use	Source
E. coli χ760	*ara*-1, *leu-1*, *azi*^{r} *ton*A^r, *lac*Y2, *pro*C119 *tsx*, *pur*E1, *gal*K2, *trp*3 *his*4, *arg*G36, *rps*L, *xyl*-1 *mtl*-1, *ilv*A6, *thr*-1, *met*-12	Marker complementation[a]	R. Curtiss III
E. coli χ985	*thr*16, *car*33, *ton*A33, *lac*Y29, *pro*C24, *tsx*63, *pur*E41, *sup*E42, *pdx*C3, *pyr*F, *trp*32, *his*53, *met*C65, rpsL97, *tte*l, *xyl*14, *ilv*277, *cyc*B2, *cyc*A1	Marker complementation[a]	R. Curtiss III
E. coli EC130	*pro*, *arg*G::Mucts62	Marker complementation[a]	N. Glansdorff
E. coli 5K	*thr*, *thi*, *hsd*R$^-$, *hsd*M$^+$	Restriction-deficient host for initial cloning and *E. coli* K12 modification of cloned DNA	J. Collins
E. coli DS410	*min*A, *min*B	Minicell production. Derivatives of DS410, containing recombinant plasmids, were obtained by transformation and selection for ampicillin resistance	J. N. Reeve

Methanosarcina barkeri MS		Source of DNA for cloning	M. P. Bryant
Methanobrevibacter smithii MS		Source of DNA for cloning	M. P. Bryant
Methanospirillum hungatei MS		Genomic DNA screened for homology to cloned *M. barkeri* MS and *M. smithii* DNAs	M. P. Bryant
Methanococcus vannielii		Genomic DNA screened for homology to cloned *M. barkeri* MS and *M. smithii* DNAs	R. S. Wolfe
Methanococcus maripaludis		Genomic DNA screened for homology to cloned *M. barkeri* MS and *M. smithii* DNAs	W. J. Jones
Methanogenium olentangyi		Genomic DNA screened for homology to cloned *M. barkeri* MS and *M. smithii* DNAs	R. E. Corder
Methanococcus deltae		Genomic DNA screened for homology to cloned *M. barkeri* MS and *M. smithii* DNAs	R. E. Corder
Halobacterium halobium		Genomic DNA screened for homology to cloned *M. barkeri* MS and *M. smithii* DNAs	J. Cassim
Bacillus subtilis 168		Genomic DNA screened for homology to cloned *M. barkeri* MS and *M. smithii* DNAs	N. H. Mendelson

[a]Complementation of *arg*G36 and *arg*G::Mucts62 (using cloned *M. barkeri* MS DNA) and *pur*E1 and *pur*E41 (using cloned *M. smithii* DNA) was obtained following amplification of the DNA in *E. coli* 5K.

plasmid DNAs, agarose gel electrophoresis using Tris-acetate buffer, and nick translation using [α-^{32}P]-dCTP (800 Ci mmol^{-1}; New England Nuclear, Boston, Mass.) were carried out as described by Davis and colleagues (1980).

DNA fragments in agarose gels were stained by submersion of the gel in a solution of 1 μg ethidium bromide ml^{-1} for 15 min. Stained gels were photographed through Kodak filters 2E plus 23A onto Polaroid type 655 film using 300 nm light transillumination.

DNA-DNA Hybridization of Dehydrated Agarose Gels

Ethidium bromide-stained agarose gels were photographed, placed in 50 mM NaOH for 1 hr and then rinsed for 30 min in water. The gels were placed on Whatman 3MM paper and thoroughly dehydrated by heating under vacuum using a Hoeffer Model SE-1140 gel drier (Hoeffer Scientific Instruments, San Francisco, Calif.). Dried gels were removed from the Whatman 3MM support and used in hybridization reactions. The dried gels were treated exactly as the nitrocellulose membrane would be used in a standard Southern blot procedure (Davis et al., 1980).

Minicell Isolation, Incorporation of L-[^{35}S]-methionine, and Analysis of Radiochemically Labeled Polypeptides by Polyacrylamide Gel Electrophoresis and Fluorography

Plasmids were introduced into the minicell-producing strain *E. coli* DS410 by transformation (Lederberg and Cohen, 1974). Minicells were prepared from these strains and allowed to incorporate L-[^{35}S]-methionine (1060 mCi mmol^{-1}; New England Nuclear, Boston, Mass.) as previously described (Reeve, 1979). Samples were analyzed by fluorography after separation by electrophoresis through 10–20% polyacrylamide gradient gels (Reeve and Shaw, 1979).

RESULTS

Isolation and Restriction Endonuclease Digestion of DNA from Methanogens

The cell envelopes of methanogens do not contain lysozyme-sensitive peptidoglycans (Kandler and König, 1978), and therefore procedures that employ lysozyme to facilitate the lysis of prokaryotic species are ineffective with methanogens. Two lysis procedures have been employed to obtain DNA from methanogens. Cryo-impacting (Smucker and Pfister, 1975) was used to lyse methanogens that have a rigid component, analogous to prokaryotic peptidoglycans, in their cell envelopes (Kandler and König, 1978; [table 1]. The addition of SDS was sufficient to lyse methanococcal species, which have proteinaceous cell envelopes (Jones et al., 1977; [table 1]). DNAs purified by phenol extraction from the cell lysates were exposed to different restriction endonucleases, and the products of these digestions were analyzed by agarose gel electrophoresis. Figure 1 shows the electrophoretic separation of DNA fragments generated by *Hind*III (tracks 1–5) and *Bam*Hl (tracks 6–10) digestion of DNAs from five different methanogens. In general, methanogen-derived DNAs were susceptible to digestion by many different restriction enzymes. Some exceptions have been observed such as the refractility of

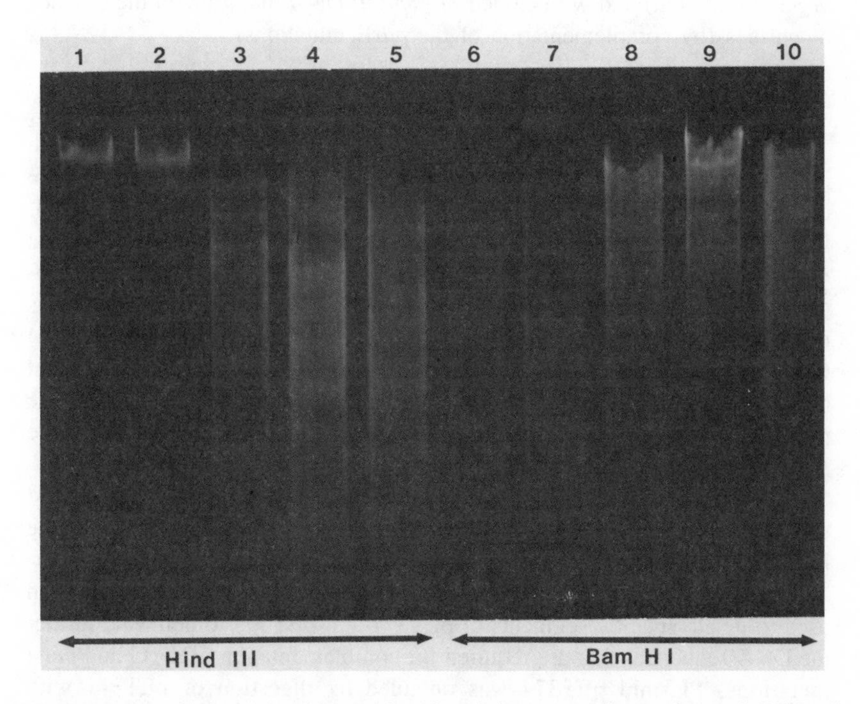

Fig. 1. Agarose gel electrophoretic separation of restriction enzyme digests of genomic DNAs of methanogens. Genomic DNA isolated from *Methanospirillum hungatei* (tracks 1 and 6); *Methanogenium olentangyi* (tracks 2 and 7); *Methanococcus deltae* (tracks 3 and 8); *Methanococcus vannielii* (tracks 4 and 9); and *Methanococcus maripaludis* (tracks 5 and 10) were digested with *Hind*III (tracks 1–5) or *Bam*H1 (tracks 6–10). DNAs from *Methanospirillum hungatei* and *Methanogenium olentangyi* are not restricted by *Hind*III, which has the recognition and cleavage sequence 5′-AAGCTT. These DNAs are also not cleaved by the enzyme *Alu*I, which recognizes and cleaves the sequence 5′AGCT, the central tetramer of the *Hind*III cleavage sequence.

DNAs from *Methanospirillum hungatei* and *Methanogenium olentangyi* to digestion with *Hind*III and *Alu*I (fig. 1).

Complementation of E. coli Auxotrophic Requirements

*Hind*III-generated fragments of DNA from *Methanosarcina barkeri* MS and *Methanobrevibacter smithii* were ligated to pBR322, and the ligation mixture was used to transform *E. coli* 5K. Transformants were selected by their ability to grow in the presence of ampicillin. Plasmid DNA was isolated from a pooled mixture of ampicillin-resistant transformants and used to transform *E. coli* χ760 and χ985. Transformants were obtained with *M. barkeri* MS DNA that grew in the absence of arginine (by complementation of the *arg*G36 mutation) and with cloned *M. smithii* DNA that grew in the absence of purines (by complementation of the *pur*E mutations).

Analysis of Recombinant Plasmids

Plasmid DNA was isolated from clones of *E. coli* χ760 capable of growth in the absence of arginine or purines and subjected to restriction endonuclease digestions. Figure 2 shows the agarose gel electrophoretic patterns produced. Plasmids pET371 and pET372 facilitate complementation of the *argG36* and *arg*G::Mu*cts*62 mutations, and plasmids pET405 and pET406 facilitate complementation of the *pur*E1 and *pur*E41 mutations. (Designation pET was reserved [2/23/81] with the Plasmid Reference Center for our use in cloning.) Plasmids pET371 and pET372 contain the same *Hind*III-generated fragment (2.9 kb) of *M. barkeri* MS DNA cloned in opposite orientations relative to the pBR322 vector DNA. Similarly, pET405 and pET406 contain the same *Hind*III-generated fragment of *M. smithii* DNA cloned in opposite orientations. Plasmids pET372 and pET406 can be obtained by *Hind*III digestion and religation of plasmids pET371 and pET405, respectively. Figure 3 shows the location of restriction enzyme cleavage sites in these plasmids.

Digestion with restriction endonucleases, followed by religation, has been used to delete specific segments of pET371 and pET405 to delineate further the DNA sequences that are required for complementation of *arg*G and *pur*E mutations. Plasmid pET374 was obtained by digestion of pET371 with *Bam*H1 and *Bgl*II, religation of the cohesive ends produced, and transformation using ampicillin resistance to select transformants (figure 3). Plasmid pET374 does not complement *arg*G mutations, indicating that the DNA deleted from pET371 in the formation of pET374 is essential for *arg*G complementation. Plasmid pET371 contains two sites for *Pst*1, one of which is located in the *M. barkeri* MS DNA. Plasmids were obtained following *Pst*1

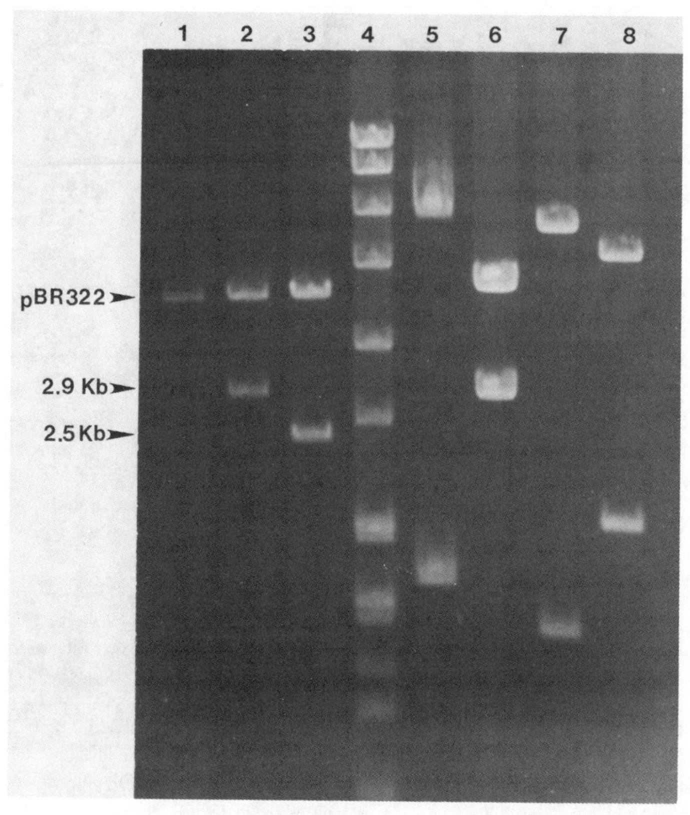

Fig. 2. Agarose gel electrophoretic separation of restriction enzyme digests of recombinant plasmid DNAs. Plasmid DNAs were prepared from cleared lysates by CsCl-ethidium bromide equilibrium density gradient centrifugation (Davis et al., 1980). The figure shows the DNA fragments produced by *Hind*III digestion of pBR322 (track 1); *Hind*III digestion of pET371 (track 2); *Hind*III digestion of pET405 (track 3); *Eco*RI digestion of the *B. subtilis* phage SPP1 used as size markers (Ratcliff et al., 1979) (track 4); *Pst*1 digestion of pET371 (track 5); *Pst*1 digestion of pET372 (track 6); *Bam*H1 digestion of pET405 (track 7) and *Bam*H1 digestion of pET406 (track 8).

digestion of pET371, religation and selection for transformants of *E. coli* χ760 that could grow in the absence of arginine. Plasmids isolated from these transformants were identical to pET371, indicating that DNA sequences between the two *Pst*1 sites were also essential for *arg*G36 complementation. (Digestion of pBR322-derived plasmids with *Pst*1 inactivates the β-lactamase

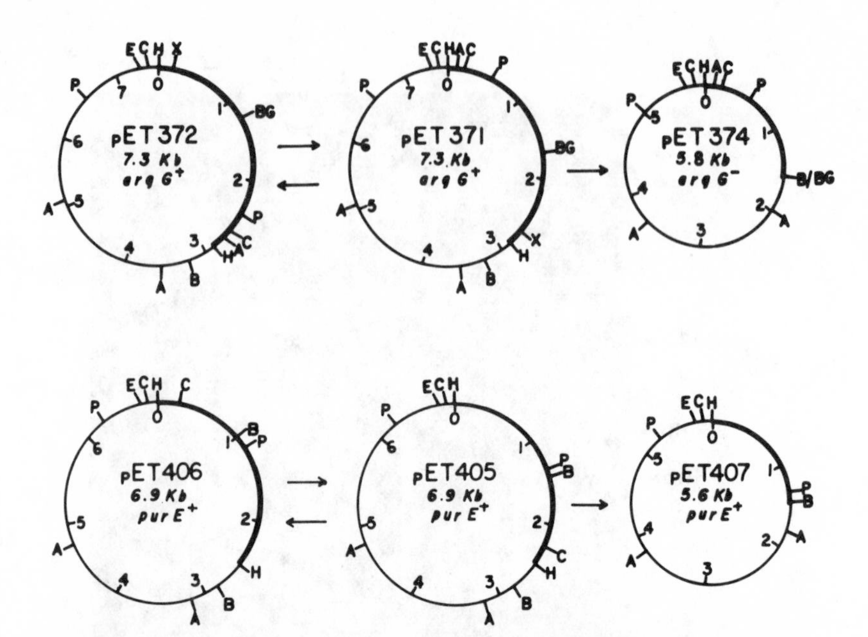

Fig. 3. Restriction maps of recombinant DNA plasmids. Plasmids pET371 and pET372 contain the same 2.9 kb fragment of *M. barkeri* MS DNA (heavy line) cloned in opposite orientations in pBR322. Plasmid pET374 was derived from pET371 by *Bam*H1 and *Bgl*II digestion and religation. Experiments (see text) indicate that *arg*G complementation requires the presence of DNA sequences extending beyond both the *Bgl*II and *Pst*1 sites in the cloned *M. barkeri* MS DNA. There are approximately 1,500 base pairs of DNA between these sites, sufficient to encode a polypeptide of approximately 50,000 molecular mass. The major polypeptide encoded by pET371 and pET372, but not by pET374, has an apparent molecular mass of 51,000 (fig. 4). Plasmids pET405 and pET406 contain the same 2.5kb fragment of *Methanobrevibacter smithii* DNA (heavy line) cloned in opposite orientations in pBR322. Plasmid pET407 was obtained by *Bam*H1 digestion of pET405 and religation. Plasmids pET405 and pET406 encode two novel polypeptides (M_r = 53,000 and 37,000) whereas pET407, which also complements *pur*E, directs the synthesis of only the 37,000 molecular mass polypeptide (fig. 4). The methanogen-derived DNA in pET405 does not have sites for the enzymes *Xor*II, *Xho*I, *Xba*I, *Sal*I, *Eco*RI, *Sma*I, *Sst*I, *Sst*II, *Bgl*II, and *Bcl*I. The methanogen-derived DNA in pET371 does not have sites for the enzymes *Bst*EII, *Bam*H1, *Bcl*1, *Eco*RI, *Kpn*1, *Sal*1, *Sma*1, *Sst*1, *Sst*II, *Xho*I, and *Xma*III. Sites for the enzymes *Eco*RI, *Hind*III, *Bam*H1, *Bgl*II, *Xba*I, *Cla*I, *Acc*I, and *Pst*1 are indicated in the figure by the letters E, H, B, Bg, X, C, A, and P, respectively.

gene, and therefore *Pst*1 digestion of pET371 followed by ampicillin selection of transformants was not appropriate for selection of deletions starting at the *Pst*1 site in pBR322).

Plasmid pET407 was obtained by digestion of pET405 with *Bam*H1, religation, and transformation using ampicillin resistance to select transformants. Plasmid pET407 not only confers ampicillin resistance but also complements *pur*E mutations, indicating that the *M. smithii* DNA segment deleted from pET405 in the formation of pET407 is not required for the complementing activity.

Polypeptide Synthesis Directed by Recombinant Plasmids in Minicells

Minicells produced by derivatives of *E. coli* DS410 carrying one of the plasmids pET371, pET372, pET374, pET405, pET406, pET407, or pBR322 were incubated in the presence of [^{35}S]-methionine to obtain synthesis of radioactively labeled, plasmid-encoded polypeptides, which were then visualized by fluorography following their separation by electrophoresis through 10–20% polyacrylamide gradient gels (Reeve, 1979). A sample of the pBR322-containing minicells was infected with coliphage T7 before addition of the [^{35}S]-methionine. The T7 gene products (Ponta et al., 1977), synthesized in radioactive form in these infected minicells, were used as molecular mass standards for estimating the molecular masses of polypeptides encoded by the recombinant plasmids. Plasmids pET371 and pET372 directed the synthesis of large amounts of a polypeptide with an apparent relative molecular mass of 51,000. This polypeptide was not synthesized in pET374-containing minicells and was presumably responsible for the complementation of *arg*G mutations (i.e., argininosuccinate synthetase activity; fig. 4). Minicells containing pET371 and pET372 also synthesized small amounts of a large number of polypeptides, all of which were smaller than the 51,000 molecular mass major product. Pulse-chase experiments were undertaken to determine if these molecules represented degradation products of the 51,000 molecular mass polypeptide. We were unable to chase radioactivity into these molecules from the major product, and it therefore seems probable that the smaller molecules represent products of events that cause premature termination of transcription and/or translation during the synthesis of the major product. Similar observations and conclusions have been made in other studies of polypeptides synthesized in plasmid-containing minicells (Meagher et al., 1977; Reeve, 1978). Plasmids pET371, pET372 and pET374 all conferred ampicillin resistance and therefore directed the synthesis of β-lactamase. The precursor and mature forms of β-lactamase synthesized in minicells containing pBR322 or the pBR322-based recombinant plasmids are indicated in figure 4. Although all the pET recombinant plasmids used in figure 4 confer resistance to high levels of ampicillin (100 μg ml^{-1}), they do not appear to

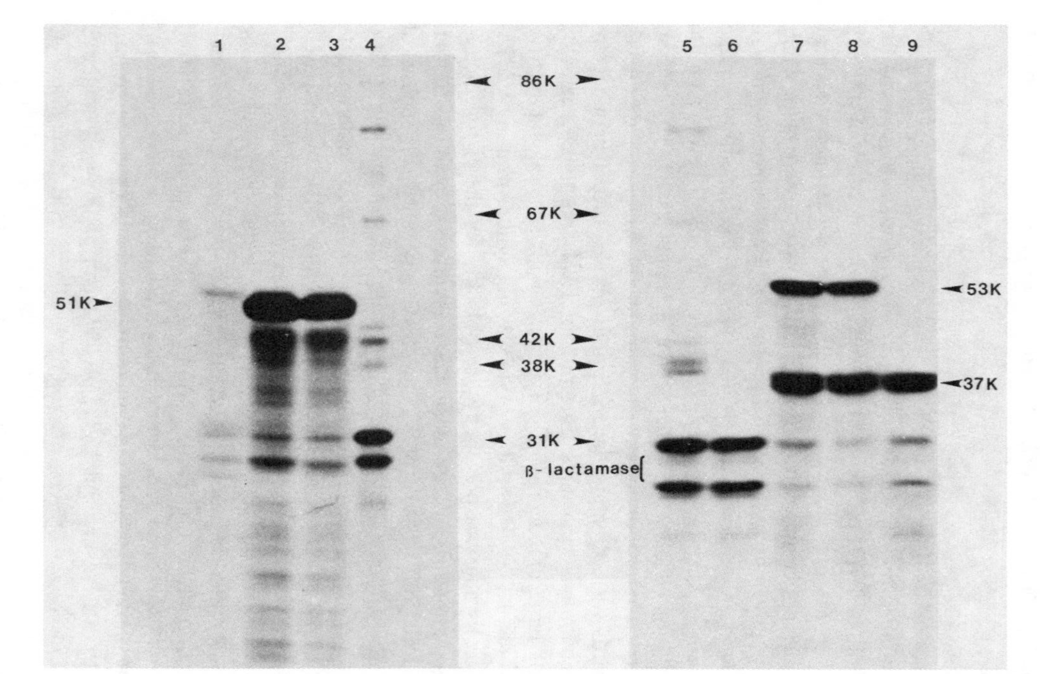

Fig. 4. Autoradiogram of the electrophoretic separation of [^{35}S]-labeled polypeptides synthesized in plasmid-containing minicells of *E. coli*. Minicells containing the different plasmids were allowed to incorporate [^{35}S]-methionine for 1 h at 37°C. Samples were prepared for electrophoresis and polypeptides contained in the samples separated by electrophoresis through a 10–20% polyacrylamide gradient gel. The gel was dried, impregnated for fluorography, and used to expose Kodak XAR-5 X-ray film to produce the autoradiogram shown. Polypeptides encoded by plasmids pET374 (track 1), pET371 (track 2), pET372 (track 3), pBR322 and coliphage T7 (tracks 4 and 5), pBR322 (track 6), pET405 (track 7), pET406 (track 8), and pET407 (track 9) are shown. The molecular masses of the T7 polypeptides, which are known precisely (Ponta et al., 1977) are indicated in the center of the figure and were used to determine the apparent molecular weights of the polypeptides encoded by the methanogen DNAs. All of the plasmids encode β-lactamase, which appears as two bands (precursor and mature form) as indicated.

direct the synthesis of as much β-lactamase as is made by pBR322-containing minicells.

Plasmids pET405 and pET406 directed the synthesis of two polypeptides (M_r = 53,000 and 37,000) that were not synthesized in pBR322-containing minicells (fig. 4). Minicells containing pET407 (a deletion derivative of pET405 that still complements *pur*E1; fig. 3) synthesized only the 37,000 molecular mass polypeptide, which therefore presumably embodies the phosphoribosylaminoimadizole carboxylase activity that is deficient in *E. coli pur*E mutants.

Hybridization Analysis of Recombinant Plasmids

It has recently been shown that the genomes of halophilic archaebacterial species contain repetitive DNA sequences (Sapienza and Doolittle, 1982). The locations of these sequences within the genomes change, suggesting that they may be transposon-like elements (Sapienza et al., 1982). We have analyzed methanogen-DNA-containing plasmids for the presence of repetitive sequences. Plasmid DNAs were made radioactive by nick translation and were used as hybridization probes to detect homology with restriction enzyme fragments present in digests of total genomic DNAs. Figure 5 shows the hybridization patterns obtained with pET371 and pET405 probes hybridized against *Hind*III and *Eco*RI digests of *M. barkeri* MS and *M. smithii* DNAs, respectively. Hybridization was obtained to *Hind*III-generated restriction fragments with electrophoretic mobilities identical to the cloned *Hind*III fragments (pET371 = 2.9 kb; pET405 = 2.5 kb). The cloned *Hind*III fragments do not contain sites for *Eco*RI (fig. 3). In agreement with this, pET371 and pET405 probes hybridized intensely to *Eco*RI fragments in the *Eco*RI digest of genomic DNAs, which were larger than the cloned *Hind*III fragments. Hybridization was also obtained, using either pET371 or pET405 probes, with additional *Hind*III and *Eco*RI fragments of genomic DNAs. The extent of hybridization was, however, less than that observed between the probes and the homologous restriction fragments. The number and sizes of the additional fragments that hybridized to the probes were consistent in several experiments, including experiments in which excesses of restriction enzymes were used in attempts to ensure complete digestions. Our results, therefore, suggest the presence of repeated sequences in the genomes of *M. barkeri* MS and *M. smithii*. The extent of sequence homologies between repeats may, however, be limited.

The cloned methanogen DNAs were also used as hybridization probes to determine if related sequences could be detected in digests of genomic DNAs from other archaebacterial and eubacterial species. Hybridization was

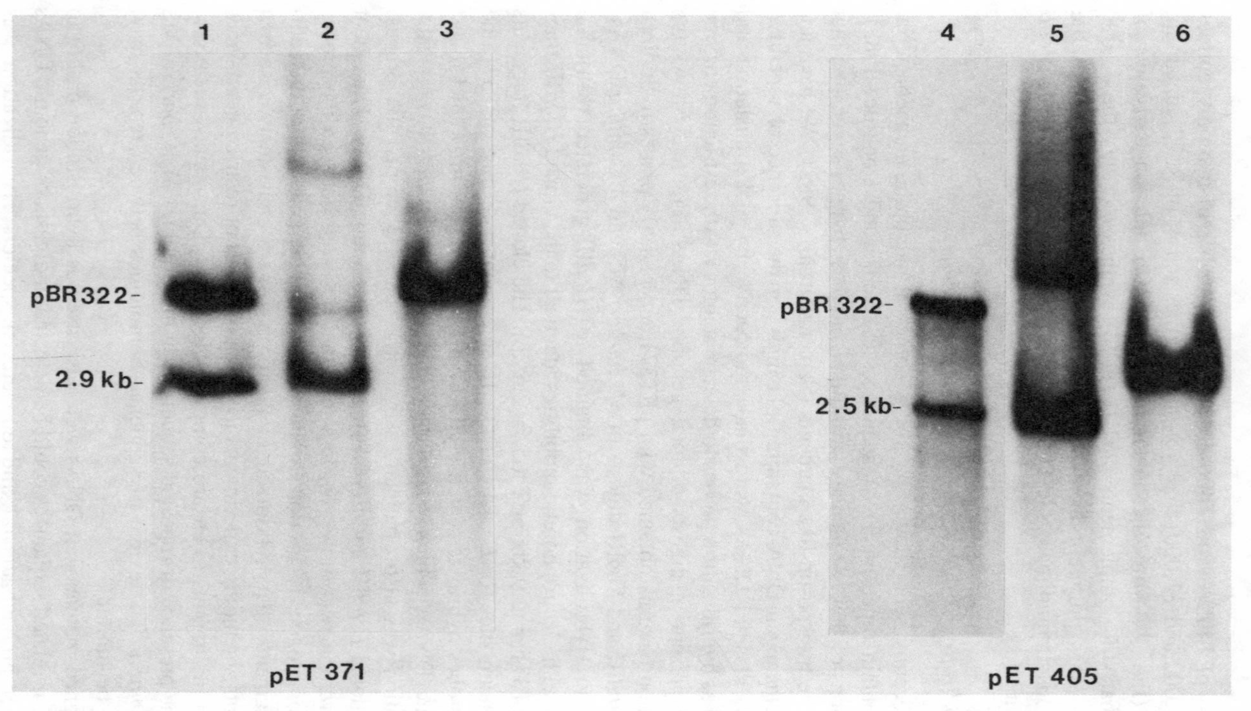

Fig. 5. Southern hybridizations between recombinant plasmids and *Hind*III or *Eco*RI digests of genomic DNAs of methanogens. Plasmids pET371 and pET405 were made radioactive by nick translation and used as probes to detect homology with *Hind*III or *Eco*RI digests of genomic DNAs. Radioactive pET371 was hybridized with *Hind*III-digested pET371 (track 1), *Hind*III-digested *M. barkeri* MS DNA (track 2), and *Eco*RI-digested *M. barkeri* MS DNA (track 3), Radioactive pET405 was hybridized with *Hind*III-digested pET405 (track 4), *Hind*III-digested *M. smithii* DNA (track 5), and *Eco*RI-digested *M. smithii* DNA (track 6).

obtained only between a cloned methanogen sequence and restriction fragments in the digest of the genome of the organism from which the DNA had been cloned. There was no detectable hybridization to sequences in the genomes of other methanogens, halophilic archaebacteria, *E. coli,* or *Bacillus subtilis* (results not shown).

DISCUSSION

Although Archaebacteria are clearly very different from Eubacteria (Balch et al., 1979), the results of this study demonstrate that DNA isolated from some archaebacterial species can be functionally expressed in *E. coli.* The observed frequency with which randomly cloned, methanogen-derived DNA complemented *E. coli* mutations under aerobic conditions was low: complementation was detected for approximately 10% of the mutations tested. Similar results were previously obtained in experiments in which eukaryotic DNAs from yeast and *Neurospora* were randomly cloned and tested for functional expression in *E. coli* (Ratzkin and Carbon 1977; Vapnek et al., 1977). The positive results support previous conclusions (Reeve et al., 1982) that methanogenic archaebacteria, like halophilic archaebacteria (Dunn et al., 1981), must employ the genetic code common to other biological species. Functional expression of methanogen-derived genes in *E. coli* also argues strongly against the presence of nonpolypeptide-encoding intervening sequences (introns) in these cloned genes. It is, of course, still possible that genes that do not function correctly in *E. coli* contain introns.

The DNAs from *M. barkeri* MS cloned in pET371 and pET372 and from *M. smithii* cloned in pET405 and pET406 are expressed regardless of their orientation relative to the vector DNA. This suggests that transcription originates within the cloned sequences. *E. coli* DNA-dependent RNA polymerase must therefore initiate transcription using methanogen-derived DNA sequences in spite of the radical differences in subunit configuration between archaebacterial and eubacterial RNA polymerases (Stetter et al., 1980). Determination of the DNA sequences used as promoters will be needed to determine if these two very different RNA-polymerases can, in fact, recognize and initiate RNA synthesis at the same DNA sequences.

Although initiation of RNA synthesis in *E. coli,* using methanogen DNA as a template, was unexpected, it is not surprising that methanogen mRNAs are translated in *E. coli.* The sequence of bases in the 16S rRNA, which is thought to hybridize to mRNA sequences and facilitate mRNA binding to ribosomes, is the same in ribosomal RNAs from *E. coli* and from archaebacteria (Steitz, 1978). Messenger RNAs transcribed from archaebacterial DNA would therefore be expected to contain ribosome binding sequences that would hybridize to *E. coli* rRNA and correctly position these mRNAs on ribosomes for subsequent translation.

The results of this study demonstrate that some genes from obligately anaerobic, archaebacterial methanogens can be functionally expressed in aerobically grown *E. coli*. We have recently grown *E. coli* under the atmospheric conditions used to grow methanogens (80% hydrogen:20% carbon dioxide; two atmospheres pressure), and therefore it may also be possible to obtain functional expression of oxygen labile methanogen-encoded enzymes in *E. coli*. Although cloning will allow analysis and manipulation of methanogen DNA, it will also be necessary to return cloned methanogen DNA to the methanogenic species from which it was obtained in order to study regulation of its expression. The development of a genetic exchange system for methanogens is crucial if advances using recombinant DNA techniques and *E. coli* are to be fully exploited. Facilities and techniques have now been developed that permit growth on plates and handling of methanogens in numbers sufficient to facilitate isolation and screening of mutants (Hook et al., 1984). Drug-resistant derivatives of *Methanococcus vannielii* so isolated are currently being used to develop a transformation system for this species.

ACKNOWLEDGMENTS

We thank Christina Morris for providing some details of the restriction maps of pET371, pET372, and pET374. This work was supported by Grant DE-AC02-81ER10945 from the Department of Energy. J. N. R. is the recipient of Research Career Development Award 1K04AG00108 from the National Institute on Aging.

LITERATURE CITED

Balch, W. E., G. E. Fox, L. J. Magrum, C. R. Woese, and R. S. Wolfe. 1979. Methanogens: re-evaluation of a unique biological group. Microbiol. Rev. 43:260–96.

Bolivar, F., and K. Backman. 1979. Plasmids of *Escherichia coli* as cloning vectors. Meth. Enzymol. 68:245–68.

Bollschweiler, C., and A. Klein. 1982. Polypeptide synthesis in *Escherichia coli* directed by cloned *Methanobrevibacter arboriphilus* DNA. Zbl. Bakt. Hyg. I. Abt. Orig. C3:101-9.

Davis, R. W., D. Botstein, and J. R. Roth. 1980. *In* Advanced bacterial genetics. Cold Spring Harbor Laboratory, Cold Spring Harbor, N.Y.

Dunn, R., J. McCoy, M. Simsek, A. Majumdar, S. H. Chang, V. L. Rajbhandary, and H. G. Khorana. 1981. The bacteriorhodopsin gene. Proc. Natl. Acad. Sci. USA 78:6744–48.

Hamilton, P. T., and J. N. Reeve. 1982. Cloning and expression of archaebacterial DNA from methanogens in *E. coli.* Abstr. Ann. Meeting Amer. Soc. Microbiol., I 89, p. 109.

Hook, L., R. E. Corder, P. T. Hamilton, J. I. Frea, and J. N. Reeve. 1984. Development of a plating system for genetic exchange studies in methanogens using a modified ultra-low oxygen chamber. *In* W. R. Strohl and O. H. Tuovinen (eds.), Microbial chemoautotrophy, pp. 275–89. Ohio State University Press, Columbus.

Jones, J. E., B. Bauers and T. C. Stadtman. 1977. *Methanococcus vannielii*: ultrastructure and sensitivity to detergents and antibiotics. J. Bacteriol. 130:1357–63.

Kandler, O., and H. König. 1978. Chemical composition of the peptidoglycan-free cell walls of methanogenic bacteria. Arch. Microbiol. 168:141–52.

Lederberg, E. M., and S. N. Cohen. 1974. Transformation of *Salmonella typhimurium* by plasmid deoxyribonucleic acid. J. Bacteriol. 199:1072–76.

Marmur, J. 1961. A procedure for the isolation of deoxyribonucleic acid from micro-organisms. J. Mol. Biol. 3:208–18.

Meagher, R. B., R. C. Tait, M. Betlach, and H. W. Boyer. 1978. Protein expression in *Escherichia coli* minicells by recombinant plasmids. Cell 10:521–36.

Miller, J. 1972. Experiments in molecular genetics. Cold Spring Harbor Laboratory Press, Cold Spring Harbor, N.Y.

Pecher, T. and A. Böck. 1981. *In vivo* susceptibility of halophilic and methanogenic organisms to protein synthesis inhibitors. FEMS Microbiol. Lett. 10:295–97.

Ponta, H., J. N. Reeve, M. Pfennig, M. Hirsch-Kaufman, M. Schweiger, and P. Herrlich. 1977. T7 development in F^+ *E. coli* cells and anucleate minicells. Nature (London) 269:440–42.

Prévot, A. R. 1980. Recherches récentes sur les bactéries méthanogènes. Bull. Inst. Pasteur 78:217–65.

Ratcliff, S. W., J. Luh, A. T. Ganesan, B. Behrens, R. Thompson, M. A. Montenegro, G. Morelli, and T. A. Trautner. 1979. The genome of *Bacillus subtilis* phage SPP1: the arrangement of restriction endonuclease generated fragments. Mol. Gen. Genet. 168:165–72.

Ratzkin, B., and J. Carbon. 1977. Functional expression of cloned yeast DNA in *Escherichia coli.* Proc. Natl. Acad. Sci. USA 74:487–91.

Reeve, J. N. 1978. Selective expression of transduced or cloned DNA in minicells containing plasmid pKB280. Nature (London) 276:728–29.

Reeve, J. N. 1979. Use of minicells for bacteriophage directed polypeptide biosynthesis. Meth. Enzymol. 68:493–503.

Reeve, J. N., and J. E. Shaw. 1979. Lambda encodes an outer membrane protein: the *lom* gene. Mol. Gen. Genet. 172:243–48.

Reeve, J. N., N. J. Trun, and P. T. Hamilton. 1982. Beginning genetics with methanogens. *In* A. Hollaender, R. D. DeMoss, S. Kaplin, J. Konisky, D. Savage, and R. S. Wolfe (eds.), Genetic engineering of micro-organisms for chemicals, pp. 233-44. Plenum Press, New York.

Sapienza, C., and W. F. Doolittle. 1982. Unusual physical organization of the *Halobacterium* genome. Nature (London) 295:384–89.

Sapienza, C., M. R. Rose, and W. F. Doolittle. 1982. High frequency genomic re-arrangements involving repeat sequence elements in the genome of *Halobacterium halobium,* an Archaebacterium. Nature (London) 299:182–85.

Smucker, R. A., and R. M. Pfister. 1975. Liquid nitrogen cryo-impacting: a new concept for cell disruption. Appl. Microbiol. 30:445–49.

Steitz, J. 1978. Methanogenic bacteria. Nature (London) 273:10.

Stetter, K. O., J. Winter and R. Hartlieb. 1980. DNA-dependent RNA polymerase of the Archaebacterium *Methanobacterium thermoautotrophicum.* Zbl. Bakt. Hyg. I. Abt. Orig. Cl:201–14.

Vapnek, D., J. A. Hautala, J. W. Jacobson, N. H. Giles, and J. R. Kushner. 1977. Expression in *Escherichia coli* K12 of the structural gene for catabolic dehydroquinase of *Neurospora crassa.* Proc. Natl. Acad. Sci. USA 74:3508–12.

Weisburg, W. G., and R. S. Tanner. 1982. Aminoglycoside sensitivity of archaebacteria. FEMS Microbiol. Lett. 14:307–10.

Woese, C. R., and G. E. Fox. 1977. Phylogenetic structure of the prokaryotic domain: the primary kingdoms. Proc. Natl. Acad. Sci. USA 74:5088–90.

Woese, C. R., L. J. Magrum, and G. E. Fox. 1978. Archaebacteria. J. Mol. Evol. 11:245–52.

Wolfe, R. S. 1979. Methanogens: a surprising microbial group. Antonie van Leeuwenhook 45:353–64.

Zeikus, J. G. 1977. The biology of methanogenic bacteria. Bacteriol. Rev. 41:514–41.

HARRY D. PECK, JR.

Physiological Diversity of the Sulfate-Reducing Bacteria

18

The sulfate-reducing bacteria have long been characterized as a small and highly specialized group of strictly anaerobic bacteria with a limited metabolic versatility. In the mid-sixties, they were classified into two genera, *Desulfovibrio* and *Desulfotomaculum* (Postgate and Campbell, 1966; Campbell and Postgate, 1965). Both groups of these bacteria accumulated acetate during growth on organic substrates plus sulfate. *Desulfovibrio* contains *c*-type cytochromes (Postgate, 1956) and a chromatophore, desulfoviridin, which produces an intense red fluorescence in the presence of alkali and ultraviolet light (365 nm) (Postgate, 1959). The formation of spores is characteristic of, and restricted to, *Desulfotomaculum*. It was generally believed that the function and mechanism of dissimilatory sulfate reduction was identical in both genera, although immunological studies suggested that the two groups were not closely related (Postgate, 1979). Two additional genera, physiologically similar to *Desulfovibrio* have been proposed, *Desulforistella* (Hvid-Hansen, 1951) and *Desulfomonas* (Moore et al., 1976). Other taxonomic observations (Rozanova and Nazina, 1976; Skyring et al., 1977; Widdel, 1980) suggest that the genus *Desulfovibrio* is in need of extensive revision. The recent isolation of two new species of *Desulfovibrio* capable of utilizing higher fatty acids up to eighteen carbons (Widdel, 1980) and the description (Widdel, 1980; Pfennig and Widdel, 1981) of five new genera of sulfate-reducing bacteria (*Desulfobacter, Desulfosarcina, Desulfonema, Desulfobulbus,* and *Desulfococcus*) that are morphologically and nutritionally diverse has completely destroyed the concept that the sulfate-reducing bacteria are a small and highly specialized group with a limited metabolic versatility. The bacteria in the seven genera only have in common their ability to utilize sulfate as a terminal electron donor and the fact that they are all strict anaerobes. On the basis of these results and a fuller understanding of the ecology of the sulfate-reducing bacteria, it has been proposed (Pfennig et al., 1981) that the dissimilatory sulfate-reducing bacteria be taxonomically treated as a single physiological-ecological group as has been done with the methanogenic and phototrophic bacteria.

A consideration of the nutrition and physiology of the new and old species of sulfate-reducing bacteria leads to some interesting observations. The sulfate-reducing bacteria can be divided into two broad groups depending on whether or not acetate is accumulated during growth with organic substrates and sulfate (Pfennig et al., 1981). The phenomenon does not appear relevant for current taxonomic schemes since acetate accumulators and nonaccumulators occur in two genera, *Desulfovibrio* and *Desulfotomaculum*. The complete oxidation of acetate to CO_2 and H_2O under anaerobic conditions was first observed by Pfennig and Biebl (1976) with *Desulfuromonas,* a sulfur-reducing bacterium, and later this capability was found in the sulfate-reducing bacteria (Widdel and Pfennig, 1977). Most of the enzymes of the citric acid cycle (Thauer, 1982) have been found in extracts of the acetate-oxidizing dissimilatory sulfate-reducing bacterium *Desulfobacter postgatei* (Widdel and Pfennig, 1981), and the citric acid cycle has been proposed as the pathway for the anaerobic oxidation of acetate with sulfate. The mechanism of net ATP formation remains a major problem, and acetate oxidation must be coupled to anaerobic electron transfer phosphorylation or perhaps the formation of an ion gradient (Hilpert and Dimroth, 1982) in order to generate the net ATP required for growth. The latter possibility must be considered in view of the intracytoplasmic membrane system described by Thauer (1982). This newly discovered ability of the sulfate-reducing bacteria for the anaerobic oxidation of acetate was suggested by the results of early workers and more recent ecological studies (Pfennig et al., 1981). In sulfate-containing environments, bacteria with this metabolic capability play a role similar to the acetate-utilizing methanogenic bacteria in low-sulfate environments as members of microbial consortia responsible for the complete degradation of complex biopolymers. In high-sulfate environments, methane is not produced from acetate, and this appears to be explained by the report that the apparent K_s for acetate (0.2 mM) of *Desulfobacter postgatei* is about tenfold lower than the apparent K_s for acetate (3 mM) of the acetate-utilizing methanogen *Methanosarcina barkeri* (Thauer, 1982).

Members of *Desulfovibrio* were the first nonphotosynthetic anaerobic bacteria in which *c*-type cytochromes were demonstrated (Postgate, 1956). Four different *c*-type cytochromes have been characterized; cytochrome c_{553} (LeGall and Bruschi, 1968), cytochrome c_3 (M_r = 13,000) (Postgate, 1956), cytochrome c_3 (M_r = 26,000) (Hatchikian et al., 1969), and nitrite reductase, a hexaheme cytochrome (Liu and Peck, 1981b). Members of *Desulfotomaculum* completely lack *c*-type cytochromes, and this characteristic had been considered distinctive for the genus. It should be noted that both *Desulfovibrio* and *Desulfotomaculum* contain *b*-type cytochromes (Campbell and Postgate, 1965; Hatchikian and LeGall, 1965). *C*-type cytochromes have been found in four of the five new genera (unknown in *Desulfococcus*), and, although these cytochromes have not been characterized (Widdel, 1980), the presence of *c*-type cytochromes can no longer be viewed as distinctive of *Desulfovibrio*.

An important taxonomic characteristic of *Desulfovibrio* has been the intense red fluorescence produced in the presence of alkali under ultraviolet light (365 nm) (Postgate, 1959), which is due to the release of siroporphyrin from bisulfite reductase (desulfoviridin) (Murphy and Siegel, 1973). Three species of *Desulfovibrio* have now been described that do not exhibit this characteristic red fluorescence and lack desulfoviridin: *D. baculatus* (Rozanova and Nazina, 1976); *D. sapovorans* (Widdel, 1980), and *D. desulfuricans* (Norway 4) (Miller and Saleh, 1964). In the case of the latter strain, desulfoviridin is replaced by a new pigment, desulforubidin, a bisulfite reductase that releases a nonfluorescent siroheme in the presence of alkali (Lee et al., 1973; Murphy and Siegel, 1973). All species of *Desulfotomaculum* have a different type of bisulfite reductase, termed P_{582} (Widdel and Pfennig, 1977; Trudinger, 1970; Skyring and Trudinger, 1973), which is nonfluorescent in the presence of alkali and ultraviolet light (Murphy and Siegel, 1973) owing to the release of siroheme rather than siroporphyrin. Desulfoviridin has been detected in extracts of *Desulfococcus multivorans* and *Desulfonema limicola*, P_{582} has been observed in extracts of *Desulfonema magnum*, and desulforubidin has been identified in extracts of *Desulfobacter postgatei* (C.-L. Liu and H. D. Peck, Jr., unpublished data). The various types of bisulfite reductase thus appear to be randomly distributed in the seven genera of sulfate-reducing bacteria and in the future will be of limited taxonomic significance.

The major ecological involvement of the sulfate-reducers is in anaerobic microbial food chains that are responsible for the complete degradation of complex biopolymers to carbon dioxide and sulfide in the presence of sulfate, and carbon dioxide and methane in the presence of sulfate. In the presence of sulfate, the sulfate-reducing bacteria serve as the terminal microorganisms responsible for the complete oxidation of the complex array of fermentation products resulting from the initial attack of other microorganisms on complex organic materials. In low-sulfate environments, the sulfate-reducing bacteria along with other types of anaerobic bacteria (Boone and Bryant, 1980; McInerney et al., 1979) participate in the process of interspecies hydrogen transfer (Bryant et al., 1977) and convert this complex assortment of fermentation products to substrates (acetate, hydrogen, and carbon dioxide) for the methanogenic bacteria. These are nutritionally complex environments that can support extensive taxonomic, nutritional, and physiological diversity among the sulfate-reducing bacteria as well as other anaerobic bacteria. In this paper dissimilatory sulfate reduction and alternate modes of growth will be discussed with emphasis on the physiological and biochemical diversity of the dissimilatory sulfate-reducing bacteria.

DISSIMILATORY SULFATE REDUCTION

The metabolic capabilities of the dissimilatory sulfate-reducing bacteria in the presence of sulfate show extensive variation both among genera and within certain genera, such as *Desulfovibrio*; and, as a physiological-ecological group, they can be considered to exhibit diverse oxidative activities. Thus, propionate, butyrate, and odd- and even-numbered fatty acids up to C_{18} are either oxidized to completion or to acetate (in some cases to acetate and propionate) (Widdel, 1980). Aromatic compounds such as cyclohexane carboxylate, benzoate, phenyl acetate, hydroxyl benzoate, 3-phenyl propionate, and hippurate can serve as electron donors for sulfate reduction in several genera and are completely oxidized to CO_2 (Widdel, 1980). Other electron donors for sulfate reduction include lactate, ethanol, propanol, methanol, pyruvate, malate, glycerol, succinate, oxamate, cysteine, and choline (Postgate, 1979). It is reasonable to expect that this list of substrates will be expanded over the next few years. There is little information available concerning the enzymology or the biochemical mechanisms involved in the oxidation of these substrates, and this aspect of the physiology of these bacteria will most certainly be a fertile area for investigation, particularly the bioenergetics of acetate oxidation, fatty acid oxidation, and the degradation of aromatic compounds.

The enzymatic reactions involved in the reduction of sulfate to sulfide by *Desulfovibrio* are summarized in equations 1-4.[1]

ATP sulfurylase:

$$ATP + SO_4^{2-} \rightarrow APS + PP_i \quad (1)$$

Inorganic pyrophosphatase:

$$PP_i + H_2O \rightarrow 2\ P_i \quad (2)$$

APS reductase:

$$APS + 2\,e^- \rightarrow AMP + SO_3^{2-}\ . \qquad (3)$$

Bisulfite reductase (desulfoviridin or desulforubidin):

$$HSO_3^- + 6\,H^+ + 6\,e^- \rightarrow HS^- + 3\,H_2O\ . \qquad (4)$$

In *Desulfotomaculum,* reaction 4 is catalyzed by a third type of bisulfite reductase that is characterized by a distinctive absorption at 582 nm in the presence of carbon monoxide. The identification of a bisulfite reductase in most of the sulfate-reducing bacteria suggests that reactions 1, 3, and 4 are common to all organisms capable of dissimilatory sulfate reduction. None of these enzymes appear to be unique to the sulfate-reducing bacteria: ATP sulfurylase is widely distributed in bacteria, plants, and animals (Peck, 1974); APS reductase is found in some of the thiobacilli (Kelly, 1982) and photosynthetic bacteria (Trüper and Fischer, 1982); and dissimilatory-type bisulfite reductases have been reported in the thiobacilli (Schedel et al., 1975), in *Methanosarcina barkeri* (Moura et al., 1982), and in the photosynthetic bacteria (Kobayashi et al., 1978; Schedel et al., 1979). In the case of *Desulfovibrio,* the utilization of two high-energy phosphates for the reduction of sulfate has made it necessary to invoke electron transfer-coupled phosphorylation as a means of providing a net production of ATP during growth on most organic substrates and sulfate (Peck, 1962). These reactions are summarized in figure 1 for the oxidation of two molecules of lactate to acetate and the reduction of one molecule of sulfate to sulfide. Two molecules of lactate are oxidized to pyruvate with the production of two molecules of hydrogen (Odom and Peck, 1981), and the two molecules of pyruvate are oxidized to acetyl phosphate and carbon dioxide by the phosphoroclastic reaction with the production of an additional two molecules of hydrogen. The acetyl phosphate is converted to ATP by the combined action of acetate and adenylate kinases, and the two high-energy phosphates are consumed in the formation of APS by ATP sulfurylase. The pyrophosphate is hydrolyzed to orthophosphate, and the APS is reduced by APS reductase to form sulfite plus AMP with the utilization of one molecule of hydrogen. The remaining three molecules of hydrogen are used in the reduction of sulfite to sulfide by bisulfite reductase. It should be noted that the two high-energy phosphates generated by substrate-level phosphorylation as acetyl phosphate are converted to ATP and consumed to the formation of APS.

The bioenergetics of dissimilatory sulfate reduction by most *Desulfotomaculum* (except for *Dt. acetoxidans* [Widdel and Pfennig, 1977]) differs

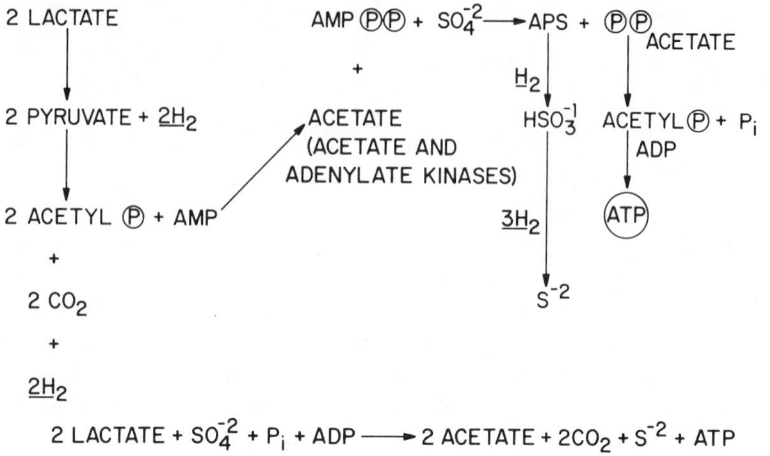

2 LACTATE → 2 PYRUVATE + $\underline{2H_2}$ → 2 ACETYL Ⓟ + AMP + 2 CO_2 + $\underline{2H_2}$

AMP ⓅⓅ + SO_4^{-2} → APS + ⓅⓅ

+ ACETATE (ACETATE AND ADENYLATE KINASES)

APS —$\underline{H_2}$→ HSO_3^{-1} —$\underline{3H_2}$→ S^{-2}

ⓅⓅ —ACETATE→ ACETYL Ⓟ + P_i —ADP→ ATP

$$2\ \text{LACTATE} + SO_4^{-2} + P_i + \text{ADP} \longrightarrow 2\ \text{ACETATE} + 2CO_2 + S^{-2} + \text{ATP}$$

Fig. 1. Enzymatic pathways and bioenergetics for the growth of *Desulfovibrio* on lactate plus sulfate (Peck and LeGall, 1982).

from that of *Desulfovibrio* such that electron transfer phosphorylation is not required for growth on organic substrates (Liu and Peck, 1981a). These bacteria lack high levels of inorganic pyrophosphatase to remove the PP_i produced during the formation of APS by ATP sulfurylase. Instead, they possess an acetate:PP_i kinase activity (equation 5) that allows them to conserve the chemical energy of the pyrophosphate bond:

$$PP_i + \text{acetate} \rightarrow \text{acetyl phosphate} + P_i\ . \quad (5)$$

The net effect of the conservation of the chemical energy of the pyrophosphate bond is to allow the bacteria to generate a net high energy phosphate at the substrate level as outlined in figure 2. The overall pathways of lactate oxidation and sulfate reduction are probably similar with the exception that the bisulfite reductase is P_{582} rather than desulfoviridin or desulforubidin. The absence of electron transfer phosphorylation was confirmed by a study of the relative growth yields of *Desulfovibrio* and *Desulfotomaculum* on lactate plus sulfate (Liu and Peck, 1981a). It thus appears that the role of sulfate in the metabolism of these desulfotomacula is that of a simple terminal electron acceptor analogous to the role of nitrate reduction to ammonia in certain clostridial fermentations (Hansan and Hall, 1975; Keith et al., 1982). Conceptually, these observations indicate an unsuspected bioenergetic diversity in

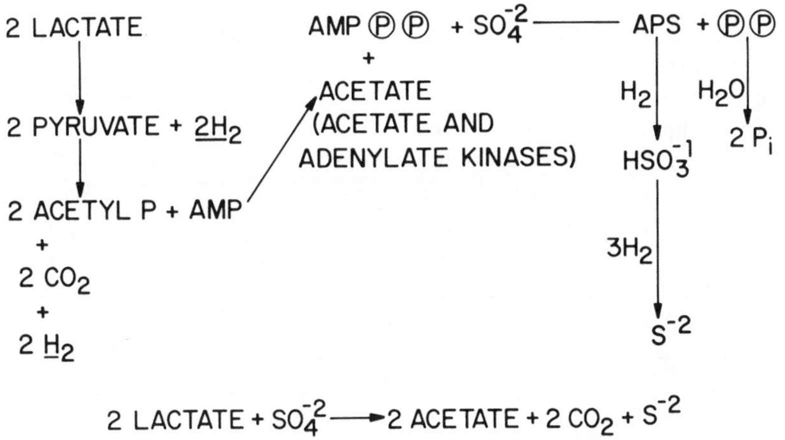

2 LACTATE → 2 PYRUVATE + $\underline{2H_2}$ → 2 ACETYL P + AMP + 2 CO_2 + 2 $\underline{H_2}$

AMP Ⓟ Ⓟ + SO_4^{-2} — APS + Ⓟ Ⓟ

\+ ACETATE (ACETATE AND ADENYLATE KINASES)

APS → (H_2) → HSO_3^{-1} → ($3H_2$) → S^{-2}

Ⓟ Ⓟ → (H_2O) → 2 P_i

$$2 \text{ LACTATE} + SO_4^{-2} \longrightarrow 2 \text{ ACETATE} + 2\, CO_2 + S^{-2}$$

Fig. 2. Enzymatic pathways and bioenergetics for the growth of *Desulfotomaculum* on lactate plus sulfate (Peck and LeGall, 1982).

the dissimilatory sulfate-reducers and allow one to treat the bioenergetics and physiology of sulfate reduction in *Desulfovibrio* and *Desulfotomaculum* as separate problems. From the standpoint of comparative biochemistry, the presence of *c*-type cytochromes (LeGall et al., 1980) and a high specific activity periplasmic hydrogenase in *Desulfovibrio* do not necessarily have any relevance for dissimilatory sulfate reduction in *Desulfotomaculum*. The possibility of other bioenergetic mechanisms involved with dissimilatory sulfate reduction cannot be excluded, and the anaerobic oxidations of acetate and possibly higher fatty acids appear to be good candidates for new bioenergetic mechanisms coupled to dissimilatory sulfate reduction.

All sulfate-reducing bacteria belonging to the genus *Desulfovibrio* have the ability to rapidly reduce sulfate with hydrogen or formate, and it was proposed many years ago that the dissimilatory sulfate-reducing bacteria are chemoautotrophs (Butlin and Adams, 1947). Later, it was demonstrated that these bacteria required a source of fixed carbon for growth (Mechalas and Rittenberg, 1960; Postgate, 1960); however, the fixation of carbon dioxide was greater than that observed with heterotrophic bacteria. Sorokin (1966) reported that a strain of *D. desulfuricans* was able to grow by the oxidation of hydrogen or formate coupled to sulfate reduction in the presence of acetate and carbon dioxide. In terms of biosynthesis of cell carbon, 70% was shown to be derived from acetate and 30% from CO_2. These results have recently been extended in studies of growth yields (Badziong and Thauer, 1978; Badziong et al., 1978; Brandis and Thauer, 1981). It was clearly demonstrated that the oxidation of hydrogen coupled to the reduction of sulfate provided sufficient energy for growth, and estimates were made for the amount of ATP produced per sulfate or thiosulfate reduced. Ribulose-bisphosphate carboxylase activity has been found in extracts of *D. vulgaris* (Alvarez and Barton, 1977), but a detailed study of the incorporation of ^{14}C-labeled acetate by *D. vulgaris* (Marburg) suggests that the carboxylase does not play a significant role in the biosynthesis of cell material (Badziong et al., 1979); however, the incorporation of $^{14}CO_2$ was not investigated. The question as to whether there exist "true" chemoautotrophic sulfate-reducing bacteria has now been resolved by the isolation of new genera that do not require fixed carbon for growth with hydrogen or formate and sulfate (Widdel, 1980; Pfennig and Widdel, 1981). Thus, *Desulfococcus multivorans, Desulfonema magnum,* and *Desulfosarcina variabilis* are capable of growth on formate plus sulfate without the addition of organic substrates, and the latter two bacteria are able to grow on hydrogen and carbon dioxide plus sulfate under the same conditions. These results establish that some sulfate-reducing bacteria are capable of the chemoautotrophic mode of growth, and it will be of considerable interest when the mechanism of carbon dioxide fixation in these new genera is determined.

GROWTH BY FERMENTATION

Postgate (1952) first showed that *D. desulfuricans* had the ability to grow in the absence of detectable sulfate on pyruvate and that it was utilized via the phosphoroclastic reaction (equation 6):

$$CH_3COCOO^- + P_i \rightarrow CH_3COOPO_3^{2-} + CO_2 + H_2 . \qquad (6)$$

Dt. nigrificans (Postgate, 1963) was later also shown to grow by the fermentation of pyruvate in a similar fashion, but not all sulfate-reducing bacteria in the genera *Desulfovibrio* and *Desulfotomaculum* possess this capacity (Postgate, 1979). From these results it was concluded that the utilization of sulfate as a terminal electron acceptor is not an obligatory mode of growth for all sulfate-reducing bacteria. From the enzymological point of view, the enzymes of dissimilatory sulfate reduction appear to be constitutive because growth with other anions of sulfur (Kobayashi et al., 1975), growth with nitrate as terminal electron acceptor (see later section), and growth with PP_i as a source of energy (Liu et al., 1982) do not have significant effects on the activities of these enzymes in crude extracts. A number of other substrates can be fermented and support the growth of the same sulfate-reducing bacteria: these include choline (Hayward and Stadtman, 1959), fumarate (Miller and Wakerley, 1966), malate (Miller et al., 1970), and lactate (Widdel and Pfennig, 1982).

Choline is fermented to yield trimethyl amine, acetate, and ethanol as shown in equation 7.

$$2(CH_3)_3\overset{+}{N}CH_2CH_2OH + H_2O \rightarrow 2(CH_3)_3\overset{+}{N}H + CH_3CH_2OH + CH_3COOH \quad (7)$$

Acetaldehyde appears to be an intermediate, and one high-energy phosphate is produced via acetyl phosphate for each two cholines degraded (Hayward, 1960). The dismutation of fumarate (and malate) yields two molecules of succinate and one molecule of acetate. For each fumarate oxidized via malate, oxalacetate and pyruvate to acetate, two fumarate molecules are reduced to succinate and the fermentation is balanced. Fumarate reductase (Hatchikian and LeGall, 1970) is found in most species of *Desulfovibrio*, and fumarate reduction to succinate with hydrogen is linked to electron transfer coupled phosphorylation (Barton et al., 1970). None of the desulfotomacula isolated to date dismutate fumarate, and they appear to lack fumarate reductase (Liu et al., 1982). The malate dehydrogenase of *Desulfovibrio* is NAD-linked (Hatchikian and LeGall, 1970) but $NADH_2$ has not been shown to be oxidized by fumarate nor by APS reductase or bisulfite reductase. *Desulfobulbus propionicus* has the ability to couple growth to the fermentation of lactate, and the mechanism appears to involve the dismutation of lactate to acetate and propionate (Widdel and Pfennig, 1982); but the enzymatic pathway of the fermentation has not yet been ascertained.

UTILIZATION OF NITRATE AS A TERMINAL ELECTRON ACCEPTOR

There have been sporadic reports in the literature that have indicated that some of the sulfate-reducing bacteria possess the ability to reduce nitrate, in addition to sulfate, as their terminal electron acceptor (Senez and Pichinoty, 1958). The difficulty of early workers in demonstrating growth with nitrate as terminal electron acceptor was probably related to two problems. First, many more bacteria reduce nitrate than sulfate, and contamination of cultures with other types of nitrate-reducing bacteria can frequently occur; and second, the reduction of sulfate generates sulfide, which is a reducing agent for the growth of the sulfate-reducing bacteria whereas the reduction of nitrate does not generate a reducing agent. A strain of *D. desulfuricans* (ATCC 27774) has been isolated that is able to reduce nitrate as its terminal electron acceptor (M. P. Bryant, personal communication), and *Desulfobulbus propionicus* has been shown to reduce nitrate to ammonia (Widdel and Pfennig, 1982). From the physiological and biochemical points of view, nitrate-grown cells of sulfate-reducing bacteria offer the advantage of being free of ferrous sulfide, and cells from mass cultures are pink rather than the usual grey-black color shown by sulfate-grown cells. In our laboratory the nitrate metabolism of *D. desulfuricans* (ATCC 27774) has been studied both with regard to its enzymology and its bioenergetics. As shown in figure 3, nitrate-grown cells of *D. desulfuricans* reduce nitrate to ammonia with the consumption of four moles of hydrogen per mole of nitrate reduced. Nitrous and nitric oxides were not detected; however, the ability of these cells to reduce nitrous and nitric oxides was not investigated. Sulfate-grown cells lacked the ability to reduce nitrate, and it was concluded that nitrate reduction was inducible in this bacterium. This was confirmed by examining various activities in extracts of both nitrate- and sulfate-grown cells; the results are summarized in table 1. Hydrogenase and the enzymes of dissimilatory sulfate reduction, ATP sulfurylase, APS reductase, bisulfite reductase, and thiosulfate reductase are present at comparable levels; however, in nitrate-grown cells, nitrate reduc-

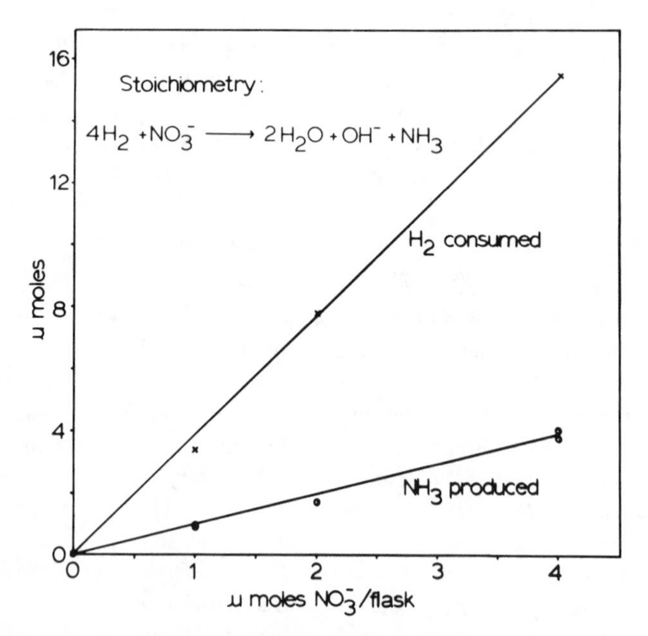

Fig. 3. Reduction of nitrate to ammonia by whole cells of *D. desulfuricans* 27774 (G. L. Dilworth and H. D. Peck, Jr., unpublished).

TABLE 1

COMPARISON OF VARIOUS ENZYMATIC ACTIVITIES IN EXTRACTS OF SULFATE- AND NITRATE-GROWN CELLS OF *D. DESULFURICANS* (27774).

Enzyme Activity	Assay	Activity[a]	
		NO_3^--Grown	SO_4^{2-}-Grown
ATP sulfurylase	Molybdolysis of ATP	7.1	5.9
APS reductase	$Fe(CN)_6^{3-}$ reduction	1.2	2.8
Bisulfite reductase	H_2 consumption	0.19	0.21
Thiosulfate reductase	H_2 consumption	0.50	0.22
Hydrogenase	H_2 consumption (benzyl viologen)	1.8	2.1
Nitrate reductase	H_2 consumption	0.012	0.000
Nitrite reductase	NO_2^- disappearance	4.8	0.35

SOURCE: G. L. Dilworth and H. D. Peck, Jr., unpublished.
[a]μmol min^{-1} mg $protein^{-1}$.

tase is present and the amount of nitrite reductase activity is elevated 5–10 fold. Some nitrite reductase activity was expected in sulfate-grown cells because bisulfite reductase exhibits nitrite reductase activity (Lee et al., 1973). The nitrate reductase is labile, and, in contrast to the enzymes of dissimilatory sulfate reduction, both the nitrate and nitrite reductases are largely membrane-bound. The nitrite reductase was purified from lactate/nitrate-grown cells and was shown to contain six *c*-type hemes per molecular mass of 65,000 (Liu and Peck, 1981b). Because bisulfite reductase is soluble and the nitrite reductase membrane-bound, it can be demonstrated that the hexaheme nitrite reductase is also present in lactate/sulfate-grown cells of this bacterium (D. J. Steenkamp and H. D. Peck, Jr., unpublished data). The ability to reduce nitrite is widely distributed in sulfate-reducing bacteria that have not been shown to grow on nitrate. However, it has not been resolved whether this activity is due to the hexaheme nitrite reductase or another type of nitrite reductase.

Studies on the bioenergetics of nitrate and nitrite reduction indicate that reduction of these anions is coupled to electron transfer phosphorylation although growth on lactate/nitrate medium should be possible solely via substrate phosphorylation because ATP is not required to reduce nitrate. A series of experiments were performed similar to those described by Peck (1960) which takes advantage of the fact that ATP is required for the reduction of sulfate but not the other anions of sulfur and nitrogen. As shown in figure 4, sulfite, thiosulfate, nitrate, and nitrite are rapidly reduced by washed cells in the presence of hydrogen; sulfate is only reduced after a long lag period (not

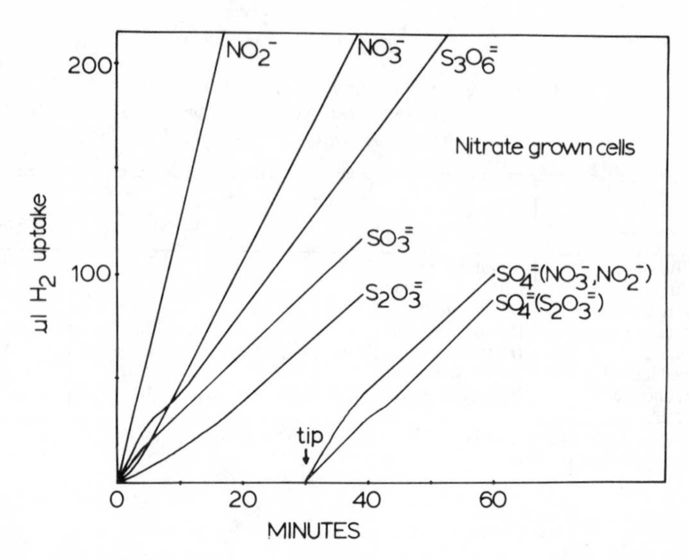

Fig. 4. Stimulation of sulfate reduction by catalytic amounts of nitrate, nitrite, and thiosulfate (G. L. Dilworth and H. D. Peck, Jr., unpublished).

shown), and this probably results from the lack of ATP for the transport and/or activation of sulfate. Addition of catalytic amounts of electron acceptors (nitrate, sulfite, nitrite, or thiosulfate) in the presence of sulfate results in the immediate restoration of sulfate reduction, and this effect has been interpreted to reflect the formation of ATP by electron transfer phosphorylation coupled to the oxidation of hydrogen and reduction of the anions. More direct evidence for the involvement of nitrite in electron transfer phosphorylation has been obtained employing lactate/nitrate-grown cells of *D. desulfuricans* (Steenkamp and Peck, 1981). Using hydrogen as reductant and nitrite as electron acceptor, proton translocation was demonstrated with $H^+/2e^-$ ratios in the range of 1.8–2.2. Proton translocation requires the presence of thiocyanate and is sensitive to uncoupling agents. This ability to couple the translocation of protons to the reduction of nitrite with hydrogen may be widespread in *Desulfovibrio* since lactate/sulfate–grown cells of *D. gigas* have the ability to reduce nitrite with hydrogen. With this organism nitrite reduction has been shown to be coupled to proton translocation and ATP synthesis (Barton et al., 1983) presumably by means of a reversible ATPase (Guarraia and Peck, 1971).

Because of the overall chemical analogy between sulfate reduction to sulfide and nitrate reduction to ammonia, it is attractive to propose that nitrate reductase plays a role equivalent to APS reductase and that nitrite reductase functions similarly to bisulfite reductase (fig. 5). This idea is supported by the fact that the levels of the enzymes of dissimilatory sulfate reduction and of *c*-type cytochromes are approximately the same in both lactate/sulfate- and lactate/nitrate-grown cells of *D. desulfuricans* (M. C. Liu, H. D. Peck, Jr., and J. LeGall, unpublished); thus, the same electron transfer chain seems to be involved with both sulfate and nitrate. The difficulties with this idea are the topography of the reductases of sulfate and nitrate metabolism (the enzymes of dissimilatory sulfate reduction are soluble and cytoplasmic whereas nitrate reductase and nitrite reductase are membrane-bound) and the fact that proton translocation appears to be directly coupled to electron transfer-associated nitrite reduction.

UTILIZATION OF INORGANIC PYROPHOSPHATE AS A SOURCE OF ENERGY FOR GROWTH

It has recently been demonstrated that three species of *Desulfotomaculum*—*Dt. orientis, Dt. ruminis,* and *Dt. nigrificans*—have the ability to utilize inorganic pyrophosphate as a source of energy for growth (Liu et al., 1982). On a minimal medium containing acetate, sulfate, yeast extract, and salts, significant growth was observed only in the presence of pyrophosphate, and

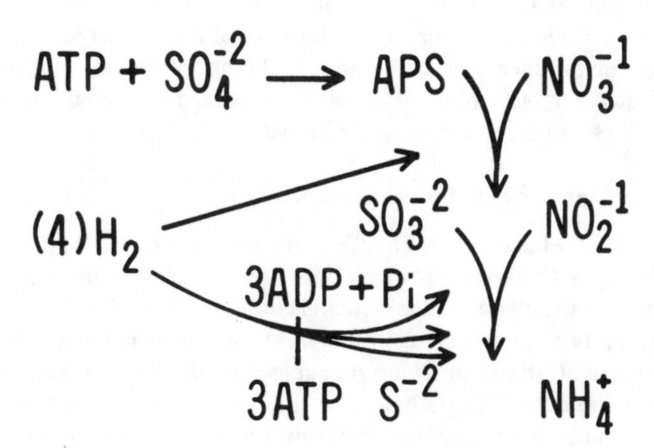

Fig. 5. Sulfate and nitrate reduction by the sulfate-reducing bacteria.

the growth response was linear up to a concentration of 0.05%. The pyrophosphate could not be replaced by orthophosphate, and, under the same growth conditions, pyrophosphate does not support the growth of *D. vulgaris* or *D. gigas*. It was suggested that acetate and yeast extract were required as a source of fixed carbon and that the presence of sulfate was required to allow the bacteria to metabolize the fixed carbon. Extracts of these three species exhibit PP_i:acetate kinase activity (equation 5) but only low levels of inorganic pyrophosphatase (Liu and Peck, 1981a). The activities of various enzymes involved in dissimilatory sulfate reduction were compared in lactate/sulfate- and PP_i-grown cells of *Dt. orientis*. ATP sulfurylase, APS reductase, bisulfite reductase (P_{582}), thiosulfate reductase, formate dehydrogenase, nitrite reductase, pyruvate dehydrogenase, pyrophosphatase, and PP_i:acetate kinase were all present at comparable levels, again indicating that these enzymes are constitutive. The major difference in the extracts was a fourfold increase in the amount of hydrogenase activity found in PP_i-grown cells. The basis for this increase in hydrogenase activity has not been determined, but it may reflect the fact that growth with PP_i is more similar to growth by interspecies hydrogen transfer than growth on lactate/sulfate media. Except for *Desulfobacter postgatei*, the possible growth of the new species of sulfate-reducing bacteria on PP_i has not been investigated. *D. postgatei* has been found to grow on PP_i, and extracts exhibit PP_i:acetate kinase activity at levels similar to those found in extracts of desulfotomacula (C.-L. Liu and H. D. Peck Jr., unpublished). A number of strictly anaerobic bacteria have been found to grow with PP_i as a source of energy (Peck et al., 1983), and PP_i enrichment cultures containing a diverse assortment of microorganisms have been obtained from anaerobic marine and freshwater environments as well as from sewage sludge and rumen fluid. Because of the bioenergetic advantage of conserving the energy of the PP_i formed during the activation of sulfate, it is anticipated that other of the new sulfate-reducing bacteria will be found to utilize PP_i as a source of energy for growth.

GROWTH IN PHOTOSYNTHETIC ASSOCIATIONS

Butlin and Postgate (1954) first demonstrated that mixed cultures of *Desulfovibrio* and *Chlorobium* can be grown in light in the presence of low concentrations of sulfate. The association constitutes a miniature sulfur cycle, and there has been a renewed interest in this type of association because of the discovery that all strains of *Chloropseudomonas ethylica* appear to be mixed cultures. *C. ethylica* (Shaposhnikov et al., 1960) was widely used in biochemical studies because it was the only green sulfur bacterium that was capable of growth on organic substrates, such as acetate and ethanol, instead of sulfide. In the early seventies, *C. ethylica* (strain 2K) was shown to be a mixed culture consisting of *Chlorobium limicola* plus at least one heterotrophic bacterium (Gray et al., 1972). It was suggested that the associated bacteria were sulfate-reducing bacteria, but they were not identified. Later, Biebl and Pfennig (1977) isolated a sulfate-reducing bacterium from cultures of *C. ethylica* (strain N_2) and identified the bacterium as *D. desulfuricans*. The bacterium was similar to the Norway strain of *D. desulfuricans* and was shown to contain desulforubidin rather than desulfoviridin as its bisulfite reductase (Fauque et al., 1979). The growth yields of *D. desulfuricans* (Essex 6) and *D. gigas* (DSM 496) grown on ethanol in association with *Chlorobium limicola* (strain 9330) have been studied by Biebl and Pfennig (1978), and the stoichiometry was reported as shown in reactions 8, 9, and 10.

Desulfovibrio:

$$34\ CH_3CH_2OH + 17\ SO_4^{2-} \rightarrow 34\ CH_3COO^- + 17\ H_2S + 34\ H_2O \quad (8)$$

Chlorobium:

$$17\ H_2S + 34\ CH_3COO^- + 28\ CO_2 + 16\ H_2O \xrightarrow{light} 24\ (C_4H_7O_3) + 17\ SO_4^{2-} \quad (9)$$

Net Reaction

$$34\ CH_3CH_2OH + 28\ CO_2 \xrightarrow{light} 24\ (C_4H_7O_3) + 18\ H_2O \quad (10)$$

Acetate cannot be oxidized by the sulfate-reducing bacterium and is quantitatively incorporated into cell material by the *Chlorobium*. The role of sulfate is catalytic, and this allows the sulfate-reducing bacterium to grow under conditions of limiting sulfate. Other cultures of *C. ethylica* were found to contain, as the green sulfur bacterium *Prosthecochloris aestuarii* and *Desulfuromonas acetoxidans,* a new physiological type of bacterium that utilizes elemental sulfur as its terminal electron acceptor and oxidizes acetate to CO_2 (Pfennig and Biebl, 1976). Some but not all species of *Desulfovibrio* can grow as sulfur-reducers (Biebl and Pfennig, 1977), but the physiological importance of this mode of growth has not been evaluated. A number of consortia involving green sulfur bacteria have been described (Pfennig and Trüper, 1974) in which a central chemoorganotropic bacterium is surrounded by synchronously dividing green sulfur bacteria. It has been suggested that the central organism is a sulfate-reducing bacterium and that the consortium is related by the oxidation and reduction of sulfur compounds.

INTERSPECIES HYDROGEN TRANSFER AND H_2 CYCLING

Sulfate-reducing bacteria belonging to the genus *Desulfovibrio* are thus far unique among the strictly anaerobic bacteria in being able to function as either hydrogen-producing or hydrogen-utilizing microorganisms in model microbial systems growing by interspecies hydrogen transfer. It was first shown that mixtures of hydrogen-utilizing methanogenic bacteria and *Desulfovibrio* had the ability to grow in low-sulfate media on ethanol plus CO_2 or lactate; the products were methane and acetate, and neither bacterium grew alone on the medium (Bryant et al., 1977). The general reactions for the growth of these bacteria on lactate are shown in equations 11 and 12:

Desulfovibrio

$$2 \text{ lactate} \rightarrow 2 \text{ acetate} + 2\, CO_2 + 4\, H_2 \tag{11}$$

H_2-utilizing methanogen

$$HCO_3^- + 4\, H_2 + H^+ \rightarrow CH_4 + 3\, H_2O\, . \tag{12}$$

Although *Desulfovibrio* does not grow significantly on lactate in the absence of sulfate, it contains all the enzymes required for the fermentation of lactate shown in equation 11; but as hydrogen accumulates, the reaction rapidly becomes thermodynamically unfavorable and growth ceases. When hydrogen is continuously removed by hydrogen-utilizing methanogenic bacteria, the thermodynamics of the overall reaction provide for a negative change in free energy and both bacteria thrive. The growth of *Desulfovibrio* in anaerobic consortia with the physiological role of hydrogen utilization is illustrated by the isolation of a new propionate-degrading bacterium *Syntrophobacter wolinii* (Boone and Bryant, 1980). This bacterium can be cultivated only with a hydrogen-utilizing methanogen or in the presence of sulfate and a *Desulfovibrio*. Products of the degradation of propionate are acetate, CO_2, and presumably hydrogen used for the reduction of either carbon dioxide or sulfate. These are the only conditions under which growth of *S. wolinii* has been observed, and it must be considered obligatory for growth by interspecies hydrogen transfer. Growth was best using the sulfate-reducing bacterium, and this may be due to the fact that the apparent K_s values for hydrogen are lower for the sulfate-reducing bacteria (1 μM) than for the hydrogen-utilizing methanogenic bacteria (6 μM) (Kristjansson et al., 1982). The reactions for the syntrophic degradation of propionate are shown in equations 13 and 14:

Syntrophobacter wolinii

$$4\ CH_3CH_2COO^- + 12\ H_2O \rightarrow 4\ CH_3COO^- + 4\ HCO_3^- + 4\ H^+ + 12\ H_2 \quad (13)$$

Desulfovibrio

$$12\ H_2 + 3\ SO_4^{2-} + 3\ H^+ \rightarrow 3\ HS^- + 12\ H_2O\ . \quad (14)$$

As in the case of lactate, the degradation of propionate to acetate, hydrogen, and carbon dioxide is thermodynamically unfavorable; but in the presence of a sulfate-reducing bacterium, the continuous removal of hydrogen changes the free energy yield and permits growth.

The physiological function of hydrogen and hydrogenase in the growth of the sulfate-reducing bacteria has been somewhat of a mystery. Clearly the oxidation of hydrogen coupled to the reduction of sulfate can serve as a source of ATP by electron transfer-coupled phosphorylation (Peck, 1960) and could support the growth of sulfate-reducing bacteria in the presence (Brandis and Thauer, 1981) and absence (Widdel, 1980) of fixed carbon. Lactate/sulfate-grown cells are able to catalyze the oxidation of hydrogen with sulfate, and they contain high levels of periplasmic hydrogenase; however, hydrogen is not usually formed as a major product during growth on organic substrates (Hatchikian et al., 1976). The presence of hydrogen does not cause a significant increase in cell yields when *Desulfovibrio* is grown on a lactate/sulfate medium (Khosrovi et al., 1971). This observation is unexpected since hydrogen oxidation with sulfate is coupled to ATP formation. During the early growth phases of *Desulfovibrio*, a phase of hydrogen production and reutilization occurs, but the physiological and biochemical ramification of this phenomenon was not obvious (Tsuji and Yagi, 1980). In addition they observed that the sparging of growing cultures with an inert gas such as argon caused a significant decrease in growth yields. Together, these observations suggested that hydrogen was involved in some manner as an intermediate in the oxidation of lactate with sulfate.

The discovery that three species of *Desulfotomaculum* conserve the chemical energy of the pyrophosphate formed during the activation of sulfate and do not require electron transfer-coupled phosphorylation for growth on organic substrates plus sulfate allows one to consider dissimilatory sulfate reduction as separate bioenergetic problems in *Desulfovibrio* and *Desulfotomaculum* (Liu and Peck, 1981a). Subsequently, Odom and Peck (1981b) proposed a hydrogen-cycling mechanism to explain electron transfer phosphorylation in *Desulfovibrio* growing on lactate plus sulfate, which is outlined in figure 6.

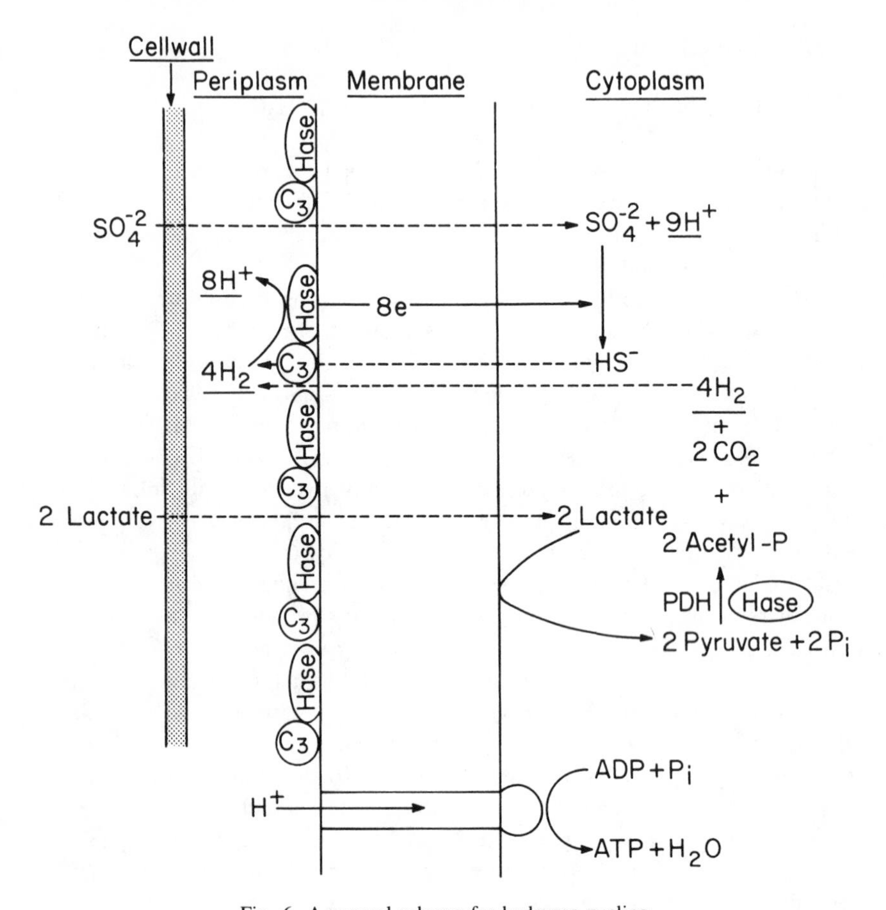

Fig. 6. A general scheme for hydrogen cycling.

For each sulfate reduced to sulfide, two molecules of lactate are oxidized to two acetate, two CO_2, and four molecules of hydrogen in the cytoplasm. The hydrogen, a permeant molecule, then diffuses across the cytoplasmic membrane and is oxidized by the periplasmic hydrogenase (Bell et al., 1974) requiring cytochrome c_3 (M_r = 13,000) as a cofactor (Bell et al., 1978). Only the electrons are transferred across the membrane to the cytoplasm, where they are used to reduce sulfate to sulfide with the consumption of eight protons. ATP for the reduction of sulfate can be viewed as being supplied by acetyl phosphate (fig. 1). The net effect of the production of eight protons in the periplasm coupled to the oxidation of hydrogen and the consumption of eight protons in the cytoplasm coupled to the reduction of sulfate is to generate a proton gradient without the participation of a typical Mitchell loop type of proton pump. The proton gradient can be utilized for the production of ATP via a reversible ATPase or to drive transport across the membrane. In the absence of sulfate, the hydrogen will diffuse into the medium, where it can be used for the reduction of carbon dioxide to methane by the methanogenic bacteria. Thus, the scheme in figure 6 explains why species of *Desulfovibrio* have the unique ability to function both as hydrogen-utilizing and hydrogen-producing microorganisms in consortia growing by means of interspecies hydrogen transfer.

The hydrogen-cycling proposal relies heavily on the ideas of Badziong and Thauer (1978) concerning the scalar production of protons during growth on hydrogen plus sulfate and is supported by the rapid production of protons coupled to the oxidation of hydrogen with a $H^+/2e^-$ of 2.0 (C.-L. Liu and H. D. Peck, Jr., unpublished). Proton translocation coupled to hydrogen oxidation with sulfite has also been reported by Kobayashi and colleagues (1982); however, there are quantitative differences between the data of the two laboratories. The evidence for the localization of enzymes and electron carriers is consistent with the hydrogen-cycling proposal (Badziong and Thauer, 1980; Odom and Peck, 1981a). The details for the cellular localization of enzymes and electron transfer component have been recently reviewed by Peck and LeGall (1982) and are summarized in figure 6. Hydrogenase, cytochrome c_{552} formate dehydrogenase, and probably cytochrome c_3 (M_r = 13,000) are clearly periplasmic. D- and L-lactate dehydrogenases and fumarate reductase are localized on the internal surface of the cytoplasmic membrane, and the soluble cytoplasmic enzymes and electron transfer components include flavodoxin, ferredoxin, pyruvate dehydrogenase, APS reductase, and bisulfite reductase, and evidence has been presented for the presence of an internal hydrogenase (Odom and Peck, 1981b). The exact localization of nitrite reductase has not been resolved, but from the available data, it is proposed to be transmembranous (Steenkamp and Peck, 1981). The most compelling experiment in support of the concept of hydrogen cycling is the observation that washed spheroplasts of lactate/sulfate-grown cells lose the ability to catalyze the oxidation of lactate with sulfate, but it can be partially restored by the addition of pure hydrogenase and cytochrome c_3 (M_r =

13,000) (Odom and Peck, 1981b). The hydrogen-cycling mechanism specifically requires that reducing equivalents from organic substrates leading to hydrogen production in the cytoplasm must not be utilized directly in the cytoplasm for dissimilatory sulfate reduction. This separation of electron transfer leading to hydrogen production from that leading to hydrogen utilization might be achieved by different electron carriers, different specificities of electron carriers, compartmentalization of the electron transfer systems, or regulation of reductases and dehydrogenases. In this regard it should be pointed out that multiple forms of hydrogenase are common in anaerobic bacteria (Chen and Blanchard, 1978; Mayhew et al., 1978; Yagi et al., 1978; Martin et al., 1980) as are multiple forms of electron carriers such as flavodoxin (D. Edmondson, personal communication), rubredoxin (Yang et al., 1980) and ferredoxin (Bruschi et al., 1976). The ferrodoxins discussed in the latter paper are composed of identical monomeric units, but ferredoxin I is a trimer and ferredoxin II a tetramer. Ferredoxin I contains four iron-four sulfur centers with an E_o' of -440 mV whereas the ferredoxin II contains three iron-three sulfur centers with an E_o' of -130 mV (Moore et al., 1978). The low-potential ferredoxin is active in the phosphoroclastic reaction but not the reduction of bisulfite with hydrogen. The high-potential ferredoxin is active in the reduction of bisulfite with hydrogen but not the phosphoroclastic reaction. In crude extracts, the four-iron ferredoxin I can be converted to the three-iron ferredoxin, and it has been proposed that the two types of ferredoxin play a important role in the separation of electron transfer sequences of *D. gigas* as shown in figure 7 (Peck and LeGall, 1982). Interconversion of the two types of ferredoxin would allow metabolic flexibility for growth on different substrates. It will be of interest to determine the extent of hydrogen cycling in the new genera of sulfate-reducing bacteria as well as its involvement in the anaerobic metabolism of other microorganisms such as the succinate-forming bacteria from the rumen (Henderson, 1980) and acetogenic bacteria such as *Acetobacterium woodii* (Winter and Wolfe, 1980).

CONCLUSION

The sulfate-reducing bactria are a taxonomically and physiologically diverse group of microorganisms that can be characterized as strict anaerobes capable of utilizing sulfate as a terminal electron acceptor. The group comprises both chemoautotrophs and chemoorganotrophs, and growth is possible by a number of alternate modes such as fermentation, nitrate reduction, pyrophosphate utilization, photosynthetic associations, and interspecies hy-

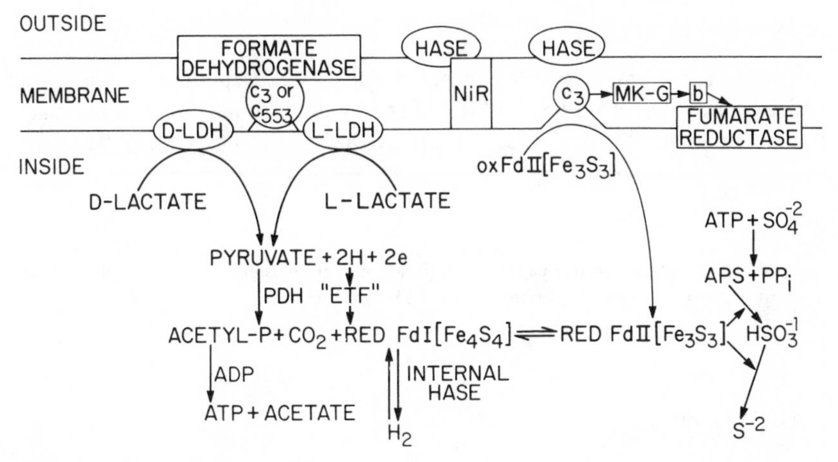

Fig. 7. Cellular localization of dehydrogenases, reductases, and electron transfer proteins in *Desulfovibrio gigas*.

drogen transfer. At least two mechanisms exist for energy-coupling to dissimilatory sulfate reduction: substrate phosphorylation in three species of *Desulfotomaculum* and hydrogen cycling in *Desulfovibrio*. It is suggested that other types of bioenergetic mechanisms may be involved in the anaerobic oxidation of acetate, the utilization of fatty acids, and the degradation of aromatic substrates. The ecological role of the dissimilatory sulfate-reducing bacteria in anaerobic high-sulfate environments is to remove the fermentation products resulting from the fermentation of complex biopolymers (Pfennig and Widdel, 1982). Under these conditions, methane is not produced because (1) many of the sulfate-reducing bacteria oxidize fermentation products to completion without the formation of extracellular acetate or hydrogen, and (2) the apparent K_s values of the sulfate-reducing bacteria for acetate and hydrogen are lower than those of the methanogenic bacteria. In low-sulfate anaerobic environments, carbon dioxide is utilized as the terminal electron acceptor by methanogenic microorganisms and reduced to methane. In this condition, interspecies hydrogen transfer becomes a predominant physiological process in the conversion of fermentation products to methane and carbon dioxide and the sulfate-reducing bacteria except for those capable of growth by interspecies hydrogen transfer are replaced by acetogenic bacteria, hydrogen-utilizing methanogens, and the acetate-utilizing methanogens.

ACKNOWLEDGMENTS

This research was supported in part by the U.S. Department of Energy under contract DE-ASO9-80 ER 10499 and by the National Science Foundation under grants PCM 8111325, PCM 8018308, and PCM 8205593.

1. Abbreviations used in this paper are: ATP, adenosinetriphosphate; APS, adenylyl sulfate (adenosine phosphosulfate); P_i, orthophosphate; PP_i, pyrophosphate.

LITERATURE CITED

Alvarez, M., and L. L. Barton. 1977. Evidence for the presence of phosphoriboisomerase and ribulose 1,5-diphosphate carboxylase in extracts of *Desulfovibrio vulgaris*. J. Bacteriol. 131:133–35.

Badziong, W., and R. K. Thauer. 1978. Growth yields and growth rates of *Desulfovibrio vulgaris* (Marburg) growing on hydrogen plus sulfate and hydrogen plus thiosulfate as the sole energy source. Arch. Microbiol. 117:209–14.

Badziong, W., and R. K. Thauer. 1980. Vectorial electron transport in *Desulfovibrio vulgaris* (Marburg) growing on hydrogen plus sulfate as sole energy source. Arch. Microbiol. 125:167–84.

Badziong, W., B. Ditter, and R. K. Thauer. 1979. Acetate and carbon dioxide assimilation by *Desulfovibrio vulgaris* (Marburg) growing on hydrogen and sulfate as sole energy source. Arch. Microbiol. 123:301–5.

Badziong, W., R. K. Thauer, and J. G. Zeikus. 1978. Isolation and characterization of *Desulfovibrio* growing on hydrogen plus sulfate as the sole energy source. Arch. Microbiol. 116:41–49.

Barton, L. L., J. LeGall, and H. D. Peck, Jr. 1970. Phosphorylation coupled to the oxidation of hydrogen with fumarate in extracts of the sulfate-reducing bacterium, *Desulfovibrio gigas*. Biochem. Biophys. Res. Commun. 41:1036–42.

Barton, L. L., J. LeGall, J. M. Odom, and H. D. Peck, Jr. 1983. Energy coupling to nitrite respiration in the sulfate-reducing bacterium *Desulfovibrio gigas*. J. Bacteriol. 153:867–71.

Bell, G. R., J. P. Lee, H. D. Peck, Jr., and J. LeGall. 1978. Reactivity of *Desulfovibrio gigas* hydrogenase toward artificial and natural electron donors. Biochimie 60:315–20.

Bell, G. R., J. LeGall, and H. D. Peck, Jr. 1974. Evidence for the periplasmic location of hydrogenase in *Desulfovibrio gigas*. J. Bacteriol. 120:994–97.

Biebl, H., and N. Pfennig. 1977. Growth of sulfate-reducing bacteria with sulfur as electron acceptor. Arch. Microbiol. 112:115–17.

Biebl, H., and N. Pfennig. 1978. Growth yields of green sulfur bacteria in mixed cultures with sulfur and sulfate reducing bacteria. Arch. Microbiol. 117:9–16.

Boone, D. R., and M. P. Bryant. 1980. Propionate-degrading bacterium, *Syntrophobacter wolinii* sp. gen. nov., from methanogenic ecosystems. Appl. Environ. Microbiol. 40:626–32.

Brandis, A., and R. K. Thauer. 1981. Growth of *Desulfovibrio* species on hydrogen and sulfate as sole energy source. J. Gen. Microbiol. 126:249–52.

Bruschi, M., E. C. Hatchikian, J. LeGall, J. J. G. Moura, and A. V. Xavier. 1976. Purification, characterization, and biological activity of three forms of ferredoxin from the sulfate reducing bacterium *Desulfobibrio gigas*. Biochim. Biophys. Acta. 449:275–84.

Bryant, M. P., L. L. Campbell, C. A. Reddy, and M. R. Crabill. 1977. Growth of *Desulfovibrio* in lactate or ethanol media low in sulfate in association with H_2-utilizing methanogenic bacteria. Appl. Environ. Microbiol. 33:1162–69.

Butlin, K. R., and J. R. Postgate. 1954. The microbial formation of sulfur in Cyrenaican Lakes. *In* J. L. Cloudsley-Thompson (ed.), Biology of deserts, pp. 112–22. Institute of Biology, London.

Butlin, K. R., and M. E. Adams. 1947. Autotrophic growth of sulfate-reducing bacteria. Nature (London) 160:154–55.

Campbell, L. L., and J. R. Postgate. 1965. Classification of the spore-forming sulfate reducing bacteria. Bacteriol. Rev. 29:359–63.

Chen, J. -S., and Blanchard, D. K. 1978. Isolation and properties of a unidirectional H_2-oxidizing hydrogenase from the strictly anaerobic N_2-fixing bacterium, *Clostridium pasteurianum* W-5. Biochem. Biophys. Res. Commun. 84:1144–50.

Fauque, G., D. Herve, and J. LeGall. 1979. Structure-function relationship in hemoproteins: the role of cytochrome c_3 in the reduction of colloidal sulfur by sulfate-reducing bacteria. Arch. Microbiol. 121:261–64.

Gray, B. H., C. F. Fowler, N. A. Nugent, and R. C. Fuller. 1972. A reevaluation of the presence of low midpoint potential cytochrome 551.5 in the green photosynthetic bacterium, *Chloropseudomonas ethylica*. Biochem. Biophys. Res. Commun. 47:322–27.

Guarraia, L. J., and H. D. Peck, Jr. 1971. Dinitrophenol-stimulated adenosine triphosphatase activity in extracts of *Desulfovibrio gigas*. J. Bacteriol. 106:890–95.

Hansan, S. M., and J. B. Hall. 1975. The physiological function of nitrate reduction in *Clostridium perfringens*. J. Gen. Microbiol. 87:120–28.

Hatchikian, E. C., and J. LeGall. 1965. Evidence for the presence of a *b*-type cytochrome in the sulfate reducing bacterium *Desulfovibrio gigas* and its role in the reduction of fumarate by molecular hydrogen. Biochim. Biophys. Acta 267:479–84.

Hatchikian, E. C., M. Bruschi, J. LeGall, and M. Dubourdieu. 1969. Cristallisation et properiètes d'un cytochrome intervenant dans la réduction du thiosulfate par *Desulfovibrio gigas*. Bull. Soc. Fr. Physiol Veg. 15:381–90.

Hatchikian, E. C., and J. LeGall. 1970. Etude du métabolisme des acides dicarboxyliques et du pyruvate chez les bactéries sulfato-réductices. II. Transport des electrons: accepteurs finaux. Ann. Inst. Pasteur 118:288–301.

Hatchikian, E. C., M. Chaigneau, and J. LeGall. 1976. Analysis of gas production by growing cultures of three species of sulfate reducing bacteria. *In* H. G. Schlegel, G. Gottschalk, and N. Pfennig (eds.), Microbial production and utilization of gases (H_2, CH_4, CO), pp. 109–18. E. Goltze KG, Göttingen.

Hayward, H. R. 1960. Anaerobic degradation of choline III acetaldehyde as an intermediate in the fermentation of choline by extracts of *Vibrio cholinicus*. J. Biol. Chem. 235:3592–96.

Hayward, H. R., and T. C. Stadtman. 1959. Anaerobic degradation of choline. 1. Fermentation of choline by an anaerobic cytochrome producing bacterium. *Vibrio cholinicus* n. sp. J. Bacteriol. 78:557–61.

Henderson, C. 1980. The influence of extracellular hydrogen on the metabolism of *Bacteroides ruminicola, Anaerovibrio lipolytica* and *Selenomonas ruminantium*. J. Gen. Microbiol. 119:485–91.

Hilpert, W., and P. Dimroth. 1982. Conversion of the chemical energy of methylmalonyl-CoA decarboxylation into a Na^+ gradient. Nature (London) 296:584–85.

Hvid-Hansen, N. 1951. Sulfate-reducing and hydrocarbon-producing bacteria in ground water. Acta Path. Microbiol. Scand. 29:314–35, 266–89.

Keith, S. M., G. T. MacFarlane, and R. A. Herbert. 1982. Dissimilatory nitrate reduction by a strain of *Clostridium butyricum* isolated from estuarine sediments. Arch. Microbiol. 132:62–66.

Kelly, D. P. 1982. Biochemistry of the chemolithotrophic oxidation of inorganic sulphur. Phil. Trans. R. Soc. Lond. B 298:499–528.

Khosrovi, B., R. MacPherson, and J. D. A. Miller. 1971. Some observations on growth and hydrogen uptake by *Desulfovibrio vulgaris*. Arch. Mikrobiol. 80:324–37.

Kobayashi, K., Y. Morisawa, T. Ishitoka, and M. Ishimoto. 1975. Biochemical studies on sulfate reducing bacteria. XIV. Enzyme levels of adenylyl sulfate reductase, inorganic pyrophosphatase, sulfite reductase, hydrogenase, and adenosine triphosphatase in cells grown on sulfate, sulfite, and thiosulfate. J. Biochem. (Tokyo) 78:1079–1145.

Kobayashi, K., Y. Seki, and M. Ishimoto. 1978. *Chromatium* sulfite reductase. I. Characterization of thiosulfate forming activity at the cell extract level. J. Biochem. (Tokyo) 84:1209–15.

Kobayashi, K., H. Hasegawa, M. Takagi, and M. Ishimoto. 1982. Proton translocation associated with sulfite reduction in a sulfate-reducing bacterium, *Desulfovibrio vulgaris*. FEBS Lett. 142:235–37.

Kristjansson, J. K., P. Schönheit, and R. K. Thauer. 1982. Different K_s values for hydrogen of methanogenic and sulfate reducing bacteria: an explanation for the apparent inhibition of methanogenesis by sulfate. Arch. Microbiol. 131:278–82.

Lee, J. P., J. LeGall, and H. D. Peck, Jr. 1973. Isolation of assimilatory- and dissimilatory-type sulfite reductases from *Desulfovibrio vulgaris*. J. Bacteriol. 115:529–42.

Lee, J.-P., C.-S. Yi, J. LeGall, and H. D. Peck, Jr. 1973. Isolation of a new pigment from *Desulfovibrio desulfuricans* (Norway 4) and its role in sulfite reduction. J. Bacteriol. 115:453–55.

LeGall, J., and M. Bruschi. 1968. Purification and properties of a new cytochrome from *Desulfovibrio desulfuricans*. *In* K. Okunuki, M. D. Kamen, and I. Sekusu (eds.), Structure and function of cytochromes, pp. 467–70. University Park Press, Baltimore.

LeGall, J., D. V. DerVartanian, and H. D. Peck, Jr. 1979. Flavoproteins, iron proteins, and hemoproteins as electron-transfer components of the sulfate-reducing bacteria. Curr. Top. Bioenerg. 9:237–65.

Liu, C.-L., and H. D. Peck, Jr. 1981a. The comparative bioenergetics of sulfate reduction in *Desulfovibrio* and *Desulfotomaculum*. J. Bacteriol. 145:966–73.

Liu, M. C., and H. D. Peck, Jr. 1981b. Isolation of a hexaheme cytochrome from *Desulfovibrio desulfuricans* and its identification as a new type of nitrite reductase. J. Biol. Chem. 256:13159–64.

Liu, C.-L., N. Hart, and H. D. Peck, Jr. 1982. Inorganic pyrophosphate: energy source for sulfate-reducing bacteria of the genus *Desulfotomaculum*. Science 217:363–64.

Martin, S. M., B. R. Glick, and W. G. Martin. 1980. Factors affecting the production of hydrogenase by *Desulfovibrio desulfuricans*. Can. J. Microbiol. 26:1209–13.

Mayhew, S. G., C. van Dijk, and H. M. van der Westen. 1978. Properties of hydrogenases from the anaerobic bacteria *Megasphaera elsdenii* and *Desulfovibrio vulgaris* (Hildenborough). *In* H. G. Schlegel and K. Schneider (eds.), Hydrogenases: their catalytic activity, structure and function, pp. 125–40. E. Goltze KG, Göttingen.

McInerney, M. J., M. P. Bryant, and N. Pfennig. 1979. Anaerobic bacterium that degrades fatty acids in syntrophic association with methanogens. Arch. Microbiol. 122:129–35.

Mechalas, B. J., and S. C. Rittenberg. 1960. Energy coupling in *Desulfovibrio desulfuricans*. J. Bacteriol. 80:501–7.

Miller, J. D. A., and A. M. Saleh. 1964. A sulphate-reducing bacterium containing cytochrome c_3 but lacking desulfoviridin. J. Gen. Microbiol. 37:419–23.

Miller, J. D. A., and D. S. Wakerley. 1966. Growth of sulphate-reducing bacteria by fumarate dismutation. J. Gen. Microbiol. 43:101–7.

Miller, J. D. A., P. M. Neuman, L. Elford, and D. S. Wakerley. 1970. Malate dismutation by *Desulfovibrio*. Arch. Mikrobiol. 71:214–19.

Moore, W. E. C., J. L. Johnson, and L. V. Holdeman. 1976. Emendation of *Bacteroidaceae* and *Butyrivibrio* and description of *Desulfomonas* gen. nov. and ten species in the genera *Desulfomonas, Butyrivibrio, Eubacterium, Clostridium* and *Ruminococcus*. Int. J. Syst. Bacteriol. 26:238–52.

Moura, J. J. G., A. V. Xavier, C. E. Hatchikian, and J. LeGall. 1978. Structural control of the redox potentials and of the physiological activity by oligomerization of ferredoxin. FEBS Lett. 89:177–79.

Moura, J. J. G., I. Moura, M. H. Santos, A. V. Xavier, and J. LeGall. 1982. Isolation of P_{590} from *Methanosarcina barkeri:* evidence for the presence of sulfite reductase activity. Biochem. Biophys. Res. Commun. 108:1002–9.

Murphy, M. J., and L. M. Siegel. 1973. Siroheme and sirohydrochlorin. The basis for a new type of porphyrin-related prosthetic group common to both assimilatory and dissimilatory sulfite reductases. J. Biol. Chem. 248:6911–19.

Odom, J. M., and H. D. Peck, Jr. 1981a. Localization of dehydrogenases, reductases, and electron transfer components in the sulfate-reducing bacterium, *Desulfovibrio gigas*. J. Bacteriol. 147:161–69.

Odom, J. M., and H. D. Peck, Jr. 1981b. Hydrogen cycling as a general mechanism for energy coupling in the sulfate reducing bacteria *Desulfovibrio* sp. FEMS Microbiol. Lett. 12:47–50.

Peck, H. D., Jr. 1960. Evidence for oxidative phosphorylation during the reduction of sulfate with hydrogen by *Desulfovibrio desulfuricans*. J. Biol. Chem. 235:2734–38.

Peck, H. D., Jr. 1962. Comparative metabolism of inorganic sulfur compounds in microorganisms. Bacteriol. Rev. 26:67–94.

Peck, H. D., Jr. 1974. Sulfation linked to ATP cleavage. *In* The enzymes, 9:651–99.

Peck, H. D., Jr. and J. LeGall. 1982. Biochemistry of dissimilatory sulphate reduction. Phil. Trans. R. Soc. (London) B 298:443–66.

Peck, H. D., Jr., C.-L. Liu, A. K. Varma, L. G. Ljungdahl, M. Szulczyuski, F. Bryant and L. Carreira. 1983. The utilization of inorganic pyrophosphate tripolyphosphate and tetrapolyphosphate as energy sources for the growth of anaerobic bacteria. *In* A. Hollaender, A. I. Laskin, and P. Rogers (eds.), The biological basis of new developments in biotechnology, pp. 317–18. Plenum Press, New York.

Pfennig, N., and H. G. Trüper. 1974. The phototrophic bacteria. *In* R. E. Buchanan and N. E. Gibbons (eds.), Bergeys manual of determinative bacteriology, pp. 24–64. Williams and Wilkins, Baltimore.

Pfennig, N., and H. Biebl. 1976. *Desulfuromonas acetoxidans* gen. nov. and sp. nov., a new anaerobic, sulfur-reducing, acetate-oxidizing bacterium. Arch. Microbiol. 110:3–12.

Pfennig, N., and F. Widdel. 1981. Ecology and physiology of some anaerobic bacteria from the microbial sulfur cycle. *In* H. Bothe and A. Trebst (eds.), Biology of inorganic nitrogen and sulfur, pp. 166–77. Springer Verlag, Berlin.

Pfennig, N., F. Widdel, and H. G. Trüper. 1981. The dissimilatory sulfate-reducing bacteria. *In* M. P. Starr, H. Stolp, H. G. Trüper, A. Balows, and H. G. Schlegel (eds.), The prokaryotes, pp. 926–40. Springer-Verlag, New York.

Pfennig, N., and F. Widdel. 1982. The bacteria of the sulphur cycle. Phil. Trans. R. Soc. London. B 298:433–42.

Postgate, J. R. 1952. Growth of sulphate-reducing bacteria in sulphate-free media. Research 5:189–90.

Postgate, J. R. 1956. Cytochrome c_3 and desulphoviridin: pigments of the anaerobe *Desulphovibrio desulphuricans*. J. Gen. Microbiol. 14:545–72.

Postgate, J. R. 1959. A diagnostic reaction of *Desulphovibrio desulphuricans*. Nature (London) 183:481–82.

Postgate, J. R. 1960. On the autotrophy of *Desulphovibrio desulphuricans*. Z. Allg. Mikrobiol. 1:53–56.

Postgate, J. R. 1963. Sulfate-free growth of *Clostridium nigrificans*. J. Bacteriol. 85:1450–51.

Postgate, J. R. 1979. The sulphate-reducing bacteria. Cambridge University Press, Cambridge.

Postgate, J. R., and L. L. Campbell. 1966. Classification of *Desulfovibrio* species, the nonsporulating sulfate-reducing bacteria. Bacteriol. Rev. 30:732–38.

Rozanova, E. P., and T. N. Nazina. 1976. A mesophilic, sulfate-reducing rod-shaped, non-spore forming bacterium. Microbiologiya 45:825–30.

Schedel, M., J. Baldensperger, and J. LeGall. 1975. Sulfur metabolism in *Thiobacillus denitrificans*. Evidence for the presence of a dissimilatory sulfite reductase. Arch. Microbiol. 105:339–41.

Schedel, M., M. Vanselow, and H. G. Trüper. 1979. Siroheme sulfite reductase isolated from *Chromatium vinosum*. Purification and investigation of some molecular and catalytic properties. Arch. Microbiol. 121:29–36.

Senez, J. C., and F. Pichinoly. 1958. Reduction de l'hydroxylamine liée à l'activitié de l'hydrogenase de *D. desulfuricans* I. Activité des cellules et des extraits. Biochim. Biophys. Acta 27:569–80.

Shaposhnikov, U. N., E. N. Kondrat'eva, and V. D. Fedorov. 1960. A new species of green sulfur bacteria. Nature (London) 187:167–68.

Skyring, G. W., and P. A. Trudinger. 1973. A comparison of the electrophoretic properties of the ATP-sulfurylases, APS-reductases, and sulfite reductases of dissimilatory sulfate reducing bacteria. Can. J. Microbiol. 19:375–80.

Skyring, G. W., H. E. Jones, and D. Goodchild. 1977. The taxonomy of some new isolates of dissimilatory sulfate-reducing bacteria. Can. J. Microbiol. 23:1415–25.

Sorokin, Yu. I. 1966. Role of carbon dioxide and acetate in biosynthesis by sulfate reducing bacteria. Nature (London) 210:551–52.

Steenkamp, D. J., and H. D. Peck, Jr. 1981. Proton translocation associated with nitrite respiration in *Desulfovibrio desulfuricans*. J. Biol. Chem. 256:5450–58.

Trudinger, P. A. 1970. Carbon monoxide reacting pigment from *Desulfotomaculum nigrificans* and its possible relevance to sulfite reduction. J. Bacteriol. 104:158–70.

Trüper, H. G., and V. Fischer. 1982. Anaerobic oxidation of sulphur compounds as electron donors for bacterial photosynthesis. Phil. Trans. R. Soc. Lond. B 298:529–42.

Thauer, R. K. 1982. Dissimilatory sulphate reduction with acetate as electron donor. Phil. Trans. R. Soc. Lond B 298:467–71.

Tsuji, K., and T. Yagi. 1980. Significance of hydrogen burst from growing cultures of *Desulfovibrio vulgaris*, Miyazaki, and the role of hydrogenase and cytochrome c_3 in energy producing system. Arch. Microbiol. 125:35–42.

Widdel, F. 1980. Anaerober Abbau von Fettsäuren und Benzoesäure durch neu isolierte Arten Sulfat-reduzierender Bakterien. Ph.D. thesis, University of Göttingen, Göttingen.

Widdel, F., and N. Pfennig. 1977. A new anaerobic, sporing, acetate oxidizing, sulfate reducing bacterium, *Desulfotomaculum* (emend.) *acetoxidans*. Arch. Microbiol. 112:119–22.

Widdel, F., and N. Pfennig. 1981. Studies on dissimilatory sulfate-reducing bacteria that decompose fatty acids. I. Isolation of new sulfate-reducing bacteria enriched with acetate from saline environments. Description of *Desulfobacter postgatei* gen. nov., sp. nov. Arch. Microbiol. 129:395–400

Widdel, F., and N. Pfennig. 1982. Studies on dissimilatory sulfate-reducing bacteria that decompose fatty acids. II. Incomplete oxidation of propionate by *Desulfobulbus propionicus* gen. nov., sp. nov. Arch. Microbiol. 131:360.

Winter, J. V., and R. S. Wolfe. 1980. Methane formation from fructose by syntrophic association of *Acetobacterium woodii* and different strains of methanogens. Arch. Microbiol. 124:73–79.

Yagi, T., A. Endo and K. Tsuji. 1978. Properties of hydrogenase from particulate fraction of *Desulfovibrio vulgaris. In* H. G. Schlegel and K. Schneider (eds.), Hydrogenases: their catalytic activity, structure, and function, pp. 107–24. E. Goltze KG, Göttingen.

Yang, S.-S., L. G. Ljungdahl, D. V. DerVartanian, and G. D. Watt. 1980. Isolation and characterization of two rubredoxins from *Clostridium thermoaceticum*. Biochim. Biophys. Acta 590:24–33.

RICHARD RADMER

Research from an Industrial Viewpoint: Martin Marietta Laboratories and Its Interactions with University Science

19

INTRODUCTION

During the past several years, academic scientists have become increasingly interested in certain areas of industrial research. There are several reasons for this renewed interest. Certainly, in these days of difficult funding, there is an increased incentive for university cooperation with other organizations, such as industry, that fund research. In support of increased industry-university cooperation in the biological sciences, I will briefly describe the operation of Martin Marietta Laboratories' Bioscience Department, and its current and possible relationships with university science. I hope that our example may serve as a model for relationships developed between other groups.

MARTIN MARIETTA CORPORATION

In order to understand the operation of the Laboratories, one must first understand the character of Martin Marietta Corporation. Martin Marietta is a large, diversified, multi-industry enterprise. The corporation is an unusual combination of commodity and high technology industries, consisting of five major companies: Cement, Construction Aggregates, Chemicals, Aluminum, and Aerospace Systems. Net sales in 1981 were approximately $3.3 billion.

The corporate laboratories reflect the diversity of the corporation, and comprise eight departments: Materials and Surface Science, Semi-conductor Physics, Chemistry, Bioscience, Aluminum, Mechanics, Ceramics, and a Center for Occupational Health and Engineering. The Laboratories also includes an Analytical Center. Martin Marietta's Environmental Center, which includes the Air Technology and Aquatic Technology groups, is colocated with the laboratories.

MARTIN MARIETTA LABORATORIES

Martin Marietta Laboratories provides science and technology support for the present and future objectives of the corporation. The means to this goal are listed in figure 1, roughly in the order of short-term applied science to long-term basic science. Examples of these items are:

1. Provide technical support services. These services include such items as waste-water analysis for the chemicals companies, and materials analyses for the Refractories Division and the Cement company.
2. Improve products and processes. For example, the Laboratories recently developed Comodye®, a computer-aided dyeing system. An interdisciplinary team of chemists and applied mathematicians developed computer models of various dye take-up rates and developed recipes for efficient, effective dyeing.
3. Develop science and technology of interest to Martin Marietta's future. The Laboratories is doing research in those emerging technologies that

MARTIN MARIETTA LABORATORIES GOAL:

TO PROVIDE SCIENCE AND TECHNOLOGY SUPPORT FOR PRESENT AND FUTURE OBJECTIVES OF MARTIN MARIETTA CORPORATION.

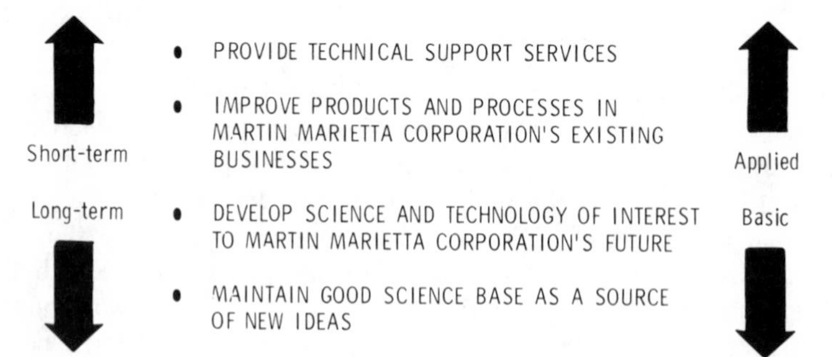

Fig. 1. Martin Marietta Laboratories' goal and the means by which the goal is achieved, listed in order from short-term applied science to long-term basic science.

may influence Martin Marietta Corporation in the foreseeable future (e.g., biotechnology in the chemicals industry; microelectronics in the aerospace industry).

4. Maintain a good science base as a source of new ideas. A primary objective of the Laboratories is getting and keeping good people. This effort is supported through contract research for outside agencies such as the National Science Foundation and the U.S. Department of Energy on projects of generic interest to Martin Marietta Corporation. Ideally, each principal investigator should spend part of his time on his own projects (say, approximately 25%). This work should keep the P.I. on the cutting edge of his discipline.

Figure 2 is a schematic diagram of the Martin Marietta Laboratories' Bioscience Department. This group functions as the "biology arm" for the corporation. As such, it has interacted with all the divisions of the corporation at one time or another. One might ask, "How is this handled, and how can the proper mix of expertise be maintained?" As shown schematically in figure 2, the Bioscience Department maintains a core program of basic photosynthesis

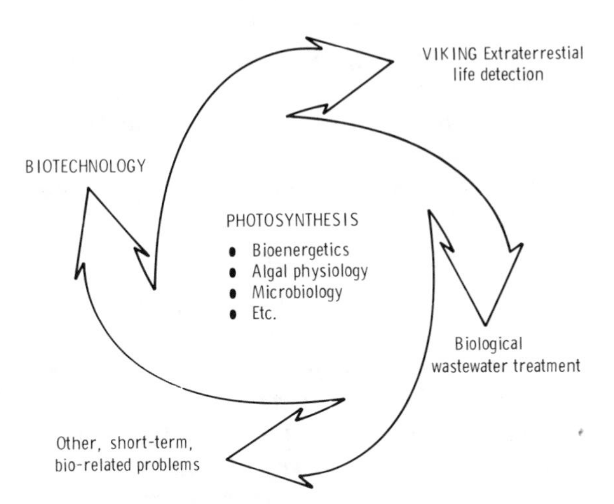

Fig. 2. Schematic diagram of Martin Marietta Laboratories' Bioscience Department. The core area of expertise is photosynthesis, from which other programs are spun off.

studies in areas such as bioenergetics, algal physiology, electron transport, and microbiology. This core program becomes the pool of expertise from which other projects of corporate-related interest can be "spun off." As corporate scientists, we apply basic science to Martin Marietta's area of interest. As basic scientists, we both add to and take from the basic science repository.

The recent projects being spun off in figure 2 include:

1. Biotechnology. We are currently developing and maintaining expertise in areas of interest to the corporation in this field. This includes projects in gene transfer, fermentation, and algal physiology.
2. Martin Marietta Corporation was the prime contractor for the Viking mission that landed on Mars. The stated goal of this monumental mission was to determine the presence or absence of life there. Although the Viking life-detection experiments were provided by NASA, and were developed under NASA auspices, Martin Marietta Corporation was ultimately responsible for making sure that they succeeded. The Laboratories' Bioscience Department worked closely with both NASA and the corporation's Denver Division to help solve the scientific and technical problems inherent in transforming laboratory procedures into automated, remotely controlled hardware. The department also designed, and had constructed, a science/hardware package that served as a backup and was the prime candidate for a future mission.
3. Biological wastewater treatment. We recently completed a project in which we helped to improve the biological wastewater treatment at one of our chemical companies.
4. Other short term biologically related projects. Our short term, one week to one month, projects are undertaken upon request to solve biology-related problems encountered within our various divisions.

INTERACTION OF THE MARTIN MARIETTA LABORATORIES' BIOSCIENCE DEPARTMENT WITH UNIVERSITY SCIENCE

Figure 3 is a diagram of how I perceive the relationship between industrial and university scientists. In both university and industry, scientists have, at the least, dual roles and must wear at the least two hats. In many respects, working in a university or in our kind of industrial laboratory is quite similar: the common denominator is research. The differences in the two sectors arise from differences in the kinds of customers that we serve; the university serves students and the public at large, whereas a laboratory such as ours serves the corporation, its divisions, and ultimately the stockholders. As shown in figure

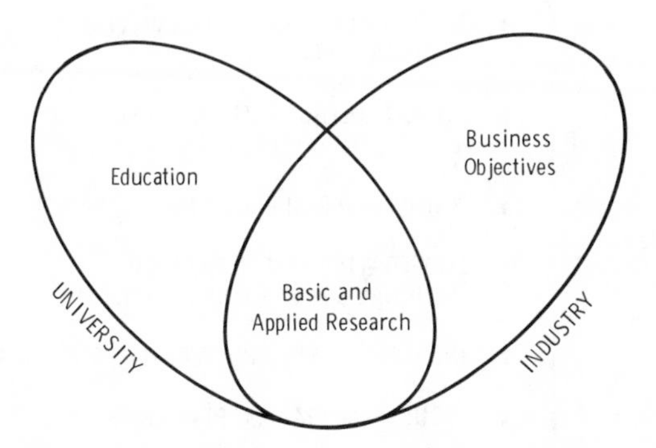

Fig. 3. Schematic diagram showing the overlap between university and industrial scientists.

3, university scientists and scientists in our laboratories share an area of basic and applied science. In the university this area is used as a jumping-off point for teaching, as well as an end in its own right. At Martin Marietta Laboratories this area is used as a jumping-off point for projects of corporate interest, as well as an end in its own right. So, in a sense, we are both basically scientists with different side jobs.

Figure 4 enumerates current and potential relationships between our laboratory and university scientists. It is ordered roughly from relationships in which scientists and institutions interact more or less equally to those in which industry has a greater burden of financial support and less involvement in the science.

1. Traditional. In this relationship university and industrial scientists interact equally and use the same peer review journals, meetings, and seminars. Some of the work that is discussed in these traditional settings may be done under the auspices of government grants rather than Martin Marietta Corporation funding.
2. Ad hoc arrangements. These include many, often location-specific arrangements that arise because of geographical proximity or common interests. For example, the University of Maryland/Baltimore County is only a mile from the Laboratories. Consequently, interaction between the Laboratories and this university is rather frequent.

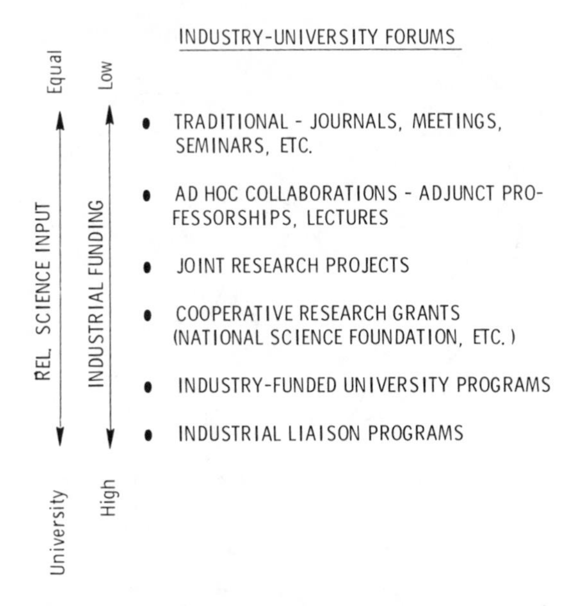

Fig. 4. Some current and potential relationships between university and industrial scientists, listed in order of increasing percentage of university science input and industrial financial support.

3. Joint research projects. Approximately one-quarter of the contract research carried on at Martin Marietta Laboratories is conducted in association with a university. There are several reasons for this:
   a. A specific university is often closer to the state-of-the-art in a given field than is our laboratory; joint projects allow our scientists to become more proficient in these areas.
   b. Joint research projects help keep our scientists in the mainstream of science development; interaction of this type is far more thorough than merely following the literature.
   c. Often one or the other of the institutions will have facilities uniquely adapted to the project.
   d. The Laboratories hopes to encourage the educational base in the university, particularly along the lines of a potential corporate interest.

   These relationships allow the Laboratories to address corporate-related problems with state-of-the-art science. Currently, the Laboratories is a prime contractor on about half of these joint research projects, and subcontractor on the other half; the Laboratories does most of the proposal work in obtaining the funds in either case.

4. Cooperative research grants. The National Science Foundation (NSF) has set aside a special pool of money to promote cooperation between universities and industries. Currently, the industrial member shares the cost of 50 percent of his part of the project. For example, for a $50,000 research project, NSF pays $25,000 and Martin Marietta pays $25,000. The university partner does not have to cost-share. Because the Laboratories is relatively small, we seldom participate in this program. The program may work better for a large laboratory attached to a high technology company, for example, Bell Laboratories.

In summary, this article is intended to convey an impression of an industrial laboratory such as Martin Marietta Laboratories.

1. University and industry laboratories such as Martin Marietta Laboratories can be quite similar; the focus of both is basic and applied science.
2. The two differ in the customers they serve and in the goals of their respective institutions.
3. There is, or can be, a good deal of cooperation between the two sectors, although this cooperation will be limited to some extent by the differing objectives of the institutions.

ACKNOWLEDGMENTS

I thank Raymond Bartlett for his substantial contributions to the preparation of this paper.

Index